Genomics, Proteomics and Vaccines

Editor

Guido Grandi
Chiron Vaccines,
Siena, Italy

John Wiley & Sons, Ltd

Other Wiley Editorial Offices

John Wiley & Sons, Inc., 111 River Street, Hoboken, NJ 07030, USA

Jossey-Bass, 989 Market Street, San Francisco, CA 94103-1741, USA

Wiley-VCH Verlag GmbH, Boschstrasse. 12, D-69469 Weinheim, Germany

John Wiley & Sons Australia Ltd, 33 Park Road, Milton, Queensland 4064, Australia

John Wiley & Sons (Asia) Pte Ltd, 2 Clementi Loop #02-01, Jin Xing Distripark, Singapore 129809

John Wiley & Sons (Canada) Ltd, 22 Worcester Road, Etobicoke, Ontario, Canada M9W 1L1

Wiley also publishes its books in a variety of electronic formats. Some content that appears in print may not be
available in electronic books.

British Library Cataloguing in Publication Data

A catalogue record for this book is available from the British Library

ISBN 0 470 85616 5

Typeset in $10\frac{1}{2}/12\frac{1}{2}$ Times by Keytec Typesetting Ltd, Bridport, Dorset, UK
Printed and bound in Great Britain by Antony Rowe, Ltd., Chippenham, Wilts
This book is printed on acid-free paper responsibly manufactured from sustainable forestry
in which at least two trees are planted for each one used for paper production.

Contents

Genomics, Proteomics and Vaccines edited by Guido Grandi
© 2004 John Wiley & Sons, Ltd   ISBN 0 470 85616 5

PART 2: TECHNOLOGIES

     and Development of Subunit Vaccines Against Bacterial
     Pathogens   223

     *Eszter Nagy, Tamás Henics, Alexander von Gabain
     and Andreas Meinke*

     10.1   Introduction   223
     10.2   Small DNA insert libraries – a tool to cover a pathogen's 'antigenome'   227
     10.3   Proper display platforms   230
     10.4   Selected human sera to provide imprints of pathogen encounters   231
     10.5   Cognate antibodies reveal the 'antigenome' of a pathogen   234
     10.6   How to retrieve from the 'antigenome' the candidate antigens for vaccine
            development   235
     10.7   Summary and discussion   237
            References   239

11   Searching the Chlamydia Genomes for New Vaccine
     Candidates   245

     Giulio Ratti, Oretta Finco and Guido Grandi

     11.1   Old problems and new perspectives for chlamydial vaccines   245
     11.2   Post-genomic approaches   250
     11.3   Genomic screening results   251
     11.4   Concluding considerations   262
            References   263

12   Proteomics and Anti-Chlamydia Vaccine Discovery   267

     *Gunna Christiansen, Svend Birkelund, Brian B. Vandahl
     and Allan C Shaw*

     12.1   Introduction   267
     12.2   Proteome analysis   269
     12.3   Proteomics as a complement for genomics   277
     12.4   Benefits that proteomics provide for vaccine development   279
            References   280

13   Proteome Analysis of Outer Membrane and Extracellular
     Proteins from *Pseudomonas aeruginosa* for Vaccine Discovery   285

     Stuart J. Cordwell and Amanda S. Nouwens

     13.1   Introduction   285

Preface

I have always been fascinated and intrigued by the history of human cultures and civilizations and how they have emerged and disappeared since 11 000 B.C., the date corresponding to the beginning of village life and the start of what geologists term the Recent Era.

In mathematical terms, the complex function describing how prosperity of a given society has varied with time is defined by several variables, often interlinked. These include the availability of domesticable plant and animal species (indispensable to trigger the passage from a hunter-gatherer to an agricultural lifestyle), climate, richness of the territory occupied and the appearance of individuals with strong personality and leadership. An additional important variable, which is often surprisingly neglected even in history text-books, is human susceptibility to microbes and infectious diseases. Indeed, epidemics caused by a variety of human pathogens have often been associated with, if not responsible for, major changes in human history. The eminent bacteriologist Hans Zinsser once wrote: 'Soldiers have rarely won wars. They more often mop up after the barrage of epidemics. And typhus with his brothers and sisters, – plague, cholera, typhoid, dysentery, – has decided more campaigns than Caesar, Hannibal, Napoleon, and the inspectors general of history'. There are several well documented examples supporting Zinsser's position and highlighting how the fate of human populations has often been dictated by epidemics.

In 430 B.C., Sparta and its Peloponnesian allies engaged in a bloody war against the Athenians, whose culture and power were at their height at that time. The Spartan invasion forced an uncounted number of villagers to find protection within Athens' city walls. As the war's great chronicler Thucydides wrote, the crowded city was soon scourged by a plague that killed a huge

Genomics, Proteomics and Vaccines edited by Guido Grandi
© 2004 John Wiley & Sons, Ltd   ISBN 0 470 85616 5

number of Athenians. The plague weakened Athens to such an extent that, despite its great naval power, it took more than thirty years to defeat Sparta, and the city never regained its political and cultural glory.

The dissolution of the great Roman empire has always been attributed to the decadence of the pagan lifestyle that led people to indulge in the pleasures of life rather than to protect the empire's borders from barbarian invasions. However, historians seldom highlight the fact that from the second century on, new diseases appeared in Europe with increasing frequency and devastation. The Oronius, Galen and Cyprian plagues (so called from the names of the chroniclers who described them), together with the accompanying famines, severely reduced the empire's population and abated its morale. In 452 A.D., when the Huns, headed by their ferocious general Attila, reached the gates of Rome, instead of entering the city, they halted and fell back. The withdrawal has often been attributed to the persuasive power of Pope Leo I, but, in fact, a severe epidemic, most likely smallpox, was raging within Rome, whose population, after subsequent epidemics, was soon to be reduced to a few thousand people. The Western Empire (in the fourth century, the Roman Empire had split into the Western Empire with its capital in Rome and the Eastern Empire with its capital in Costantinople) finally crashed in 476 A.D.

During the following century, the great eastern emperor Justinian was reconquering western territories. By 542 A.D., he had taken back much of North Africa, Sicily and part of Spain. However, in the middle of his glorious campaigns the first indisputably reported bubonic plague broke out, causing one of the worst population crashes in human history. The Byzantine historian Procopius described it as 'a pestilence by which the whole human race came near to being annihilated'. When the bout of plague ended, 40 percent of the people in Constantinople had died. Plague returned frequently until 590 A.D. and localized outbreaks occurred for another 150 years, halving human populations in many parts of the world. The Western Empire was never conquered again.

In the 14th century, Europe was experiencing a period of cultural growth and expansion. The population continued to grow; Crusades and trade routes put Europe in contact with the Middle East, Arabia and China, favouring increasing literacy and promoting what would have been later called Renaissance. This flourishing expansion was dramatically slowed down by the worst disaster in human history, the second bubonic plague pandemic, the Black Death. It was brought into Europe in the summer of 1347 by Genoese traders who contracted it in the Crimean city port of Kaffa when it was under siege by Janibeg, Khan of Tartars (the legend says that before withdrawing because plague had started killing the Tartars, Janibeg ordered plague-infected corpses to be catapulted into Kaffa, thus spreading the epidemic within the city walls). The disease rapidly reached all the main cities in Europe and it was estimated that up to one half of the people in Europe, North Africa and Asia perished during the long epidemics. All aspects of European life, from art to commerce, were severely affected and the post-plague labour shortage was one of the main reasons for European involvement in the slave trade.

The conquest of the Americas is usually attributed to the braveness, and sometime ferocity, of the Spanish *conquistadores* and of the European immigrants who succeeded in defeating the aggressive native populations. However, when Cortés attacked with his 300 soldiers the Aztec capital Tenochtitlán, he found the city, originally inhabited by 300,000 people, slaughtered by smallpox brought in by one of his African slaves who was affected by a virus strain to which Cortés' soldiers were immune. In his battle report, Cortés wrote, 'A man could not set his foot down unless on the corpse of an Indian'. Less than fifty years after Cortés arrived in Mexico, of the original 30 million people, only 3 million survived, 18 million having been killed by smallpox alone. Similarly, 80% of the Inca population in Peru disappeared in the 200 years following the Spanish invasion.

In North America, the natives suffered similar slaughter by pathogens brought into their territories by European invaders. As in Latin America, it was mainly smallpox and measles that kept killing Amerindians, reducing by about 90% the original population of 100 million. As a result, Europeans succeeded in imposing their languages, religions, political power and what the historian Alfred Crosby called ecological imperialism, whereby most of the original ecosystems were completely europeanized by imported plant and animal species.

Until the middle of the last century, infectious diseases were the first cause of death in humans; they continued to influence human cultures and civilizations and favoured the expansion of stronger, more immune-protected populations at the expense of weaker, immune-susceptible ones.

In the second half of the 20th century, two extraordinary revolutions occurred, which had no precedents in history in terms of prolongation and amelioration of human life: the large scale availability of antibacterial drugs and the practice of mass vaccination. For the first time, bacterial infections could be effectively defeated and deadly viral and bacterial diseases such as smallpox, polio, rabies, tetanus and diphtheria could be prevented. In the developed countries, where antibiotics and vaccines were discovered and readily available, mortality rates from infectious diseases rapidly declined and lost their first ranking in the cause-of-death list, being overtaken by cancer and cardiovascular diseases. Smallpox global vaccination eradicated this killer virus, which now exists in only a few, well-contained laboratories, and anti-polio and anti-measles vaccines are expected to have similar effects within the next few years.

The effectiveness of antibiotics and vaccines has been so impressive that for sometime it created the false illusion that humans would soon live in a healthy, pathogen-free environment, inhabited only by harmless, friendly microorganisms. However, this is far from being the case. During the last several million years of existence on earth, microorganisms have learnt how to adapt to rapid and often drastic environmental changes. The relative simplicity of their genetic material, together with the extraordinary high number of individuals belonging to each species, allow them, through spontaneous mutations and acquisition of exogenous genetic material, to test different solutions for better survival in a changing environment. This 'learning process' led bacteria to acquire several sophisticated and effective machineries to cope with antibacterial compounds. Today, scientists and physicians look at the rapid expansion of antibiotic resistance as a serious threat to the future of human health.

Vaccination is an extremely effective way to protect animals and humans against bacterial and viral infections. Exploiting the mechanisms of natural immune defences that, despite the insurgence of disease, often allow individuals to survive primary infections and to be protected against subsequent challenges from the same pathogens, vaccines have the invaluable advantage over antibiotics of being able to prevent illness and, in the long term, have the potential to eliminate the pathogen (as has been the case for smallpox). At present, different vaccine formulations against 25 different pathogens are available on the market; if all national health and political authorities would take concerted action for global vaccination campaigns, many of these pathogens could be eliminated from earth forever.

However, similarly to antibiotics, vaccines have to deal with the extraordinary capacity of microorganisms to cope with the threat of a changing environment. In particular, pathogens have developed the fascinating ability to escape the surveillance of our immune system by using two major mechanisms. On the one hand, many pathogens are capable of hiding themselves from host responses. Typical examples are extracellular bacterial pathogens, which surround themselves with protective layers of different chemical nature, and HIV (human immunodeficiency virus), whose ability to invade and destroy CD4+ T cells, key players of our immune system, makes it so deadly and vaccines so difficult to develop. On the other hand, pathogens often develop the capacity of infecting different hosts. This is a particularly clever strategy to survive and proliferate: by jumping from one host to another, they can use one host to accumulate sufficient mutations to allow them to escape the immune defences developed by the other hosts during previous infections. A paradigmatic example is the influenza virus, which, thanks to its capacity to infect several animal species, from birds to pigs, happily pays humans a visit every winter.

The two mechanisms of pathogen defence explain why vaccines against several pathogens such as HIV, HCV (hepatitis C virus), *Group A Streptococcus, Staphylococcus, Plasmodium falciparum* and *Schistosomes* (just to mention a few) are difficult to develop and currently not available. They also explain the appearance of new species of animal and human pathogens, which, like HIV and the recent coronavirus that causes SARS (severe acute respiratory syndrome), are generally highly infective and virulent. Without doubt, the exponential growth of both human and domesticated animal populations will promote the outbreak of new infective species in the future.

In conclusion, the attractive perspective of our living in a pathogen-free environment is destined to remain a utopia. More practically, our own health and that of future generations will depend on our ability to find new, rapid, effective strategies for vaccine development to fight both existing pathogens and those that will inevitably emerge in the future.

The aim of this book is to describe in a comprehensive manner and to provide up-to-date examples of an emerging strategy that has the potential to replace the existing approaches to vaccine discovery.

Traditionally, vaccines have been developed exploiting two technologies. According to the first, pathogens or attenuated, non-pathogenic derivatives are grown under appropriate laboratory conditions (synthetic liquid cultures, cell cultures, tissue cultures, etc.), collected, inactivated and formulated. This technology, tested for the first time by Edward Jenner more than 200 years ago and subsequently rationalized by Louis Pasteur, is very efficacious but suffers from two major drawbacks. Firstly, there are pathogens that are difficult or almost impossible to cultivate in a scalable setting (a typical example is *Plasmodium falciparum*, the aetiological agent of malaria, which can grow only in the mosquito salivary glands). Secondly, today regulatory authorities require extremely high safety and quality standards for all new vaccine formulations and obtaining approval to distribute vaccines based on live and inactivated organisms is becoming more and more difficult.

The second technology is based on the utilization of isolated components of the pathogen, which alone are capable of eliciting protective immune responses against the intact pathogen. The great advantages of this technology are that there is no risk that vaccines can provoke the disease (it is well known that the polio vaccine based on an attenuated, live strain was extremely effective but caused one case of polio per one million vaccinees) and that, if recombinant forms of the selected components are utilized, the pathogen need not be cultivated. The major disadvantage of this technology is that the identification of the few protective components from the pool of molecules present in the pathogen is usually complex and time consuming.

The new strategy for vaccine discovery described in this book, which is based on the exploitation of genomic and post-genomic technologies, provides a solution to shorten the time required to identify vaccine candidates, while enhancing the probability of success.

Genomics and post-genomics reached a watershed in 2001, with the publication of the complete sequence of the human genome. While this impressive technological achievement is already producing an astonishing amount of invaluable scientific discoveries, extensive exploitation of the human genome sequence for practical applications will require a few more years to emerge. Genomic and post-genomic technologies applied to viral and bacterial pathogens are not only almost equally important from a scientific perspective, but also have the potential to be translated more rapidly into useful products and processes.

As illustrated in this book, the characterization of bacterial genomes, transcriptomes and proteomes offers the opportunity to select the few proteins likely to elicit protective immunity from the plethora of molecules constituting a given pathogen. Once identified, the selected proteins are subjected to high throughput cloning, expression and purification, and finally tested in appropriate correlate-of-protection assays.

The first successful example of this approach was published in *Science* in 2000 by Pizza, Scarlato and co-workers, who showed how, starting from the knowledge of the genome sequence of the bacterial pathogen *Neisseria meningitidis*, vaccine candidates never described before could be identified. Following the *Neisseria* paradigm, officially named 'Reverse Vaccinology', several other examples are appearing in the scientific literature, illustrating the general applicability of this approach to bacterial vaccine discovery.

Therefore, it seemed an appropriate time to publish a book describing in detail what 'Reverse Vaccinology' is all about, the most relevant technologies involved, and some significant successful examples of its application.

The book is organized in three sections. The first introduces the reader to the history of vaccinology and provides information on how vaccines are expected to evolve in the future. The section includes a chapter detailing how genomes, transcriptomes and proteomes can complement each other in the search for new vaccines.

The second section is dedicated to the description of relevant technologies, including genome sequencing and analysis, DNA microarrays, 2D electrophoresis and 2D chromatography, mass spectrometry and high-throughput protein expression and purification. All chapters include many references, which can be utilized by the reader for a deeper comprehension of each technology.

Finally, some relevant examples of the exploitation of genomics and post-genomics in vaccine discovery are presented, including previously unpublished data (as of August 2003). Further chapters in this section offer useful descriptions of, and several references on, the biology and pathogenesis of clinically important bacterial pathogens.

Guido Grandi
Siena, August 10, 2003

Many of the concepts contained in the Preface have been inspired by four books, which have also been used as sources of historical data. I warmly recommend the reading of these books, whose references are given below.

Jared Diamond *Guns, Germs and Steel – The Fate of Human Societies*, W. W. Norton & Co. New York, London

Arno Karlen *Man and Microbes – Disease and Plagues in History of Modern Times*, Simon & Schuster

Sheldon Watts *Epidemics and History. Disease, Power and Imperialism*, Yale University Press, New Haven and London

Norman F. Cantor *In the Wake of Plague – The Black Death and the World it Made*, Perennial.

List of Contributors

Pierre F. Baldi School of Information and Computer Science, Institute for Genomics and Bioinformatics, University of California, Irvine, CA 92697, USA

Giuliano Bensi Chiron Vaccines, Via Fiorentina 1, 53100 Siena, Italy

Pierre-Alain Binz Swiss Institute of Bioinformatics, Proteome Informatics Group, CMU, Michel Servet 1, 1211 Geneva, Switzerland

Svend Birkelund Department of Medical Microbiology and Immunology, Faculty of Health Sciences, University of Aarhus, University Park 240, DK-8000 Aarhus C, Denmark

Gunna Christiansen Department of Medical Microbiology and Immunology, Faculty of Health Sciences, University of Aarhus, University Park 240, DK-8000 Aarhus C, Denmark

Stuart J. Cordwell Australian Proteome Analysis Facility, Sydney, Australia 2109

Tamara Feldblyum The Institute for Genomic Research (TIGR), 9712 Medical Center Drive, Rockville, MD 20850, USA

Oretta Finco Chiron Vaccines, Via Fiorentina 1, 53100 Siena, Italy

Guido Grandi Chiron Vaccines, Via Fiorentina 1, 53100 Siena, Italy

Mahmoud Hamdan Computational, Analytical and Structural Sciences, Discovery Research, GlaxoSmithKline, Via Fleming 4, Verona, Italy

Genomics, Proteomics and Vaccines edited by Guido Grandi
© 2004 John Wiley & Sons, Ltd   ISBN 0 470 85616 5

G. Wesley Hatfield Department of Microbiology and Molecular Genetics, College of Medicine, University of California, Irvine, CA 92697, USA

Tamás Henics Intercell AG, Campus Vienna Biocentre 6, 1030 Vienna, Austria

She-pin Hung Department of Microbiology and Molecular Genetics, College of Medicine, University of California, Irvine, CA 92697, USA

Andreas Kreusch Genomics Institute of the Novartis Research Foundation, 10675 John Jay Hopkins Drive, San Diego, CA 92121, USA

Maria Lattanzi Chiron Vaccines, Via Fiorentina 1, 53100 Siena, Italy

Scott A. Lesley Genomics Institute of the Novartis Research Foundation, 10675 John Jay Hopkins Drive, San Diego, CA 92121, USA

Domenico Maione Chiron Vaccines, Via Fiorentina 1, 53100 Siena, Italy

Vega Masignani Chiron Vaccines, Via Fiorentina 1, 53100 Siena, Italy

Andreas Meinke Intercell AG, Campus Vienna Biocentre 6, 1030 Vienna, Austria

Eszter Nagy Intercell AG, Campus Vienna Biocentre 6, 1030 Vienna, Austria

Amanda S. Nouwens Australian Proteome Analysis Facility, Sydney 2109, Australia

Mariagrazia Pizza Chiron Vaccines, Via Fiorentina 1, 53100 Siena, Italy

Rino Rappuoli Chiron Vaccines, Via Fiorentina 1, 53100 Siena, Italy

Giulio Ratti Chiron Vaccines, Via Fiorentina 1, 53100 Siena, Italy

Frederic Reymond DiagnoSwiss SA, c/o CIMO SA, Route de l'Ile-au-Bois 2, CH-1870 Monthey, Switzerland

Pier Giorgio Righetti Department of Agricultural and Industrial Biotechnologies, University of Verona, Strada Le Grazie No. 15, Verona, Italy

Immaculada Margarit y Ros Chiron Vaccines, Via Fiorentina 1, 53100 Siena, Italy

Joël S. Rossier DiagnoSwiss SA, c/o CIMO SA, Route de l'Ile-au-Bois 2, CH-1870 Monthey, Switzerland

Davide Serruto Chiron Vaccines, Via Fiorentina 1, 53100 Siena, Italy

Allan C. Shaw Department of Medical Microbiology and Immunology, Faculty of Health Sciences, University of Aarhus, University Park 240, DK-8000 Aarhus C, Denmark

Suman Sundaresh School of Information and Computer Science, Institute for Genomics and Bioinformatics, University of California, Irvine, CA 92697, USA

John L. Telford Chiron Vaccines, Via Fiorentina 1, 53100 Siena, Italy

Hervé Tettelin The Institute for Genomic Research (TIGR), 9712 Medical Center Drive, Rockville, MD 20850, USA

Brian B. Vandahl Department of Medical Microbiology and Immunology, Faculty of Health Sciences, University of Aarhus, University Park 240, DK-8000 Aarhus C, Denmark

Alexander von Gabain Intercell AG, Campus Vienna Biocentre 6, 1030 Vienna, Austria

Part 1
Introduction

1

Vaccination: Past, Present and Future

Maria Lattanzi and Rino Rappuoli

1.1 Introduction

Infectious diseases are one of the most terrible enemies mankind has faced during its whole history. They have changed human fate and the course of history and have influenced national economics more than any war.

There are many examples that can be cited to assess the consequences of infectious diseases on the economic development of societies. Paradigmatic is what happened in Siena in 1348, when plague struck with unprecedented cruelty. In the 14th century, Siena, a city located on the way from Rome to France and Northern Europe, had one of the most powerful economies of its time. The wealthy inhabitants planned to build the largest cathedral in history whose construction started in 1338. Unfortunately, after only 11 years, in 1347, the plague (also called the 'Black Death') appeared in Northern Europe and killed 30 per cent of the European population. In May 1348 the epidemic spread to Siena where more than two-thirds of the inhabitants died within three months, including most of the masons and carvers who were building the cathedral. Today, the remainder of the unfinished original construction is still visible and can be considered 'a monument to infectious diseases'.

In 1300, vaccines were not available, and Siena did not have the choice of preventing the plague. Only recently has mankind managed to partially control infectious diseases thanks to the discovery and development of antibiotics and vaccines. The new technologies to develop vaccines available today have the power to achieve even much more spectacular results in the fight against infectious diseases.

Genomics, Proteomics and Vaccines edited by Guido Grandi
© 2004 John Wiley & Sons, Ltd   ISBN 0 470 85616 5

1.2   Vaccination: the past

The observation that people who survive an infectious disease do not relapse into the same disease again is extremely old. The historian Thucydides, describing the Peloponnesian War, reports that during the plague in Athens in 430 B.C. it was a common practice to use those who had recovered from the disease to take care of the sick, because 'the same man was never attacked twice' (Silverstein, 1989). Many authors have reported this observation since then, among them Procopius (541 A.D.), Fracastoro (1483–1553) and Alessandro Manzoni (1785–1873). Describing the terrible plague that killed half of the population in Milan in 1630, Manzoni reported that 'those who had recovered were quite a privileged class' because they could walk without fear anywhere, while the rest of the population were hiding themselves, fearing the constant risk of infection and death (Manzoni, 1972).

Nowadays, naturally acquired immunity to diseases, the basic principle of vaccinology, is common knowledge. However, practices of inducing artificial immunity by deliberate infection of healthy people are also very old. Variolation, the transfer of infected material from a smallpox lesion to healthy people to make them resistant to subsequent exposures to this deadly disease, was used in 590 A.D. in Asia. However, it had probably been practiced a long time before then. During the Middle Ages, the practice was spread to southwest Asia, the Indian subcontinent, North Africa and Turkey. The first half of the 18th century saw an extensive diffusion of variolation in Europe, and especially in the United Kingdom the reports of the miracles of variolation to the English Royal Society by Emanuele Timoni (family physician of the British ambassador in Costantinopolis) and Jacob Pylarini (Venetian Consul in Smyrna). A few years later, Lady Mary Montagou, wife of the British ambassador in Turkey, promoted this practice so efficiently that the sons of the Royal Family were variolated in 1722.

The practice was nevertheless a desperate undertaking, because up to four per cent of those variolated developed a severe, fatal form of the disease. However, this was much lower than the 20–30 per cent fatality rate from natural small pox.

A more successful approach came in 1796 with Edward Jenner, an English physician. During his practice in the countryside, he had noticed that farmers exposed to infected materials from cows did not develop the disease but acquired immunity to smallpox. Jenner decided to use the less dangerous material derived from the bovine (vaccinus) lesions to 'vaccinate' a boy (James Phipps) and showed that he was immune to a subsequent challenge with smallpox. The scientific approach to vaccination came only a century later, when Louis Pasteur introduced the concept that infectious diseases were caused by micro-organisms. Using this empiric approach he developed the first vaccine against rabies, which on 6 June, 1885, was successfully used to inoculate Joseph Meister, an Alsatian boy bitten by a rabid dog (Silverstein, 1989; Ada, 1993). Large-scale vaccination was adopted only following the discovery of a safe and reproducible way to inactivate toxins and pathogens with formaldehyde treatment, performed by Glenny and Hopkins in 1923 and Ramon in 1924, and of the stable attenuation of pathogens by serial passage *in vitro*.

From 1920 to 1980, these simple, basic technologies were used to develop vaccines that controlled many infectious diseases (Table 1.1). The achievements of this period have been remarkable. The introduction of routine mass vaccination is responsible for the eradication of smallpox virus in 1977, and poliomyelitis is expected to be eradicated by 2005. Moreover, the incidence of

Table 1.1 Vaccines introduced in routine immunization programs before the 1980s

Disease	Year
Diphtheria	1939
Tetanus	1949
Pertussis (whole cell)	1953
IPV	1954
Influenza	1958
OPV	1961
Measles	1963
Mumps	1966
Rubella	1969
Meningococcus (capsular polysaccharide)	1969

Table 1.2 Mass vaccination reduced by more than 97 per cent the incidence of nine devastating diseases from the US and eliminated two of them (smallpox and poliomyelitis)

Disease	Max. N of cases (year)	N of cases in 2001	Reduction (%)
Smallpox	48 164 (1901–1904)	0	100
Poliomyelitis	21 269 (1952)	0	100
Diphtheria	206 939 (1921)	2	99.99
Measles	894 134 (1941)	96	99.9
Rubella	57 686 (1969)	19	99.78
Mumps	152 209 (1968)	216	99.86
Pertussis	265 269 (1934)	4788	98.20
H. influenzae type b	20 000 (1992)	242	98.79
Tetanus	1 560 (1923)	26	98.44

(Source: Center for Disease Control. (1998) Impact of vaccines universally recommended for children – United States, 1990–1998. *Morb. Mortal. Wkly Rep.* **48**, 577–581.)

seven other threatening diseases (diphtheria, measles, rubella, mumps, pertussis, *Haemophilus influenzae* type b – Hib – and tetanus) has dropped by more than 97 per cent in those countries where the relative vaccines have been introduced (Table 1.2).

1.3   Vaccination: the present

In the mid-1970s, the technologies necessary for vaccine development had been stretched to the extreme and most of the vaccines that could be produced with the available technologies had been developed.

In the last two or three decades, the emergence of new diseases such as AIDS, hepatitis C, and Lyme disease, the re-emergence of diseases thought to be controlled and the dramatic spreading of resistance to antimicrobial agents by several pathogens (Table 1.3) made people realize that the available technologies for vaccine development had limitations and modern approaches were needed.

Table 1.3   New and emerging pathogens identified in the past 20 years

Period	Agent
1980–1990	HTLV
	HIV
	Staphylococcal toxins
	Escherichia coli O157
	Borrelia burgdorferi
	Helicobacter pylori
	Human herpesvirus type 6
	Ehrlichia chalffeensis
1990–2000	Hepatitis C
	Vibrio cholerae O139
	Bartonella spp.
	Hantavirus
	Nipah virus
	West Nile virus
	TT virus
	SARS coronavirus

Therefore, from the beginning of the 1980s, taking advantage of the better understanding of the immune system and the emergence of the recombinant DNA technologies, new vaccines, most of which were based on specific pathogen components, were developed.

1.3.1 Conjugate vaccines

The observation that mutants of encapsulated bacteria without the capsule are non-pathogenic, since they are highly sensitive to serum complement, led to the use of the capsular polysaccharide as the antigen of choice for the development of subunit vaccines.

Purified capsular polysaccharides have been used to develop vaccines against *Neisseria meningitidis* group A (MenA), group C (MenC) group Y (MenY) and group W135 (MenW135) in the late 1960s (Jodar *et al.*, 2002; Morley and Pollard, 2002), against 23 serotypes of *S. pneumoniae* in the early 1990s (Wuorimaa and Kayhty, 2002), against Hib in the 1980s (Ward and Zangwill 1999) and against *Salmonella typhi* using the Vi polysaccharide (Plotkin and Bouveret-Le Cam, 1995). However, the use of all of these vaccines is unsatisfactory. Capsular polysaccharides are T-independent antigens that induce transient antibody responses (mainly of IgM and IgG2 isotypes) in individuals above 18 years of age. Their immunogenicity and efficacy is usually very poor or absent in infants. Furthermore, polysaccharides do not induce any immuno-logical memory and repeated immunizations are unable to increase antibody titers and, in some cases, provoke tolerance in adults (Granoff *et al.*, 1998).

These drawbacks were solved by covalently linking the sugar to a carrier protein (conjugation). This procedure converts T-independent antigens into T-dependent antigens by providing a source of appropriate T-cell epitopes (present in the carrier protein). This technology has now been successfully applied to develop very efficacious vaccines against several capsulated bacteria (Table 1.4).

Table 1.4   Vaccines licensed after 1980

Disease	Year
Hepatitis B (HBV)	1986
Haemophilus influenzae (conjugate, Hib)	1990
Salmonella typhi (live-attenuated; capsular polysaccharide)	1991
Hepatitis A	1994
Varicella	1995
Acellular pertussis (aP)	1996
Influenza (adjuvanted)	1997
Lyme disease	1998
Meningococcus (conjugate)	2000
Pneumococcus (conjugate)	2000

Hib

The first conjugate vaccine developed was the one against Hib (Ward and Zangwill, 1999). The vaccine was introduced for routine immunization in the United States in December 1987 and afterwards in several countries worldwide; even many developing countries are implementing this vaccination in their national programmes. The Hib conjugate vaccine, used to immunize infants from the age of two months, has contributed to reduce dramatically the incidence of invasive Hib diseases in all those countries that introduced vaccination within a few years (Peltola, 2000). In the US, annual cases of Hib meningitis among children under 5 years of age decreased from more than 10 000 cases to less than 200 cases within a 10 year period after licensure of Hib conjugate vaccines (Bisgard *et al.*, 1998), a decrease that has also been maintained in the following years (Anonymous, 2002). In addition to preventing Hib invasive disease, several studies have shown that conjugate vaccines are effective in reducing nasopharyngeal colonization (Takala *et al.*, 1991; Murphy *et al.*, 1993; Barbour *et al.*, 1995; Forleo-Neto *et al.*, 1999), and therefore that they may confer protection to populations not targeted for immunization through herd immunity (Barbour, 1996). Conversely, reduction of Hib carriage may open ecological niches for *H. influenzae* non-type-b strains and therefore potentially increase the risk of colonization and invasive disease by these strains (Lipsitch, 1997, 1999). However, the risk attributable to serotype replacement is small in comparison with the large reduction in Hib meningitis due to immunization (Anonymous, 2002; Ribeiro *et al.*, 2003).

The recent report of a few cases of vaccine failures in the United Kingdom (Garner and Weston, 2003) highlights the need to maintain surveillance as the use of conjugate vaccines expands worldwide.

S. pneumoniae

The development of pneumococcal conjugate vaccine began with monovalent formulation (Schneerson *et al.*, 1986), followed by a gradual increase in the valency. Multivalent conjugate vaccines consist of a mixture of individual monovalent polysaccharide protein conjugates. Several carrier proteins are utilized, including *H. influenzae* protein D (PD), *N. meningitidis* outer membrane protein complex (OPMC), non-toxic variant of diphtheria toxin 197 (CRM), diphtheria (D) and tetanus (T) toxoid (Wuorimaa and Kayhty, 2002). The first conjugate vaccine licensed for human use, initially in the USA on 17 February 2000, and subsequently in Europe, contains polysaccharides from seven serotypes (4, 6B, 9V, 14, 18C, 19F and 23F) conjugated to the CRM_{197} as carrier protein by reductive amination. This vaccine has been shown to be safe and highly immunogenic in all age groups (Eskola, 2000) and to prime memory responses in early life (Ahman *et al.*, 1998). Significantly, it exhibited 57 per cent efficacy in preventing acute otitis media due to the *S. pneumoniae* strains covered by the vaccine (although protection varied widely among serotypes) (Eskola *et al.*, 2001), and 73 per cent efficacy in preventing clinically diagnosed and radiologically confirmed pneumonia in infants (Black *et al.*, 2000).

N. meningitidis

At the end of the 1980s a conjugate vaccine against meningococcus C began its development and, after many clinical trials also in children and toddlers, it was shown to be safe, immunogenic and able to induce immunological memory (Costantino *et al.*, 1992; Anderson *et al.*, 1994; Twumasi *et al.*, 1995; Lieberman *et al.*, 1996; Fairley *et al.*, 1996; Leach *et al.*, 1997; Borrow *et al.*, 2000).

Conjugated vaccines against serogroup C meningococcus were developed by several manufacturers and licensed (between October 1999 and August 2000) in the UK, where they have been used in a countrywide vaccination initiative. From November 1999 onward, all the population between 2 months and 18 years of age was vaccinated against meningococcus C. The results were impressive: only a few months after the beginning of the vaccination campaign the cases of serogroup C meningococcus meningitis practically disappeared among the vaccinated population. The efficacy has been calculated to be 91.5 per cent in infants receiving three doses of vaccine and 89.3 per cent in toddlers receiving one dose. Moreover, the vaccine provides some evidence of herd immunity, as there is also a reduction of meningococcus C disease in unvaccinated individuals, ranging from 34 per cent in 9–14 year olds to 61 per cent in 15–17 year olds (Balmer, Borrow and Miller, 2002).

The good results obtained in the UK and the increase in disease incidence in the year 2000 were determining factors in the decision taken by the Spanish health authorities to include this vaccine (licensed in Spain in August 2000) in the routine vaccination schedule at 2, 4 and 6 months of age, and to carry out a nationwide mass vaccination campaign during October–December 2000 in children aged between 2 months and 6 years (Ministerio de Sanidad, 2000). The effectiveness of the vaccination in this age group during 2001, and the first 28 weeks of 2002 was 100 per cent (94.27–100 per cent) (Salleras, Dom'ýnguez and Cardiñosa, 2003a, 2003b).

In conclusion, the conjugate vaccine against serogroup C meningococcus has already shown to be very effective in preventing disease. The success with serogroup C meningococcus suggests that similar results will be obtained with conjugates of serogroups A, Y and W135, which are now being tested in several preclinical and clinical studies. The proof of principle obtained with the conjugate vaccine against meningococcus C strongly suggests that within a few years licensed vaccines against four out of five life-threatening meningococcal serogroups will be available (Rappuoli, 2001a).

1.3.2   Recombinant DNA technology for subunit vaccines

According to conventional approaches for vaccine development, the pathogen is first studied to identify the factors capable of eliciting protective immunity, and the identified factors are then purified from large-scale cultures of the pathogen. This approach has severe limitations not only because of safety issues, but also because several pathogens are very difficult, sometimes impossible, to grow at industrial level.

The advent of recombinant DNA technologies has greatly facilitated the development of new vaccines. Once the protective antigens of the pathogen are identified, the corresponding coding genes can be cloned in other organisms, which become the vaccine factories.

This approach has generated two very efficacious recombinant vaccines: the hepatitis B vaccine and the acellular vaccine against *Bordetella pertussis*.

HBV vaccine

The first recombinant subunit vaccine developed was the one against hepatitis B (Valenzuela *et al.*, 1982). It was already known that protection from the disease was based on the presence of antibodies against the surface antigen of the virus (the HBsAg). The development of the vaccine was largely successful because the HBsAg circulated in the bloodstream in large quantities (Mahoney and Kane, 1999). The vaccine, produced in yeast and mammalian cells, is highly effective: it induces seroprotective levels of antibodies (i.e. >10 mIU/ml) in more than 95 per cent of subjects and prevents the development of chronic hepatitis in at least 75 per cent of vaccinated infants born from HBeAg-positive mothers (Poovorawan *et al.*, 1992). On the basis of disease burden and the safety and efficacy of the vaccine, the WHO recommended that by the end of the 20th century HBV vaccine be integrated into routine infant and child-hood immunization programs for all countries (Kao and Chen, 2002). By June 2001 129 countries included hepatitis B vaccine in their routine immunization programmes, and by the end of 2002 41 out of the 51 countries of the WHO European Region had implemented universal hepatitis B immunization (Van

Damme and Vorsters, 2002). The efficacy of universal immunization has been shown in different countries, with striking reductions of the prevalence of HBV carriage in children. Most importantly, the HBV vaccine can be considered the first successful anti-cancer vaccine, as 20 years of mass vaccination has clearly reduced the incidence of hepatocellular carcinoma in children, at least in Taiwan (Kao and Chen, 2002).

Acellular vaccine against B. pertussis

A vaccine composed of the whole, inactivated *B. pertussis* cells has been available for mass vaccination since the late 1940s (Kendrick *et al.*, 1947). This vaccine was very efficacious in preventing the disease, but the presence of severe adverse reactions, although never proved to be a consequence of vaccination, caused a drop in vaccine uptake in the 1970s, and stressed the need for a new and safer vaccine. As pertussis toxin represents one of the major virulence factors of *B. pertussis*, several researchers developed acellular pertussis vaccines containing chemically inactivated, purified PT. However, it is recognized that chemical treatment of PT with formaldehyde and glutaralde-hyde is associated with significant reversion rates (National Bacteriological Laboratory, Sweden, 1988). To overcome these problems, Pizza and co-workers applied genetic engineering to develop a mutant PT, which had all the antigenic properties but lacked the toxic effects: this genetically detoxified pertussis toxin has been the first recombinant bacterial subunit vaccine developed on the market (Pizza *et al.*, 1989).

The safety and immunogenicity of genetically inactivated PT was tested in clinical trials, both in adult volunteers and in infants and children, as a monovalent mutant PT alone (Podda *et al.*, 1990, 1992), in association with FHA and pertactin (Podda *et al.*, 1991; Podda 1993), and also with FHA and pertactin in association with diphtheria and tetanus toxoids (DTaP) (Italian Multicentre Group, 1994). These trials showed that the different acellular vaccine formulations containing the non-toxic PT mutant were extremely safe, and much safer than whole-cell pertussis vaccines. Furthermore, all formulations induced high titres of anti-PT neutralizing antibodies and very strong antigen-specific T-cell proliferative responses. Interestingly enough, five to six years after the primary immunization schedule, this vaccine still exhibited an efficacy of about 80 per cent, and both antigen-specific antibody and CD4+ T-cell responses were still detectable at significant levels (Di Tommaso *et al.*, 1997; Salmaso *et al.*, 2001).

1.4   Vaccination: the future

1.4.1   'Reverse vaccinology'

The genomic era allows scientists to change the paradigm to vaccine discovery. A new genomic approach, now called 'reverse vaccinology', has been applied for the first time to tackle *N. meningitidis* group B (MenB) vaccine development.

The most obvious approach to the development of a vaccine against MenB would be to exploit its capsular polysaccharide. Unfortunately, a polysaccharide-based approach cannot be used for this strain, because the MenB capsular polysaccharide is identical to a widely distributed human carbohydrate (α(2-8)N-acetyl neuraminic acid or polysialic acid), which is a self-antigen and therefore is a poor immunogen in humans. Furthermore, use of this polysaccharide in a vaccine may elicit autoantibodies (Finne *et al.*, 1987; Hayrinen, 1995). An alternative approach to vaccine development is based on surface-exposed proteins contained in outer membrane vesicles (OMVs). These vaccines have been shown both to elicit serum bactericidal antibody responses and to protect against developing meningococcal disease in clinical trials (Tappero *et al.*, 1999). However, the applicability of these vaccines is limited by the sequence variability of major protein antigens, which therefore are able to elicit an immune response only against homologous strains, especially below the age of five (Poolman, 1995).

To circumvent these problems, researchers at Chiron had developed a genome-based approach which is described in detail in other chapters of this book.

Very briefly, Chiron's scientists used the geneome sequence determined in collaboration with the Institute for Genomic research (TIGR, Washington, DC) to predict 600 putative surface antigens. Three hundred and fifty of these were expressed in *Escherichia coli*, purified and analysed for their capacity of eliciting bactericidal antibodies in the mouse model. Some of the antigens were very bactericidal against most of the clinical isolates so far tested and are currently under evaluation in humans.

Following the success of MenB vaccine discovery, other vaccines are currently under investigation using reverse vaccinology including vaccines against *S. pneumoniae* (Wizemann *et al.*, 2001), *Streptococcus agalactiae* (Tettelin *et al.*, 2002), *Staphylococcus aureus* (Etz *et al.*, 2002), *Porphyromonas gingivalis* (Ross *et al.*, 2001), *Chlamydia pneumoniae*, (Montigiani *et al.*, 2002) and the parasite *Plasmodium falciparum* (Smooker *et al.*, 2000).

Very recently, the concept of reverse vaccinology has been applied to viruses as well, and all encoded proteins are being considered as potential antigens.

Promising results with HIV early proteins such as Tat, Rev, Pol, etc. show that this approach may provide concrete new weapons in the fight against AIDS (Osterhaus *et al.*, 1999; Cafaro *et al.*, 2000; Pauza *et al.*, 2000).

1.4.2 Improved delivery: mucosal vaccines

With a few exceptions, most vaccines are currently delivered by intramuscular injection. Development of mucosally delivered vaccines has many advantages: not only for the increased compliance among recipients, but also because it should stimulate an immune response at the mucosal sites, the main entrance of most pathogens, responses that are usually not involved by systemic vaccines. The mucosal surface is enormous: the human body has a total mucosal surface of 400 m^2, an area equivalent to that of a tennis court. Mucosal surfaces represent a critical component of the mammalian immunologic repertoire. Most external mucosal surfaces are replete with organized follicles and scattered antigen-reactive or sensitized lymphoid elements, including B cells, T lympho-cytes, T-cell subsets, plasma cells, and a variety of other cellular elements involved in the induction and maintenance of immune response. The major antibody isotype in external secretions is secretory immunoglobulin A (SIgA). It is, however, interesting that the major effector cells in mucosal surfaces are not IgA B cells, but T lymphocytes of CD4+ as well as CD8+ phenotypes. It is estimated that T lymphocytes may represent up to 80 per cent of the entire mucosal lymphoid cell population (Ogra, Faden and Welliver, 2001).

In order to stimulate immune response at the mucosal site, protective antigens need to be presented with special adjuvants. The heat-labile toxin of *E. coli* (LT) and the cholera toxin (CT) are the strongest mucosal adjuvants known so far. These toxins consist of two subunits: the A subunit that contains the enzymatic activity, and the B subunit that binds to the GM$_1$ ganglioside and to other glycolipids. After internalization into the cell, the A subunit of the toxin is proteolytically cleaved, binds to NAD and transfers the ADP-ribose group to the α subunit of Gs, a GTP-binding protein, which regulates the activity of adenylate cyclase (Holmes, 1997). This enzyme becomes permanently activated and causes accumulation of cAMP and secretion of electrolytes and water, which is the pathogenesis of the watery diarrhoea caused by these pathogens in infected people (Levine *et al.*, 1983; Field, Rao and Chang, 1989).

Site-directed mutagenesis has been widely used to produce non-toxic mutants while still maintaining their strong mucosal adjuvancy: with this approach more than 50 different mutants have been generated (Tsuji *et al.*, 1990; Burnette *et al.*, 1991; Lobet, Cluff and Cieplak, 1991; Tsuji *et al.*, 1991; Pizza *et al.*, 1994; Dickinson and Clements, 1995; Douce *et al.*, 1995; Fontana *et al.*,

1995; De Haan *et al.*, 1996; Douce *et al.*, 1997; Yamamoto *et al.*, 1997; Douce *et al.*, 1998; Giuliani *et al.*, 1998). Among them, the most promising non-toxic derivatives are LTK63 (containing a serine-to-lysine substitution in position 63 of the A subunit) that has totally lost its enzymatic activity and its toxic properties even at very high doses, and LTK72 (with alanine-to-arginine substitution in position 72 of the A subunit), which retains some residual enzymatic and toxic activity, although at several orders of magnitude lower than those observed with wild-type toxin (Giuliani *et al.*, 1998). Both molecules are able to elicit a mucosal immune response to co-administered synthetic peptides and bacterial and viral antigens in many animal models such as mice, rabbits, guinea pigs and minipigs (Pizza *et al.*, 2001). The mucosal adjuvanticity of these mutants was potentiated when the vaccine and the adjuvant were delivered intranasally together with nanoparticles favouring the uptake of the vaccine at the mucosal level (Baudner *et al.*, 2002). Immunological studies showed that the mutants prime CD4+ T cells and also CD8+ CTL specific for the co-administered antigen, regardless of the administration route (Simmons *et al.*, 1999). Interestingly, it was shown that LTK63, the mutant devoid of enzymatic activity, enhanced Th1 and Th2 responses to co-administered antigens. In contrast, LTR72, the partially detoxified mutant, enhanced Th2 response, but suppressed Th1 response. LTK63 enhanced IL-12 and TNF-a production and NFkB translocation, whereas LT or LTR72 failed to activate NFkB but stimulated cAMP production (Ryan *et al.*, 2000).

The clinical experience with these molecules is still limited. However, the LTK63 mutant is now being tested as an intranasal adjuvant in a phase I clinical trial together with a subunit influenza vaccine.

1.5 Conclusion: the intangible value of vaccination

Today, we have the technology to make vaccines against most infectious diseases and in theory we could free mankind from most of them. However, despite the enormous technological progress, today vaccines are not easier to develop because of the increase in GMP and clinical trials standards that have soared investments in vaccine development to the highest levels ever seen. Nowadays, licensing a new vaccine requires an investment of up to 850 million dollars (EFPIA, 2002). Moreover, the vaccine business has a relatively low profitability and therefore industrial interest in vaccines is limited.

Worldwide, vaccine sales are estimated to be approximately US$ 6.5 billion, representing only two per cent of the global pharmaceutical market, roughly equalling the sales of one successful drug (Gréco, 2002).

Indeed, the economics of vaccines have become the major obstacle to vaccine development.

Today, the economic value of vaccines is calculated by cost-effectiveness studies. These studies have clearly shown that the costs for prophylactic treatments are substantially lower than the cumulative costs of therapeutic approaches which include costs for drugs, hospitalization, loss of working days etc. For instance, it has been calculated that for every single US$ spent on mumps–measles–rubella (MMR) vaccine, more than US$ 21 are saved in direct medical care cost, while the diphtheria, tetanus and acellular pertussis (DTaP) vaccine saves US$ 24 for every single US$ spent in vaccination costs (Ehreth, 2003).

Although the economical benefit of vaccination as calculated by the current methodologies is already enormous and would largely justify the expansion of vaccination practice, the question is whether economists are using the right approach to calculate the economical value of prevention. Indeed the benefits of vaccines go far beyond the cost saved to treat the diseases. What is the value of being alive? What is the value of being healthy? What is the value of the lost opportunity for economic growth? If the above intangible values were included in a cost–benefit analysis, the present cost–benefit ratio would grow by a factor between 10 and 100. Such calculation provides the economical and social rationale to convince industry and national and international authorities to invest in vaccine research and development (Rappuoli, Miller and Falkow, 2002; Masignani, Lattanzi and Rappuoli, 2003).

References

Ada, G. L. (1993) Vaccines. In Paul, W. E. (Ed.). *Fundamental Immunology* 3rd edn. Raven, New York, 1309–1352.

Ahman, H. *et al.* (1998) *Streptococcus pneumoniae* capsular polysaccharide–diphtheria toxoid conjugate vaccine is immunogenic in early infancy and able to induce immunologic memory. *Pediatr. Infect. Dis. J.* **17**, 211–216.

Anderson, E. L., Bowers, T., Mink, C. M., Kennedy, D. J., Belshe, R. B., Harakeh, H., Pais, L., Holder, P., and Carlone, G. M. (1994) Safety and immunogenicity of meningococcal A and C polysaccharide conjugate vaccine in adults. *Infect. Immun.* **62**(8) 3391–3395.

Anonymous. (2002) Progress toward elimination of *Haemophilus influenzae* type b invasive disease among infants and children – United States, 1998–2000. *MMWR* **51**, 234–237.

Balmer, P., Borrow, R., and Miller, E. (2002) Impact of meningococcal C conjugate vaccine in the UK. *J. Med. Microbiol.* **51**, 717–722.

Barbour, M. L. (1996) Conjugate vaccines and the carriage of *Haemophilus influenzae* type b. *Emerg. Infect. Dis.* **2**, 176–182.

Barbour, M. L., Mayon-White, R. T., Coles, C., Crook, D. W., and Moxon, E. R. (1995) The impact of conjugate vaccine on carriage of *Haemophilus influenzae* type b. *J. Infect. Dis.* **171**, 93–98.

Baudner, B. C., Balland, O., Giuliani, M. M. *et al.* (2002) Enhancement of protective efficacy following intranasal immunization with vaccine plus a nontoxic LTK63 mutant delivered with nanoparticles. *Infect. Immun.* **70**, 4785–4790.

Bisgard, K. M., Kao, A., Leake, J., Strebel, P. M., Perkins, B. A., and Wharton, M. (1998) *Haemophilus influenzae* invasive disease in the United States, 1994–1995: near disappearance of a vaccine-preventable childhood disease. *Emerg. Infect. Dis.* **4**, 229–237.

Black, S., Shinefield, H., Fireman, B. *et al.* (2000) Efficacy, safety, and immunogenicity of heptavalent pneumococcal conjugate vaccine in children. Northern California Kaiser Permanente Vaccine Study Center Group. *Pediatr. Infect. Dis. J.* **19**, 187–195.

Borrow, R., Fox, A. J., Richmond, P. C., Clark, S., Sadler, F., Findlow, J., Moris, R., Begg, N. T., and Cartwright, K. A. (2000) Induction of immunological memory in UK infants by a meningococcal AC conjugate vaccine. *Epidemiol. Infect.* **124**(3), 427–432.

Burnette, W. N., Mar, V. L., Platter, B. W., Schlotterbeck, J. D., McGinley, M. D., Stoney, K. S., Rodhe, M. F., and Kaslow, H. R. (1991) Site-specific mutagenesis of the catalytic subunit of cholera toxin: substituting lysine for arginine 7 causes loss of activity. *Infect. Immun.* **59**, 4266–4270.

Cafaro, A., Caputo, A., Maggiorella, M. T. *et al.* (2000) SHIV89.6P pathogenicity in cynomolgus monkeys and control of viral replication and disease onset by human immunodeficiency virus type 1 Tat vaccine. *J. Med. Primatol.* **29**(3/4), 193–208.

Costantino, P., Viti, S., Podda, A., Velmonte, M. A., Nencioni, L., and Rappuoli, R. (1992) Development and phase I clinical testing of a conjugate vaccine against meningococcus A and C. *Vaccine* **10**(10), 691–698.

De Haan, L., Verweij, W. R., Feil, I. K., Lijnema, T. H., Hol, W. G., Agsteribbe, E., and Wilschut, J. (1996) Mutants of the *Escherichia coli* heat labile enterotoxin with reduced ADP-ribosylation activity or no activity retain the immunogenic properties of the native holotoxin. *Infect. Immun.* **64**, 5413–5416.

Di Tommaso, A., Bartalini, M., Peppoloni, S., Podda, A., Rappuoli, R., and de Magistris, M. T. (1997) Acellular pertussis vaccines containing genetically detoxified pertussis toxin induce long-lasting humoral and cellular responses in adults. *Vaccine* **15**, 1218–1224.

Dickinson, B. L., and Clements, J. D. (1995) Dissociation of *Escherichia coli* heat-labile enterotoxin adjuvanticity from ADP-ribosyltransferase activity. *Infect. Immun.* **63**, 1617–1623.

Douce, G., Fontana, M. R., Pizza, M., Rappuoli, R., and Dougan, G. (1997) Intranasal immunogenicity and adjuvanticity of site-directed mutant derivatives of cholera toxin. *Infect. Immun.* **65**, 2821–2828.

Douce, G., Giuliani, M. M., Giannelli, V., Pizza, M., Rappuoli, R., and Dougan, G. (1998) Mucosal immunogenicity of genetically detoxified derivatives of heat labile toxin from *Escherichia coli*. *Vaccine* **16**(11/12), 1065–1073.

Douce, G., Turcotte, C., Cropley, I., Roberts, M., Pizza, M., Domenghini, M., Rappuoli, R., and Dougan, G. (1995) Mutants of *Escherichia coli* heat-labile toxin lacking ADP-ribosyltransferase activity act as nontoxic, mucosal adjuvants. *Proc. Natl. Acad. Sci. USA* **92**, 1644–1648.

Ehreth, J. (2003) The global value of vaccination. *Vaccine* **21**, 596–600.

Eskola, J. (2000) Immunogenicity of pneumococcal conjugate vaccines. *Pediatr. Infect. Dis. J.* **19**, 388–393.

Eskola, J., Kilpi, T., Palmu, A. *et al.* (2001) Efficacy of a pneumococcal conjugate vaccine against acute otitis media. *N. Engl. J. Med.* **344**, 403–409.

Etz, H., Minh, D. B., Henics, T. *et al.* (2002) Identification of in vivo expressed vaccine candidate antigens from *Staphylococcus aureus*. *Proc. Natl. Acad. Sci. USA* **99**(10), 6573–6578.

European Federation of Pharmaceutical Industries and Associations (EFPIA). (2002) 2001–2002, the Year in Review.

Fairley, C. K., Begg, N., Borrow, R., Fox, A. J., Jones, D. M., and Cartwright, K. (1996) Conjugate meningococcal serogroup A and C vaccine: reactogenicity and immunogenicity in United Kingdom infants. *J. Infect. Dis.* **174**(6), 1360–1363.

Field, M., Rao, M. C., and Chang, E. B. (1989) Intestinal electrolyte transport and diarrhoeal disease. 2. *N. Engl. J. Med.* **321**, 800–806.

Finne, J., Bitter-Suermann, D., Goridis, C., and Finne, U. (1987) An IgG monoclonal antibody to group B meningococci cross-reacts with developmentally regulated polysialic acid units of glycoproteins in neural and extraneural tissues. *J. Immunol.* **138**(12), 4402–4407.

Fontana, M. R., Manetti, R., Giannelli, V., Magagnoli, C., Marchini, A., Domenighini, M., Rappuoli, R., and Pizza, M. (1995) Construction of nontoxic derivatives of cholera toxin and characterization of the immunological response against the A subunit. *Infect. Immun.* **63**, 2356–2360.

Forleo-Neto, E., de Oliveira, C. F., Maluf, E. M. *et al.* (1999) Decreased point prevalence of *Haemophilus influenzae* type b (Hib) oropharyngeal colonization by mass immunization of Brazilian children less than 5 years old with Hib polyribosyl-ribitol phosphate polysaccharide–tetanus toxoid conjugate vaccine in combination with diphtheria–tetanus toxoids–pertussis vaccine. *J. Infect. Dis.* **180**, 1153–1158.

Garner, D., and Weston, V. (2003) Effectiveness of vaccination for *Haemophilus influenzae* type b. *Lancet* **361**, 395–396

Giuliani, M. M., Del Giudice, G., Giannelli, V., Douce, G., Dougan, G., Rappuoli, R., and Pizza, M. (1998) Mucosal adjuvanticity and immunogenicity of LTR72, a novel mutant of *Escherichia coli* heat-labile enterotoxin with partial knockout of ADP-ribosyltransferase activity. *J. Exp. Med.* **187**, 1123–1132.

Glenny, A. T., and Hopkins, B. E. (1923) Diphtheria toxoid as an immunizing agent. *Br. J. Exp. Pathol.* **4**, 283–288.

Granoff, D. M., Gupta, R. K., Belshe, R. B., and Anderson, E. L. (1998) Induction of immunologic refractoriness in adults by meningococcal C polysaccharide vaccination. *J. Infect. Dis.* **178**, 870–874.

Gréco, M. (2002) The future of vaccines: an industrial perspective. *Vaccine* **20**, S101–S103.

Hayrinen, J., Jennings, H., Raff, H. V., Rougon, G., Hanai, N., Gerardy-Schahn, R., and and Finne, J. (1995) Antibodies to polysialic acid and its N-propyl derivative: binding properties and interaction with human embryonal brain glycopeptides. *J. Infect. Dis.* **171**(6), 1481–1490.

Holmes, R. K. (1997) Heat-labile enterotoxins (*Escherichia coli*). In Rappuoli, R., Montecucco, C. (Eds.). *Guidebook to Protein Toxins and Their Use in Cell Biology.* Oxford University Press, Oxford, 30–33.

Italian Multicentre Group for the study of Recombinant Acellular Pertussis Vaccine, Podda, A., Carapella de Luca, E., Contu, B., Furlan, R., Maida, A., Moiraghi, A., Stramare, D., Titone, L., Uxa, F., Di Pisa, F., Peppoloni, S., Nencioni, L., and Rappuoli, R. (1994) Comparative study of a whole-cell pertussis vaccine and a recombinant acellular pertussis vaccine. *J. Pediatr.* **124**, 921–926.

Jodar, L., Feavers, I. M., Salisbury, D., and Granoff, D. M. (2002) Development of vaccines against meningococcal disease. *Lancet* **359**, 1499–1508.

Kao, J. H., and Chen, D. S. (2002) Global control of hepatitis B virus infection. *Lancet Infect. Dis.* **2**(7), 395-403.

Kendrick, P. L., Eldering, G., and Dixon, M. K. (1947) Mouse protection tests in the study of pertussis vaccine: a comparative study using intracerebral route for challenge. *Am. J. Publ. Health* **37**, 803.

Leach, A., Twumasi, P. A., Kumah, S., Banya, W. S., Jaffar, S., Forrest, B. D., Granoff, D. M., LiButti, D. E., Carlone, G. M., Pais, L. B., Broome, C. V., and Greenwood, B. M. (1997) Induction of immunologic memory in Gambian children by vaccination in infancy with a group A plus group C meningococcal polysaccharide–protein conjugate vaccine. *J. Infect. Dis.* **175**(1), 200–204.

Levine, M. M., Kaper, J. B., Black, R. E., and Clements, M. L. (1983) New knowledge on pathogenesis of bacterial enteric infections as applied to vaccine development. *Microbiol. Rev.* **47**, 510–550.

Lieberman, J. M., Chiu, S. S., Wong, V. K., Partidge, S., Chang, S. J., Chiu, C. Y., Cheesling, L. L., Carlone, G. M., and Ward, J. I. Safety and immunogenicity of a serogroups A:C Neisseria meningitidis oligosaccharide–protein conjugate vaccine in young children. A randomized controlled trial. *JAMA* **275**(19), 1499–1503.

Lipsitch, M. (1997) Vaccination against colonizing bacteria with multiple serotypes. *Proc. Natl. Acad. Sci. USA* **94**, 6571–6576.

Lipsitch, M. (1999) Bacterial vaccines and serotype replacement: lessons from *Haemophilus influenzae* and prospects for *Streptococcus pneumoniae. Emerg. Infect. Dis.* **5**, 336–345.

Lobet, Y., Cluff, C. W., and Cieplak, W. Jr. (1991) Effect of site-directed mutagenic alterations on ADP-ribosyltransferase activity of the A subunit of Escherichia coli heat-labile enterotoxin. *Infect. Immun.* **59**, 2870–2879.

Mahoney, F. J., and Kane, M. (1999) Hepatitis B vaccine. In Plotkin, S. A., and Orenstein, W. A. (Eds.). Vaccines 3rd edn. Saunders, Philadelphia, PA, 158–182.

Manzoni, A. (1972) *The Betrothed*. Transl. Penman, B. London: Penguin.

Masignani, V., Lattanzi, M., and Rappuoli, R. (2003) The value of vaccines. Vaccine 21 S2/110–S2/113.

Ministerio de Sanidad y Consumo. (2000) Calendario de vacunación 2001. *Biol. Epidemiol. Sem.* **8**, 265–76.

Montigiani, S., Falugi, F., Scarselli, M. *et al.* (2002) Genomic approach for analysis of surface proteins in *Chlamydia pneumoniae*. *Infect. Immun.* **70**(1), 368–379.

Morley, S. L., and Pollard, A. J. (2002) Vaccine prevention of meningococcal disease, coming soon? *Vaccine* **20**, 666–687.

Murphy, T. V., Pastor, P., Medley, F., Osterholm, M. T., and Granoff, D. M. (1993) Decreased *Haemophilus* colonization in children vaccinated with *Haemophilus influenzae* type b conjugate vaccine. *J. Pediatr.* **122**, 517–523.

National Bacteriological Laboratory, Sweden (1988) A clinical trial of acellular pertussis vaccines in Sweden. *Technical Report*. Stockholm.

Ogra, P. L., Faden, H., and Welliver, R. C. (2001) Vaccination strategies for mucosal immune responses. *Clin. Microbiol. Rev.* **14**(2), 430–445.

Osterhaus, A. D., van Baalen, C. A., Gruters, R. A. *et al.* (1999) Vaccination with Rev and Tat against AIDS. *Vaccine* **17**(20/21), 2713–2714.

Pauza, C. D., Trivedi, P., Wallace, M. *et al.* (2000) Vaccination with Tat toxoid attenuates disease in simian/HIV-challenged macaques. *Proc. Natl. Acad. Sci. USA* **97**(7), 3515.

Peltola, H. (2000) Worldwide *Haemophilus influenzae* type b disease at the beginning of the 21st century: global analysis of the disease burden 25 years after the use of the polysaccharide vaccine and a decade after the advant of conjugates. *Clin. Microbiol. Rev.* **13**, 302–317.

Pizza, M., Covacci, A., Bartoloni, A., Perugini, M., Nencioni, L., de Magistris, M. T., Villa, L., Nucci, D., Manetti, R., Bugnoli, M., Giovannoni, F., Olivieri, R., Barbieri, J. T., Sato, H., and Rappuoli, R. (1989) Mutants of pertussis toxin suitable for vaccine development. *Science* **246**, 497–500.

Pizza, M., Fontana, M. R., Giuliani, M. M., Domenighini, M., Magagnali, C., Giannelli, V., Nucci, D., Hol, W., Manetti, R., and Rappuoli, R. (1994) A genetically detoxified derivative of heat-labile *Escherichia coli* enterotoxin induces neutralizing antibodies against the A subunit. *J. Exp. Med.* **6**, 2147–2153.

Pizza, M., Giuliani, M. M., Fontana, M. R. *et al.* (2001) Mucosal vaccines: non toxic derivatives of LT and CT as mucosal adjuvants. *Vaccine* **19**, 2534–2541.

Pizza, M., Scarlato, V., Masignani, V., Giuliani, M. M., Aricò, B., Comanducci, M., Jennings, G. T., Baldi, L., Bartolini, E., Capecchi, B., Galeotti, C. L., Luzzi, E., Manetti, R., Marchetti, E., Mora, M., Nuti, S., Ratti, G., Santini, L., Savino, S., Scarselli, M., Storni, E., Zuo, P., Broeker, M., Hundt, E., Knapp, B., Blair, E., Mason, T., Tettelin, H., Hood, D. W., Jeffries, A. C., Saunders, N. J., Granoff, D. M.,

Venter, J. C., Moxon, E. R., Grandi, G., and Rappuoli, R. (2000) Identification of vaccine candidates against serogroup B meningococcus by whole-genome sequencing. *Science* **287**, 1816–1820.

Plotkin, S. A., and Bouveret-Le Cam, N. (1995) A new typhoid vaccine composed of the Vi capsular polysaccharide. *Arch. Intern. Med.* **155**, 2293–2299.

Podda, A., Carapella de Luca, E., Titone, L., Casadei, A. M., Cascio, A., Bartalini, M., Volpini, G., Peppoloni, S., Marsili, I., Nencioni, L., and Rappuoli, R. (1993) Immunogenicity of an acellular pertussis vaccine composed of genetically inactivated pertussis toxin combined with filamentous hemagglutinin and pertactin in infants and children. *J. Pediatr.* **123**, 81–84.

Podda, A., Carapella de Luca, E., Titone, L., Casadei, A. M., Cascio, A., Peppoloni, S., Volpini, G., Marsili, I., Nencioni, L., and Rappuoli, R. (1992) Acellular pertussis vaccine composed of genetically inactivated pertussis toxin: safety and immunogenicity in 12- to 24- and 2- to 4-month-old children. *J. Pediatr.* **120**, 680–685.

Podda, A., Nencioni, L., de Magistris, M. T., Di Tommaso, A., Bossu, P., Nuti, S., Pileri, P., Peppoloni, S., Bugnoli, M., Ruggiero, P., Marsili, I., D'Errico, A., Tagliabue, A., and Rappuoli, R. (1990) Metabolic, humoral, and cellular responses in adult volunteers immunized with the genetically inactivated pertussis toxin mutant PT-9K/129G. *J. Exp. Med.* **172**, 861–868.

Podda, A., Nencioni, L., Marsili, I., Peppoloni, S., Volpini, G., Donati, D., Di Tommaso, A., de Magistris, M. T., and Rappuoli, R. (1991) Phase I clinical trial of an acellular pertussis vaccine composed of genetically detoxified pertussis toxin combined with FHA and 69 kDa. *Vaccine* **9**, 741–745.

Poolman, J. T. (1995) Development of a meningococcal vaccine. *Infect. Agents Dis.* **1**, 13–28.

Poovorawan, Y., Sanpavat, S., Pongpunlert *et al.* (1992) Long term efficacy of hepatitis B vaccine in infants born to hepatitis B e antigen-positive mothers. *Pediatr. Infect. Dis. J.* **11**, 816–821.

Ramon, G. (1924) Sur la toxine et sur l'anatoxine diphtheriques. *Ann. Inst. Pasteur* **38**, 1–10.

Rappuoli, R. (2001a) Conjugates and reverse vaccinology to eliminate bacterial meningitis. *Vaccine* **19**, 2319–2322.

Rappuoli, R. (2001b) Reverse vaccinology, a genome-based approach to vaccine development. *Vaccine* **19**, 2688–2691.

Rappuoli, R., Miller, H. I., and Falkow, S. (2002) The intangible value of vaccination. *Science* **297**, 937–939.

Ribeiro, G. S., Reis, J. N., Cordeiro, S. M., Lima, J. B. T., Gouveia, E. L., Petersen, M., Salgado, K., Silva, H. R., Zanella, R. C., Almeida, S. C. G., Brandileone, M. C., Reis, M. G., and Ko, A. I. (2003) Prevention of *Haemophilus influenzae* type b (Hib) meningitis and emergence of serotype replacement with type a strains after introduction of Hib immunization in Brazil. *J. Inf. Dis.* **187**, 109–116.

Ross, B. C., Czajkowski, L., Hocking, D. *et al.* (2001) Identification of vaccine candidate antigens from a genomic analysis of *Porphyromonas gingivalis*. *Vaccine* **19**(30), 4135–4142.

Ryan, E. J., McNeela, E., Pizza, M., Rappuoli, R., O'Neill, L., and Mills, K. H. (2000) Modulation of innate and acquired immune responses by *Escherichia coli* heat-labile toxin: distinct pro- and anti-inflammatory effects of the nontoxic AB complex and the enzyme activity. *J. Immunol.* **165**(10), 5750–5759.

Salleras, L., Dom'ýnguez, A., and Cardeñosa, N. (2003a) Dramatic decline of serogroup C meningococcal disease in Catalonia (Spain) after a mass vaccination campaign with meningococcal C conjugated vaccine. *Vaccine* **21**, 729–733.

Salleras, L., Dom'ýnguez, A., and Cardeñosa, N. (2003b) Impact of mass vaccination with polysaccharide conjugate vaccine against serogroup C meningococcal disease in Spain. *Vaccine* **21**, 725–728.

Salmaso, S., Mastrantonio, P., Tozzi, A. E. *et al.* (2001) Sustained efficacy during the first 6 years of life of 3-component acellular pertussis vaccines administered in infancy: the Italian experience. *Pediatrics* **108**, E81.

Schaible, U. E., and Kaufmann, S. H. E. (2000) CD1 molecules and CD1-dependent T cells in bacterial infections: a link from innate to acquired immunity? *Semin. Immunol.* **12**, 527–535.

Schneerson, R., Robbins, J., Parke, J. Jr. *et al.* (1986) Quantitative and qualitative analyses of serum antibodies elicited in adults by *Haemophilus influenzae* type b and pneumococcus type 6A capsular polysaccharide–tetanus toxoid conjugates. *Infect. Immun.* **52**, 519–528.

Silverstein, A. M. (1989) *A History of Immunology*. San Diego: Academic Press.

Simmons, C. P., Mastroeni, P., Fowler, R. *et al.* (1999) MHC class I-restricted cytotoxic lymphocyte responses induced by enterotoxin-based mucosal adjuvants. *J. Immunol.* **163**, 6502–6510.

Smooker, P. M., Setiady, Y. Y., Rainczuk, A., and Spithill, T. W. (2000) Expression library immunization protects mice against a challenge with virulent rodent malaria. *Vaccine* **18**(23), 2533–2540.

Takala, A. K., Eskola, J., Leinonen, M. *et al.* (1991) Reduction of oropharyngeal carriage of *Haemophilus influenzae* type b (Hib) in children immunized with an Hib conjugate vaccine. *J. Infect. Dis.* **164**, 982–986.

Tappero, J. W., Lagos, R., Ballesteros, A. M., Plikaytis, B., Williams, D., Dykes, J., Gheesling, L. L., Carlone, G. M., Hoiby, E. A., Holst, J., Nokleby, H., Rosenqvist, E., Sierra, G., Campa, C., Sotolongo, F., Vega, J., Garcia, J., Herrera, P., Poolman, J. T., and Perkins, B. A. (1999) Immunogenicity of 2 serogroup B outer-membrane protein meningococcal vaccines: a randomised controlled trial in Chile. *J. Am. Med. Assoc.* **281**, 1520–1527.

Tettelin, H., Masignani, V., Cieslewicz, M. J., Eisen, J. A., Peterson, S., Wessels, M. R. Paulsen, I. T., Nelson, K. E., Margarit, I., Read, T. D., Madoff, L. C., Wolf, A. M., Beanan, M. J., Brinkac, L. M., Daugherty, S. C., DeBoy, R. T., Durkin, A. S., Kolonay, J. F., Madupu, R., Lewis, M. R., Radune, D., Fedorova, N. B., Scanlan, D., Khouri, H., Mulligan, S., Carty, H. A., Cline, R. T., Van Aken, S. E., Gill, J., Scarselli, M., Mora, M., Iacobini, E. T., Brettoni, C., Galli, G., Mariani, M., Vegni, F., Maione, D., Rinaudo, D., Rappuoli, R., Telford, J. L., Kasper, D. L., Grandi, G., and Fraser, C. M. (2002) Complete genome sequence and comparative genomic analysis of an emerging human pathogen, serotype V *Streptococcus agalactiae. Proc. Natl. Acad. Sci. USA* **99**, 12391–12396.

Tsuji, T., Inoue, T., Miyama, A., and Noda., N. (1991) Glutamic acid-112 of the A subunit of heat-labile enterotoxin from enterotoxigenic *Escherichia coli* is important for ADP-ribosyltransferase activity. *FEBS Lett.* **291**, 319–321.

Tsuji, T., Inoue, T., Miyama, A., Okamoto, K., Honda, T., and Miwatani, T. (1990) A single amino acid substitution in the A subunit of *Escherichia coli* enterotoxin results in a loss of its toxic activity. *J. Biol. Chem.* **36**, 22520–22525.

Twumasi, P. A. Jr., Kumah, S., Leach, A., Ceesay, S. J., O'Dempsey, T. J., Todd, J., Broome, C. V., Carlone, C. V., Pais, G. M., Holder, P. K. *et al.* (1995) A trial of a group A plus group C meningococcal polysaccharide-protein conjugate vaccine in African infants. *J. Infect. Dis.* **171**(3), 632–638.

Valenzuela, P., Medina, A., Rutter, W. J. *et al.* (1982) Synthesis and assembly of hepatitis B virus surface antigen particles in yeast. *Nature* **298**, 347–350.

Van Damme, P., and Vorsters, A. (2002) Hepatitis B control in Europe by universal vaccination programmes: the situation in 2001. *J. Med. Virol.* **67**, 433–439.

Ward, J. I., and Zangwill, K. M. (1999) *Haemophilus influenzae* vaccines. In Plotkin, S. A., Orenstein, W. A. (Eds.). *Vaccines* 3rd edn., Saunders, Philadelphia, PA, 183–221.

Wizemann, T. M., Heinrichs, J. H., Adamou, J. E. *et al.* (2001) Use of a whole genome approach to identify vaccine molecules affording protection against *Streptococcus pneumoniae* infection. *Infect. Immun.* **69**(3), 1593–1598.

Wuorimaa, T., and Kayhty, H. (2002) Current state of pneumococcal vaccines. *Scand. J. Immunol.* **56**, 111–129.

Yamamoto, S., Kiyono, H., Yamamoto, M., Imaoka, K., Yamamoto, M., Fujihashi, K., Van Ginkel, F. W., Noda, M., Takeda, Y., and McGhee, J. R. (1997) A nontoxic mutant of cholera toxin elicits Th2-type responses for enhanced mucosal immunity. *Proc. Natl. Acad. Sci. USA* **94**, 5267–5272.

2

Bioinformatics, DNA Microarrays and Proteomics in Vaccine Discovery: Competing or Complementary Technologies?

Guido Grandi, PhD

2.1 Introduction

Vaccinology is the sector of biomedical applications that seems to have benefited least from the unprecedented technological growth of the last few decades. A quick glance at the list of vaccines currently available on the market (Grandi, 2001) illustrates how most vaccine formulations rely on production technologies developed several decades ago. Approximately 60 per cent of commercialized products contain either killed or attenuated micro-organisms, a practice experimented more than 200 hundred years ago by Edward Jenner and subsequently rationalized and expanded by Luis Pasteur. Another 30 per cent is represented by vaccines constituted by pathogen-derived components proved to elicit protective immunity. They include toxins from *Clostridium tetani* and *Corynebacterium diphtheriae*, whose chemical detoxification process was first reported by Glenny and Hopkins in 1923, and glycoconjugate vaccines (such as the *Haemophilus influenzae* and meningococcus vaccines) whose efficacy was described in the 1920s (Avery and Goebel, 1929). Only 10 per cent of marketed vaccines are

Genomics, Proteomics and Vaccines edited by Guido Grandi
© 2004 John Wiley & Sons, Ltd   ISBN 0 470 85616 5

based on components obtained with recombinant DNA technologies, and among them the anti-hepatitis B vaccine is the only one for which more traditional approaches fail to provide effective alternatives (for example, the anti-*Bordetella pertussis* recombinant vaccine, despite the fact it is probably the most efficacious vaccine available (Greco *et al.*, 1996), represents only a negligible fraction of today's anti-whooping cough vaccine market, largely dominated by non-recombinant formulations).

However, new strategies for vaccine development are now urgently needed for two reasons. First, there are several infectious diseases for which traditional approaches have failed (Grandi, 2001). Second, regulatory authorities now require vaccines to meet extremely stringent safety and physiochemical characterization standards. Indeed, several of the vaccines developed over the past century would never have met the standards required today for registration, and the only reason they are still on the market is that the many years of vaccination in humans have demonstrated their safety and high benefit-to-risk ratio.

In the last two decades, some exciting approaches have been proposed and extensively investigated as possible alternatives to traditional vaccines. They include peptide-based vaccines (BenMohamed, Wechsler and Nesbuin, 2002), anti-idiotipic antibody-based vaccines (Magliani *et al.*, 2003) and DNA vaccines (Liu, 2003). However, despite the fact that the 'proof of concept' of their efficacy has been largely demonstrated in different animal models against numerous infectious diseases, and several clinical studies have been performed in humans, none of them have reached the market yet, and it is highly unlikely we shall see one approved in the near future.

This chapter has been written with the aim of concentrating in a few pages the whole message of this book, namely that the 'nomics' technologies (genomics, transcriptomics, proteomics) coupled to the high-throughput protein expression and purification technologies represent a concrete, exciting approach to vaccine development. In particular, I shall describe how bioinformatics, DNA microarrays and proteomics, starting from the knowledge of the genome sequence of a given bacterial pathogen, can lead to the selection of vaccine candidates in a very competitive time frame. Using as 'case study' the global genome approach to anti-meningococcus B vaccine developed in our laboratories, I shall then compare the three strategies, showing how they complement rather than overlap one another. Finally, I shall propose a global 'nomics' approach to vaccine candidate identification, which combines the three strategies, offering the highest chances of identifying the best candidates in an efficient and rational manner.

2.2   From genome sequence to vaccine discovery

One way our immune system protects us against pathogenic bacteria is by recognizing a few components of the pathogen and mounting specific responses against them capable of neutralizing the infective agent. Unfortunately, this adaptive immune response takes some time to become fully effective, and this is why primary exposures to pathogens often lead to the insurgency of disease. The key to the development of effective vaccines is the identification of the pathogen components used by the immune system to neutralize infections. Their administration would in fact educate the immune system to fight the pathogens from the very moment they enter our organism, thus preventing them from causing disease.

It turns out that many of the bacterial components eliciting protective immune responses are represented by protein antigens. This observation leads to the consideration that the knowledge of all protein components of a given bacterial pathogen would greatly facilitate vaccine candidate identification. In fact, with the list of total proteins available, vaccine identification would 'simply' be a question of (1) defining a correlate-of-protection assay amenable to the screening of a large number of candidates, (2) finding ways to select, within the list of proteins, those likely to become vaccine candidates, thus reducing the proteins to be assayed to a manageable number, (3) developing high-throughput expression and purification strategies to produce the recombinant forms of all selected proteins, and (4) screening the recombinant proteins with the correlate-of-protection assay until the antigens eliciting a protective response against the pathogen are unravelled.

Despite its conceptual simplicity, until recently vaccine discovery starting from the knowledge of whole genes/proteins had been impractical because of the tremendous technological challenges this approach implies. The genomic and post-genomic revolution has now drastically changed the situation. Genome sequencing and analysis are so advanced that for a bacterial genome to be fully sequenced is more a question of days rather than weeks, and its annotation is usually completed in a few months (Chapter 3). Genome annotation not only defines the list of total proteins and groups them into different families on the basis of their predicted function, but also provides information on the possible localization of the proteins within the bacterial cell. Since, for extracellular pathogens many of the antigens eliciting a protective immune response are secreted and/or membrane associated (these proteins have a high probability of coming into contact with our immune system), *in silico* analysis represents a first important strategy for the selection of possible vaccine candidates (Figure 2.1). A second antigen selection criterion is provided by the analysis of bacterial transcriptome using DNA microarrays. This technology provides a

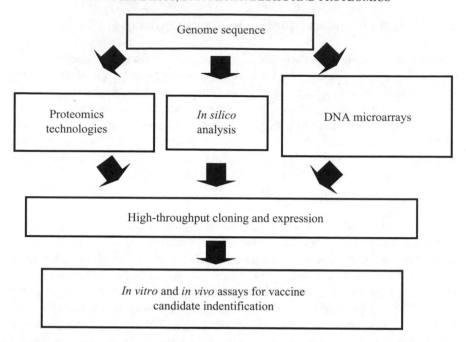

Figure 2.1 The 'nomics' approach to the identification of new vaccine candidates. The genes that potentially encode the vaccine antigens are selected using three different approaches: (1) *in silico* analysis to predict genes encoding secreted and surface-associated antigens and virulence factors; (2) proteomics analysis of membrane preparations and (3) DNA microarrays to identify highly expressed genes and *in vivo* up-regulated genes. After selection, the genes are cloned and expressed in heterologous systems. The recombinant antigens are then purified and used in the *in vitro/in vivo* assays for candidate identification

picture of the overall transcriptional activity of a given pathogen and allows the comparison of gene expression under different growth/environmental conditions (Chapter 4). Particularly, by analysing the transcription profile in bacteria interacting with their hosts, we can obtain indications of genes specifically regulated during infection. These genes encode proteins of obvious potential interest for vaccine development. Finally, a third approach to vaccine candidate selection is based on proteomic technologies (Chapters 5 and 6). These technologies allow a sufficiently detailed qualitative and quantitative characterization of the proteins present in the different cellular components of a given pathogen.

Using the three strategies mentioned above, selected pools of genes/proteins of potential interest for vaccine development are eventually identified.

Altogether, the pools usually include 20–30 per cent of all genes present in bacterial genomes.

Once identified, the group of candidates needs to be expressed and purified to be tested in an appropriate correlate-of-protection assay. As described in Chapter 7, the high-throughput expression and purification systems now available allow the isolation of sufficient quantities of a large number of recombinant proteins in a relatively short period of time. In our laboratories, three skilled scientists (one molecular biologist and two biochemists) have a working capacity of 100 proteins per month, meaning that the expression and purification of all selected candidates is usually completed by a relatively small group of scientists in a few months.

The last stretch of the pathway to vaccine discovery is the screening assay. This is usually the most time-consuming and critical step of the entire process for two main reasons. First, there is very little room for automation. The assays are often based on the immunization of animals with each purified protein. Animal survival after challenge with adequate doses of pathogen, and/or quality of immune responses elicited (production of antigen-specific T cells and antibodies), is used to define the potentiality of each antigen to become a vaccine. These types of assay usually require many animals and several months to be completed. In general, the length of the primary screening strictly depends on the number of animals a research team can handle simultaneously, and normally one year is needed for completion. The second critical aspect of the screening step is the quality of the assay used, which should mimic as far as possible what occurs in humans in terms of both disease and correlate of protection. Inadequate selection of the screening assay can severely jeopardize the success of the entire vaccine project.

Despite the bottleneck represented by the screening step, a well organized team can move through the entire process in 12–18 months. This is quite an astonishing achievement considering that, with more conventional approaches, similar research times are necessary, and often not sufficient, just to evaluate the potentiality of single candidates. In fact, according to traditional strategies, candidates are selected among the most immunogenic antigens as inferred, for example, by sera and/or blood cells analyses of infected/covalescent patients. Once identified, antigens are purified from the pathogen and finally tested with the appropriate assay. Overall, this is quite a complex process, particularly as far as the identification of immunogenic antigens and their subsequent purification from the infective agent are concerned. The beauty of the 'nomics' approach is that indications of the immunogenic properties of candidates are not required. This offers two advantages. First, no laborious experiments are required for the identification and purification of antigens. Second and most important, no reductive bias is imposed upfront to candidate screening. It is in fact not necessarily true that immunogenic antigens are the only ones to elicit protective immunity.

To illustrate the power of the 'nomics' approach, I shall now present the data recently obtained in our laboratories on the identification of vaccine candidates against *Neisseria meningitidis* group B (*Meningococcus B* (MenB)). In particular, I shall describe the results obtained with each of the three strategies for antigen selection (*in silico* analysis, transcriptome analysis and proteome analysis), and then analyse them in terms of efficiency and potential synergies.

2.3   A case study: the anti-meningococcus B vaccine

Meningococcal meningitis and sepsis are caused by *Neisseria meningitidis*, a Gram-negative, capsulated bacterium, classified into five major pathogenic serogroups (A, B, C, Y and W135) on the basis of the chemical composition of their capsular polysaccharides (Gotschlich, Liu and Artenstein, 1969; Gotschlich, Goldschneider and Artenstein, 1969) (see Chapter 8 for more details on classification, meningococcal disease and epidemiology). Very effective vaccines based on capsular polysaccharides against meningococcus C are already on the market and anti-meningococcus A/C/Y/W polyvalent vaccines are expected to be launched in the near future. However, there are no vaccines available for the prevention of MenB disease, responsible for a large proportion (from 32 to 80 per cent) of all meningococcal infections (Schotten *et al.*, 1993). The use of capsular polysaccharides is not feasible since in MenB its chemical composition is identical to a widely distributed human carbohydrate ($\alpha(2\rightarrow8)$ N-acetyl neuraminic acid or polysialic acid), which, being a self-antigen, is a poor immunogen in humans. Furthermore, the use of this polysaccharide in a vaccine may elicit autoantibodies (Hayninen *et al.*, 1995; Finne *et al.*, 1987). An alternative approach to vaccine development is based on the use of surface-exposed proteins, which in some instances have been shown to elicit protective bactericidal antibodies (Poolman, 1995; Martin, Cadieux, Hamel and Brodeur, 1997). However, many of the major surface protein antigens show sequence and antigenic variability, thus failing to confer protection against heterologous strains. Therefore, the challenge for anti-MenB vaccine research is the identification of highly conserved antigens eliciting protective immune responses (bactericidal antibodies) against a broad range of MenB isolates.

2.3.1   The 'in silico' approach

In an attempt to identify highly conserved, protective protein antigens, we followed the strategy thoroughly described in Chapter 8. Very briefly, the MenB genome sequence (Tettelin *et al.*, 2000) was subjected to computer analysis to identify genes potentially encoding surface-exposed or exported proteins. Of the 650 proteins predicted to move from the cytoplasm to the periphery of the cells, approximately 50 per cent were successfully expressed in *Escherichia coli* as either His-tag or glutathione S-transferase (GST) fusion proteins (Pizza *et al.*, 2000). The recombinant proteins were purified and used to immunize mice and the immune sera were tested for bactericidal activity, an assay that strongly correlates with protection in humans (Goldschneider, Gotschlich and Artenstein, 1989). Twenty-eight sera turned out to be bactericidal. To test sequence conservation of the protective antigens, 34 different N. meningitidis clinical isolates were selected and the nucleotide sequences of the corresponding genes of all protective antigens were compared. This kind of analysis allowed the identification of five highly conserved antigens. Most importantly, the sera against these five antigens and, in particular, combinations of these sera were capable of killing most of the meningococcal strains so far utilized in the complement-mediated bactericidal assay, thus making the antigens particularly promising for vaccine formulations. Phase I clinical studies are currently in progress to establish the ability of these antigens to induce bactericidal antibodies in humans.

2.3.2   The transcriptome analysis approach

The first step toward meningococcal infection is the colonization of the nasopharyngeal mucosa after bacterial invasion through the respiratory tract. For reasons still poorly understood, in a few cases the pathogen can penetrate the mucosal epithelium and gain access to the circulation. In individuals lacking humoral immunity to meningococci, proliferation of the organisms in the blood stream leads to septicaemia, characterized by circulatory collapse, multiple organ failure and extensive coagulopathy. In addition, from the blood stream the pathogen often reaches the subarachnoid compartment and initiates meningitis (Nassif *et al.*, 1999).

To elucidate the molecular events leading to bacterial colonization, we used DNA microarray technology to follow the changes in gene expression profiles when *N. meningitidis* interacts with human epithelial cells (Grifantini *et al.*, 2002). Bacteria were incubated with human epithelial cells and cell-adhering bacteria were recovered and total RNA was purified at different times. In a parallel experiment, RNA was prepared from bacteria grown in the absence of epithelial cells. After labelling with different fluorochromes, the two RNA samples were mixed together and used in hybridization experiments on DNA microarrays carrying the entire collection of PCR-amplified MenB genes (2152 genes). Relative emission intensities of the two fluorochromes after laser excitation of each spot (gene) were used to assess transcription activity of each gene under the two experimental conditions analysed. From these kinds of experiment, we found that adhesion to epithelial cells altered the expression of several genes, including adhesion genes, host–pathogen cross-talk genes (genes encoding transport machineries for ammonium, amino acids, chloride, sulphate and iron), amino acids and selenocysteine biosynthesis genes, DNA metabolism genes (methylases, nucleases, transposases, helicases and ligases) and hypothetical genes (107 genes).

Twelve proteins whose transcription was found to be particularly activated during adhesion were expressed in *E. coli*, purified and used to produce antisera in mice. Sera were finally tested for their capacity to mediate complement-dependent killing of MenB. Five of these sera (against the products of the hypothetical genes NMB0315 and NMB1119, the adhesin MafA, the MIP-related protein and N-acetylglutamate synthetase) showed substantial bactericidal activity against the different meningococcal strains.

These data show how DNA microarray analysis can be successfully exploited to identify new protective antigens against MenB. Being particularly expressed during the infection process, these antigens are expected to be ideal candidates for vaccine formulations.

2.3.3   The proteome approach

The most direct way to characterize meningococcal surface proteins, which, as already pointed out, are the most likely vaccine candidates, is to collect bacterial membranes and find ways to separate and identify protein components. Nowadays, proteomic technologies permit this with a sufficient degree of accuracy.

Table 2.1   Classification of membrane-associated proteins identified by 2DE/mass spectrometry. Proteins are grouped in 17 major functional families (bold characters) within which some relevant subfamilies are indicated. The 'Identified' and 'Annotated' columns refer to the number of proteins belonging to each family identified by 2DE/mass spectrometry, and annotated from the genome sequence (Tettelin *et al.*, 2000), respectively. The total number of identified proteins is 250. The sum of the proteins listed in the 'Identified' column exceeds this number since some proteins have been grouped in more than one family

	Identified	Annotated	%
Amino acid biosynthesis	24	78	31
Aspartate family	6	17	35
Pyruvate family	8	12	67
Biosynthesis of cofactors, prosthetic groups and carriers	9	81	11
Cell envelope	23	120	19
Cellular processes	29	132	22
Cell division	5	18	28
Chemotaxis and motility	3	3	100
Detoxification	3	10	30
DNA transformation	7	16	44
Toxin production and resistance	2	25	8
Pathogenesis	8	55	15
Central intermediary metabolism	5	40	13
DNA metabolism	7	98	7
Energy metabolism	46	162	28
ATP−proton motive force interconversion	4	8	50
Fermentation	5	12	42
Glycolysis/gluconeogenesis	9	15	60
Pentose phosphate pathway	3	6	50
Pyruvate dehydrogenase	3	3	100
TCA cycle	10	18	56
Fatty acid and phospholipid metabolism	4	31	13
Hypothetical proteins	13	327	4
Protein fate	15	70	21
Protein synthesis	47	123	38
Ribosomal proteins: synthesis and modification	20	58	34
Translation factors	5	12	42
tRNA aminoacylation	19	29	66
Purines, pyrimidines, nucleosides and nucleotides	12	41	29
Purine ribonucleotide biosynthesis	9	15	60
Pyrimidine ribonucleotide biosynthesis	2	9	22
Regulatory functions	6	52	12
Transcription	8	37	22
DNA-dependent RNA polymerase	3	4	75
Transport and binding proteins	15	125	12
Unknown function	7	132	5
No family (hypothetical proteins)	8	–	

In our laboratories, membrane protein characterization was carried out by resolving enriched preparations of membrane proteins on two-dimensional (2D) gels and subsequently identifying the separated proteins by matrix-assisted laser desorption/ionization time-of-flight mass spectrometry (MALDI-TOF).

From this kind of analysis, 250 proteins were identified (Ferrari *et al.*, manuscript in preparation). Unexpectedly, the list includes several proteins usually considered to be restricted to the cytoplasmic milieu (Table 2.1). For example, 56 per cent of the tricarboxylic acid cycle enzymes (10 out of 18) and 60 per cent of the enzymes of the glycolysis pathway (9 out of 15) were visible on the 2D map of the meningococcal membranes. Considering that 2D electrophoresis allows the visualization of only a fraction of all proteins present in a complex mixture (Gygi *et al.*, 2000; Futcher *et al.*, 1999), it is reasonable to assume that these two fundamental metabolic processes can take place also at the membrane level. A proportion of the 250 proteins found in the membrane fractions may well represent experimental artifacts due to sample contamination with cytoplasmic proteins. Although particular care was taken in sample preparation (cells were collected at the beginning of the exponential growth phase and membrane fractions were extensively washed with alkaline buffer), this possibility cannot be ruled out. To address this issue, we are systematically analyzing 2D gels of both membrane and soluble (cytoplasmic) fractions. Presence, absence or relative intensities of each protein in the gel of the two preparations can then be used to assign protein localization with a certain degree of reliability. Preliminary data suggest that 10–20 per cent of the proteins predicted to be restricted to the cytoplasmic milieu by computer analysis (Figure 2.4) are in fact membrane associated (Ferrari *et al.*, unpublished).

'Cytoplasmic' proteins can be not only membrane associated but also surface exposed. This was demonstrated by immunizing mice with the purified recombinant form of newly discovered membrane proteins and by using the mouse sera to stain the surface of meningococcal cells. Typical examples of this analysis are given in Figure 2.2, which shows how antibodies against the ribosomal subunit S7, the elongation factor TufA and DsbA, the thiol–disulphide interchange protein, recognized intact meningococcal cells with high efficiency.

In conclusion, the proteomic analysis of bacterial membranes is an important step toward the thorough resolution of membrane protein components. On the one hand, it provides experimental support to *in silico* predictions of protein compartmentalization. On the other hand, it sheds light on new aspects of the topological organization of the membrane, helping future refinements of the algorithms used for predicting protein location.

From a vaccine viewpoint, the implementation of the list of membrane

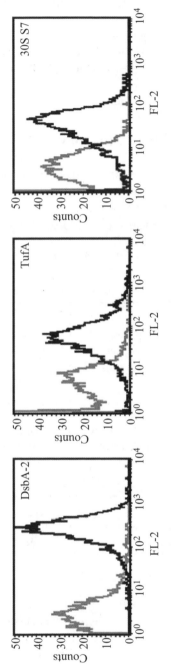

Figure 2.2 Identification of new meningococcal surface-exposed proteins. Proteomic analysis of membrane protein preparations of meningococcus B led to the identification of several proteins unexpectedly found associated with the membranes (see Section 3.3). Some of them are not only membrane associated but also surface exposed. This was demonstrated by immunizing mice with the recombinant form of the proteins and by testing the capacity of the corresponding sera to recognize the bacteria. The figure reports the FACS analysis of bacteria stained with sera against the ribosomal protein S7, DsbA (thiol–disulphide interchange protein) and the elongation factor TufA. Grey lines, bacteria stained with pre-immune sera; black lines, bacteria stained with sera against each recombinant protein

proteins offers the opportunity to test additional antigens for their capacity to elicit a protective immune response.

In a preliminary screening, we have selected 20 membrane proteins that have been expressed in *E. coli* and tested in the bactericidal assay. Two of the selected proteins, the products of NMB1313 and NMB1302 genes, induced antibodies capable of killing meningococcus B (Ferrari *et al.*, unpublished).

2.4   Comparison of the three approaches

In the previous section, we saw how the three antigen selection approaches can be successfully exploited to identify vaccine candidates. An interesting exercise is now to try to compare the strategies and see to what extent they overlap and/or complement one another. Advantages and disadvantages of each selection approach are summarized in Table 2.2.

Table 2.2   Comparison of the three 'nomics' approaches for antigen selection

	Advantages	Drawbacks
In silico analysis	• Very rapid. No experimental work needed for gene selection • Prediction of surface antigens and virulence factors	• Relay on algorithms not fully optimized yet • Lack of information on gene expression
DNA microarrays	• Semi-quantitative data on gene expression levels • Identification of genes likely to be involved in virulence	• No indications of protein localization • Labour intensive and expensive • Relatively high quantities of bacterial cells needed for the analysis (10^8–10^9)
Proteomics	• Quantitative data on antigen expression • Identification of membrane associated antigens	• Labour intensive and very expensive • Large numbers of bacterial cells needed for the analysis

In silico selection of membrane proteins has the great advantage that no experimental work is needed once the genome is fully sequenced. It is therefore relatively rapid and often provides extremely useful indications on protein function. However, it has two major drawbacks. First, it is currently unable to predict whether, when and to what extent proteins are expressed. From a vaccine discovery viewpoint, this limitation translates into an unnecessary extra work load, since proteins that are poorly expressed in the pathogen are expected to be bad vaccine candidates and therefore they could be excluded from the high-throughput screening. This is clearly illustrated in Figure 2.3, where the transcription activity of all MenB genes during the exponential growth phase is

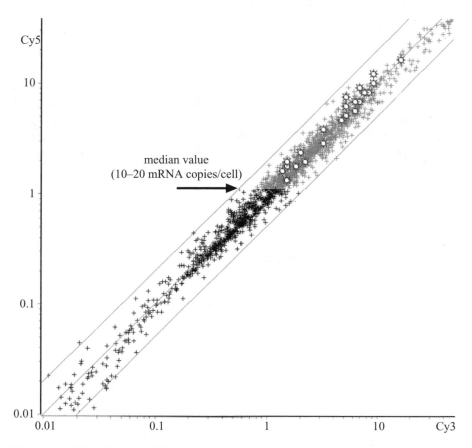

Figure 2.3 Identification of highly expressed genes in Meningococcus B by DNA micro-arrays. Bacteria were grown in GC medium and when the culture reached an OD600 value of 0.3 cells were collected and RNA purified (Grifantini *et al.*, 2002). RNA was labelled with either Cy3 or Cy5 and hybridized with the whole collection of meningococcal genes (2152) spotted on a glass slide (Grifantini *et al.*, 2002). Conventionally, highly expressed genes are here considered those genes (grey spots) giving relative fluorescence signals above the median value of all the signals of the slide. Asterisks correspond to the fluorescence intensities of those genes selected by the *in silico* analysis approach (see Section 3.1) and proved to be vaccine candidates

analysed. Under these experimental conditions, 50 per cent of the genes give a fluorescent signal above the median value, corresponding to an expression level higher than 5–10 mRNA copies per cell. If we now look at the 28 protective antigens identified by using the *in silico* analysis approach (see section 3.1), only a minor fraction is expressed at a level below the threshold median value (and these antigens were the least promising of the list and ultimately excluded from the final vaccine formulation), indicating that the *a priori* exclusion of the poorly expressed genes would have avoided the cloning, expression, purification and screening of approximately 200 (30 per cent of the 650 selected by the computer) useless proteins! The second drawback of the *in silico* approach is that it relies on algorithms, which are not yet fully refined. For example (Figure 2.4), of the 250 membrane proteins identified by 2DE/mass spectrometry (see section 3.3), only 66 per cent would be predicted in the membrane compartment by at least one of the seven most reliable algorithms currently available (Psort, SignalP, DAS, SOSUI, TMHMM, TopPre2 and TmPre) (Ferrari *et al.*, manuscript in preparation). Therefore, potential vaccine candidates could be lost if antigen selection were solely based on computer prediction.

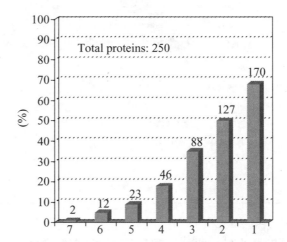

Figure 2.4 Reliability of current algorithms in predicting protein localization. The 250 membrane-associated proteins identified by proteomic analysis (see Section 3.3) were analysed by seven algorithms used to predict protein localization: Psort, SignalP, DAS, SOSUI, TMHMM, TopPre2 and TmPre (Ferrari *et al.*, manuscript in preparation). As shown in the figure, only 68 per cent of the 250 proteins would be assigned to the membrane compartment by at least one of the algorithms utilized. Interestingly, only two proteins (the pilin E, and the B subunit of ATP synthase F0) are predicted to be membrane associated by all algorithms

Antigen selection based on transcription analysis has the advantage that it provides a semi-quantitative indication of gene expression. Furthermore, if appropriate experimental conditions are adopted, indications can be gained on genes specifically expressed during infection (see section 3.2). These genes, which might be involved in virulence and pathogenicity, are of particular interest for vaccine development. However, the approach does not provide information on protein localization, and therefore, if all genes expressed above a selected threshold level and those over-expressed under *in vivo*-simulating conditions were to enter the high-throughput pipeline, a considerable amount of unnecessary extra work would be done. Furthermore, transcriptome analysis is technically demanding and requires a relatively high number of bacteria to be carried out. This could be a serous limitation, particularly when bacteria recovered from animals or humans are to be analysed.

Finally, proteome analysis is, theoretically speaking, the best selection criterion one could use. It has the great advantage of providing direct qualitative and quantitative information on membrane proteins. However, practically speaking, it poses several limitations. First of all, only a fraction of all protein components can be identified, particularly when 2DE coupled to mass spectrometry is used (Gygi *et al.*, 2000). Second, large numbers of bacterial cells are needed, limiting the analysis to the *in vitro* conditions. Third, the technology is laborious, expensive and time consuming and requires skilled personnel.

2.5   Conclusions: a 'nomics' approach to vaccine discovery

The advantages and disadvantages of the three approaches strongly indicate that none of them can be used alone without losing candidates and/or being inefficient. Retrospectively, our MenB studies indicate that transcription analysis of infection-regulated genes (section 3.2) and characterization of the membrane subproteome (section 3.3) led to the identification of vaccine candidates the *in silico* approach alone would have missed. Conversely, proteome analysis would have identified only five of the 28 candidates selected by *in silico* analysis. Probably, the most effective approach is the selection of highly transcribed and *in vivo* up-regulated genes by DNA microarrays. In fact, in addition to providing new candidates (section 3.2), practically none of the antigens selected both *in silico* and by proteomics would have been lost. However, transcription analysis is poorly selective, since the list of candidates generated may include as much as 60 per cent of the whole protein list. Therefore, without applying an additional selection filter, the approach would be highly inefficient.

We would like to conclude this chapter by trying to propose a general strategy for antigen selection that, by properly combining the three approaches, should guarantee the identification of most vaccine candidates in a highly efficient manner. Figure 2.5 schematizes the proposed 'nomics' flowchart for vaccine discovery. The strategy starts from the identification of highly expressed genes by DNA microarray. Practically speaking, the pathogen under investigation is grown in liquid culture and the transcription activity of all its genes is analysed. Having fixed an arbitrary threshold for gene expression (on the basis of our MenB studies, I tentatively propose 5–10 mRNA copies/gene/cell), a first pool of candidates accounting for up to 50–60 per cent of the whole protein list is obtained. To identify, within this large pool, the most interesting proteins for vaccine development, *in silico* analysis is applied to select only the putative secreted and membrane-associated proteins. We strongly recommend utilizing more than one algorithm for membrane-protein prediction, since each of the available algorithms has its own pros and cons. This process should reduce the original pool to a more manageable and reliable size of approximately 400 genes (20–30 per cent of the genome). However, since existing algorithms are not infallible, proteome analysis of membrane preparation is necessary to include those proteins that are erroneously predicted to be cytoplasmic by *in silico* analysis. Furthermore, proteome analysis allows the rescue of those proteins which, despite being poorly represented at RNA level and therefore excluded from the pool by the first selection criterion, appear to be sufficiently abundant in the membrane. In this context, it is worth mentioning here that the amount of protein in the cell does not always correlate with the level of its corresponding RNA (Futcher *et al.*, 1999). Overall, the proteomic approach should add 50–100 new candidates to the pool. Finally, as we have clearly shown in meningococcus (Grifantini *et al.*, 2002), bacterial interaction with the host induces a profound remodelling of its membrane-protein composition. Therefore, if experimentally feasible, it would be extremely useful to follow the gene expression profile under conditions mimicking natural infection. Not only would these experiments rescue those genes highly expressed *in vivo* but originally excluded from the pool of candidates because they were poorly expressed under '*in vitro*' conditions, but they would also provide useful information on proteins playing specific roles in the infection process that are therefore particularly interesting from a vaccine viewpoint.

At the end of this selection process, the pool of best candidates should include for a pathogen carrying 2000–3000 genes, approximately 550–600 genes, which are ready to move to the high-throughput expression step and to the screening assay.

I should like to point out that, while this proposed flowchart for antigen

THE 'NOMICS' FLOWCHART FOR VACCINE DISCOVERY

1. Identify 'highly' expressed genes by DNA microarrays

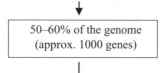

50–60% of the genome
(approx. 1000 genes)

2. Remove non-membrane-associated proteins as defined by *in silico* analysis
(use several algorithms)

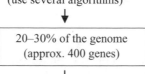

20–30% of the genome
(approx. 400 genes)

3. Add new membrane-associated proteins as defined by membrane proteome analysis

25–35% of the genome
(approx. 500 genes)

4. Add *in vivo* up-regulated genes as defined by DNA microarrays

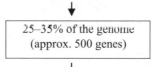

30–35% of the genome
(approx. 600 genes)

Figure 2.5 A proposed 'nomics' approach for vaccine candidate identification. A DNA microarray is first used to identify highly transcribed genes (>5 mRNA copies/cell) by analysing the gene transcription profile from bacteria grown in liquid culture. Highly transcribed genes are then subjected to *in silico* analysis to predict their localization, and membrane-associated proteins are selected. Most of the existing vaccine candidates are expected to be included in this pool of selected genes. However, a few candidates belonging to the following classes would be lost: (1) membrane-associated proteins erroneously predicted as 'cytoplasmic'; (2) relatively abundant membrane proteins with low RNA levels (protein concentration does not always correlate with RNA expression); (3) proteins poorly expressed when bacteria are grown in liquid culture but up-regulated under *in vivo*-simulating conditions. To rescue candidates belonging to classes (1) and (2), membrane protein identification is carried out by using proteomic technologies and newly identified membrane-associated proteins are added to the pool of candidates. Finally, DNA microarray technology is used to identify new *in vivo* up-regulated antigens belonging to class (3)

selection is, in my opinion, currently the most effective one, it may well become obsolete in a few years' time. The tremendous progress in the understanding of the physiology and system biology of bacterial pathogens expected to occur in the near future, thanks to the application of the 'nomics' technologies, would probably refine the algorithms of bioinformatics to such an extent that extremely reliable predictions of vaccine candidates would simply require a mouse click on your personal computer.

References

Avery, O. T., and Goebel, W. F. (1929) Chemo-immunological studies on conjugated carbohydrate–proteins. II. Immunological specificity of synthetic sugar–proteins. *J. Exp. Med.* **50**, 521–533.

BenMohamed, L., Wechsler, S. L., and Nesburn, A. B. (2002) Lipopeptide vaccines: yesterday, today and tomorrow. *Lancet Infect. Dis.* **2**, 425–431.

Finne, J., Bitter-Suermann, D., Goridis, C., and Finne, U. (1987) An IgG monoclonal antibody to group B meningococci cross-reacts with developmentally regulated polysialic acid units of glycoproteins in neural and extraneural tissues. *J. Immunol.* **138**, 4402–4407.

Futcher, B., Latter, G. I., Monardo, P., McLaughlin, C. S., and Garrels, J. I. (1999) A sampling of the yeast proteome. *Mol. Cell. Biol.* **19**, 7357–7368.

Glenny, A. T., and Hopkins, B. E. (1923) Diphtheria toxoid as an immunizing agent. *Br. J. Exp. Pathol.* **4**, 283–287.

Goldschneider, I, Gotschlich, E. C., and Artenstein, M. S. (1969) Human immunity to meningococcus. I. The role of humoral antibodies. *J. Exp. Med.* **129**, 1307–1326.

Gotschlich, E. C., Goldschneider, I., and Artenstein, M. S. (1969) Human immunity to the meningococcus. IV. Immunogenicity of group A and group C meningococcal polysaccharides in human volunteers. *J. Exp. Med.* **129**, 1367–1384.

Gotschlich, E. C., Liu, T. Y., and Artenstein, M. S. (1969) Human immunity to the meningococcus. 3. Preparation and immunochemical properties of the group A, group B and group C meningococcal polysaccharides. *J. Exp. Med.* **129**, 1349–1365.

Grandi, G. (2001) Antibacterial vaccine design using genomics and proteomics. *Trends Biotechnol.* **19**, 181–188.

Greco D., Salmaso, S., Mastrantonio, P., Giuliano, M., Tozzi, A. E., Anemona, A., Ciofi degli Atti, M. L., Giammanco, A., Panei, P., Blackwelder, W. C., Klein, D. L., and Wassilak, G. F. (1996) A controlled trial of two acellular vaccines and one whole-cell vaccine against pertussis. Progetto Pertosse Working Group. *N. Engl. J. Med.* **334**, 341–349.

Grifantini, R., Bartolini, E., Muzzi, A., Draghi, M., Frigimelica, E., Berger, J., Ratti, J. G., Petracca, R., Galli, G., Agnusdei, M., Giuliani, M. M., Santini, L., Brunelli, N., Tettelin, H., Rappuoli, R., Randazzo, F., and Grandi, G. (2002) Previously unrecognized vaccine candidates against group B meningococcus identified by DNA micro-arrays. *Nature Biotech.* **20**, 914–921.

Gygi, S. P., Corthals, G. L., Zhang, Y., Rochon, Y., and Aebersold, R. (2000) Evaluation of two-dimensional gel electrophoresis-based proteome analysis technology. *Proc. Natl. Acad. Sci. USA* **97**, 9390–9395.

Hayrinen, J., Jennings, H., Raff, H. V., Rougon, G., Hanai, N., Gerardy-Schahn, R., and Finne, J. (1995) Antibodies to polysialic acid and its N-propyl derivative: binding properties and interaction with human embryonal brain glycopeptides. *J. Infect. Dis.* **171**, 1481–1490.

Liu, M. A. (2003) DNA vaccines: a review. *J. Int. Med.* **253**, 402–410.

Magliani, W., Conti, S., Salati, A., Arseni, S., Ravanetti, L., Frazzi, R., and Polinelli L. (2003) Biotechnological approaches to the production of idiotypic vaccines and antiidiotypic antibiotics. *Curr. Pharm. Biotechnol.* **4**, 91–97.

Martin, D., Cadieux, N., Hamel, J., and Brodeur, B. R. (1997) Highly conserved Neisseria meningitidis surface protein confers protection against experimental infection. *J. Exp. Med.* **185**, 1173.

Nassif, X., Pujol, C., Morand, P., and Eugène. E. (1999) Interactions of pathogenic *Neisseria* with host cells. Is it possible to assemble the puzzle? *Mol. Microbiol.* **32**, 1124–1132.

Pizza, M., Scarlato, V., Masignani, V., Giuliani, M. M., Aricò, B., Comanducci, M., Jennings, G. T., Baldi, L., Bartolini, E., Capecchi, B., Galeotti, C. L., Luzzi, E., Manetti, R., Marchetti, E., Mora, M., Nuti, S., Ratti, G., Santini, L., Savino, S., Scarselli, M., Storni, E., Zuo, P., Broecker, M., Hundt, E., Knapp, B., Blair, E., Mason, T., Tettelin, H., Hood, D. W., Jeffries, A. C., Saunders, N. J., Granoff, D. M., Venter, J. C., Moxon, E. R., Grandi, G., and Rappuoli R. (2000) Identification of vaccine candidates against serogroup B meningococcus by whole-genome sequencing. *Science*, **287**, 1816–1820.

Poolman, J. T. (1995) Development of a meningococcal vaccine. *Infect. Agents Dis.* **4**, 13–28.

Scholten, R. J., Bijlmer, H. A., Poolman, J. T., Kuipers, B., Caugant, D. A., Van Alphen, L., Dankert, J., and Valkenburg, H. A. (1993) Meningococcal disease in The Netherlands, 1958–1990: a steady increase in the incidence since 1982 partially caused by new serotypes and subtypes of *Neisseria meningitidis. Clin. Infect. Dis.* **16**, 237–246.

Tettelin, H., Saunders, N. J., Nelson, K. E., Jeffries, A. C., Heidelberg, J., Eisen, J. A., Ketchum, K. A., Hood, D. W., Dodson, R. J., Nelson, W. C., Gwinn, M. L., DeBoy, R., Peterson, J. D., Hickey, E. K., Haft, D. H., Salzberg, S. L., White, O., Fleischmann, R. D., Dougherty, B. A., Mason, T., Ciecko, A., Parksey, D. S., Blair, E., Cittone, H., Clark, E. B., Cotton, M. D., Utterback, T. R., Khouri, H., Qin, H., Vamathevan, J., Gill, J., Scarlato, V., Masignani, V., Pizza, M., Grandi, G., Sun, L., Smith, H. O., Fraser, C. M., Moxon, E. R., Rappuoli, R., and Venter, J. C. (2000) Complete genome sequence of *Neisseria meningitidis* serotype B strain MC58. *Science*, **287**, 1809–1815.

Part 2
Technologies

3

Genome Sequencing and Analysis

Hervé Tettelin and Tamara Feldblyum

3.1 Introduction

The potential of genomics-based approaches to expand our understanding of the biology of pathogens was summarized in an editorial by Dr. Barry Bloom, current Director of the Harvard School of Public Health, in the journal *Nature*, following publication of the second complete microbial genome sequence in 1995 (Bloom, 1995). In it he stated 'The power and cost effectiveness of modern genome sequencing technology mean that complete genome sequences of 25 of the major bacterial and parasitic pathogens could be available within 5 years. For about $100 million we could buy the sequence of every virulence determinant, every protein antigen, and every drug target. It would represent for each pathogen a one-time investment from which the information derived would be available to all scientists for all time. We could then think about a new post-genomic era of microbe biology'.

We have, in fact, surpassed this goal, and genomics efforts in laboratories around the world have already delivered the complete sequence of more than 80 microbial species including major bacterial, fungal, and parasitic pathogens of human, animals, and plants (see http://www.tigr.org/tigr-scripts/CMR2/CMRHomePage.spl).

Genomics, Proteomics and Vaccines edited by Guido Grandi
© 2004 John Wiley & Sons, Ltd   ISBN 0 470 85616 5

3.2   Genome sequencing

The relatively small genome size of bacterial human pathogens (0.5–10 Mb) makes them perfectly suited for the whole-genome shotgun sequencing strategy (reviewed by Fraser and Fleischmann 1997 and Frangeul *et al.*, 1999). The random shotgun strategy has proven to be robust, and has been successful when applied to genomes with differing characteristics. These characteristics include variations in genome size, base compositions from very low to very high GC%, and the presence of various repeat elements, IS elements and multiple chromosomal molecules and plasmids. This method, first applied to generate the genome sequence of *Haemophilus influenzae* (Fleischmann *et al.*, 1995) involves the steps shown in Figure 3.1 and described below.

3.2.1   Library construction and template preparation

The creation of a random library of cloned chromosomal DNA fragments is essential to a successful genome sequencing project. Where possible, a random library is best constructed from mechanically sheared fragments since fragmentation of DNA by enzymatic cleavage introduces bias due to the slightly non-random genomic distribution of restriction enzyme sites. Plasmid libraries are most likely to be completely random if they have relatively small inserts in a narrow size range in order to minimize differences in growth rate and to minimize *in vitro* rearrangement. Considerable savings of effort can be achieved if the inserts are at least twice the average sequence read length, allowing each template to be sequenced from both ends without redundancy. Production of such mated sequence pairs is also of critical importance for assembly of shotgun data sets, as the presence of mates with one read in one contig and the other in a different contig allows the order and orientation of contigs to be determined and the sizes of intervening gaps to be estimated.

It is generally desirable to produce the bulk of shotgun sequences from two libraries of different insert lengths, typically 2 kb and 10 kb. This offers a compromise between the randomness/clone integrity of small inserts and the superior repeat spanning and intermediate-range linking capability of the 10 kb inserts.

After shearing, the genomic DNA is end repaired with kinase and T4 DNA polymerase treatments, and size selected by electrophoresis on 1 per cent low-melting-point agarose. One of the methods employed to improve genomic library quality is based on the ligation of *Bst*XI adapters (Invitrogen) to the sheared DNA. The DNA is then purified by gel electrophoresis to remove

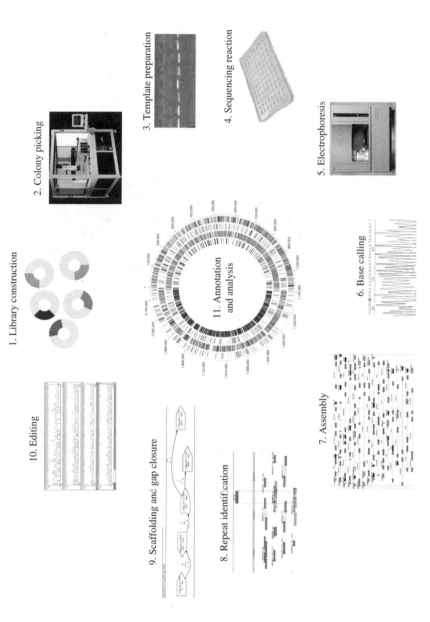

Figure 3.1 Description of the main steps required for the whole-genome shotgun sequencing of a bacterial pathogen and the analysis of the genome data

excess adapters, and the fragments, now with 3'-CACA overhangs, are inserted into a *Bst*XI-linearized plasmid vector with 3'-TGTG overhangs. The adapter ligation strategy reduces the chimeric and no-insert rate of the plasmid libraries to very low levels: insert DNA cannot self-ligate to form chimeras due to the presence of the non-compatible 4 bp overhangs and, likewise, vector DNA cannot self-ligate to form plasmid without insert. Ligation of 20 ng of linear vector with one molar equivalent of adapted insert DNA typically produces libraries containing approximately 50 million clones. Libraries containing 100 million independent clones or more are easily obtained simply by scaling the final ligation reaction.

It is sometimes useful to generate light sequence coverage from other types of library for species whose genomic DNA is unstable in *Escherichia coli*. One method to circumvent the problem is to use 50 kb libraries constructed as described above but digested with a restriction enzyme (*Bgl*II) that does not cleave the vector, but generally cleaves several times within the 50 kb insert. A *Bam*HI kanamycin resistance cassette is added and ligation is carried out at 37 °C in the continual presence of *Bgl*II. As *Bgl*II–*Bgl*II ligations occur they are continually cleaved, whereas *Bam*HI–*Bgl*II ligations are not cleaved. A high yield of internally deleted circular library molecules is obtained in which the residual insert ends are separated by the kanamycin resistance element (Figure 3.2). In addition to enabling selection of deleted plasmids, the inserted resistance element prevents the formation of chimeric recombinants by separating the two insert ends. The internally deleted libraries, when plated on agar containing ampicillin/carbenicillin and kanamycin (15 µg/ml), produce relatively uniform large colonies and can be propagated and sequenced as a small-insert plasmid library. This method offers markedly enhanced repeat spanning and long-range linking capability, and is a practical means of scaffolding across unclonable regions of genomic DNA.

Several methods for DNA template isolation from genomic libraries exist. At TIGR, the alkaline lysis procedure (Sambrook, Fritsch and Maniatis, 1989) modified for processing in 384-well plates and the TempliPhi method (Amersham Biosciences) proved to be the most cost effective and robust methods that are amenable to automation in high-throughput production environments.

High-purity plasmid DNA is prepared from liquid bacterial growth. This can be performed by hand or using a DNA purification robotic work-station such as the one from Thermo CRS. In the alkaline lysis method, bacterial cells are lysed, cell debris is removed by centrifugation and plasmid DNA is recovered from the cleared lysate by isopropanol precipitation. The DNA precipitate is washed with 70 per cent ethanol, dried and re-suspended in 10 mM Tris HCl buffer containing a trace of blue dextran. The typical yield of the plasmid DNA per clone from this method is approximately 600–800 ng, providing sufficient

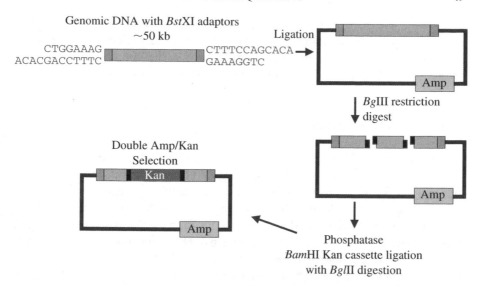

Genomic DNA with *Bst*XI adaptors
~50 kb

CTGGAAAG
ACACGACCTTTC

CTTTCCAGCACA
GAAAGGTC

Ligation

Amp

*Bgl*II restriction
digest

Amp

Phosphatase
*Bam*HI Kan cassette ligation
with *Bgl*II digestion

Double Amp/Kan
Selection

Kan

Amp

Figure 3.2   Construction of a 50 kb linking clone library for shotgun sequencing. After ligation of 50 kb inserts into the vector, the inserts are digested with a restriction enzyme (*Bgl*II) that does not cleave the vector, but generally cleaves several times within the 50 kb inserts. A *Bam*HI kanamycin resistance cassette is added and ligation is carried out at 37 °C in the continual presence of *Bgl*II. As *Bgl*II–*Bgl*II ligations occur they are continually cleaved, whereas *Bam*HI–*Bgl*II ligations are not cleaved. A high yield of internally deleted circular library molecules is obtained in which the residual insert ends are separated by the kanamycin resistance element. Amp, ampicillin selection marker; Kan, kanamycin selection marker

DNA quantity for at least four sequencing reactions per template. At TIGR currently, 30 000 templates can be prepared, in a 12 hour period using the robotic work-station. The TempliPhi DNA sequencing template amplification kit is based on a rolling circle type of amplification of DNA. The kit contains the Phi29 DNA polymerase, random hexamer primers, dNTPs, and reaction buffer. Isothermal DNA amplification at 30 °C for 4–6 hours yields 1–3 µg of DNA, sufficient for multiple sequencing reactions.

Template production quality is monitored by assessing the cell growth and by sampling DNA from each production lot. Library quality is assessed in three steps. First, plasmid inserts are excised by restriction digest and checked for the presence of insert and appropriate size. Second, a few clones (e.g. 96 or 384) are sequenced and the success at generating high-quality sequences measured. The sequences are also searched against public databases to verify that they originate from the organism of interest. Third, a larger number of sequences (e.g. 1000 per Mb of genome size) are generated, assembled and verified for randomness and proper sampling of repeats across the genome. If these steps are passed without problems, high-throughput sequencing is conducted.

3.2.2 High-throughput sequencing

In the last years, high-throughput sequencing technologies have transitioned from gel-plate-based approaches (e.g. manual sequencing with S^{35} radioactivity, ABI 373, ABI PRISM 377, Li-cor IR^2 DNA sequencer) to capillary sequencers (e.g. ABI Prism 3100, 3700 and 3730xl DNA analysers, Megabace 1000 and 5000). Methods and instruments for preparing and sequencing DNA were reviewed recently (Meldrum, 2000a,b). The transition to capillary machines eliminated problems of lane tracking inherent to gel plate technologies, and the automation of capillary machines allowed a dramatic increase in throughput together with providing longer sequence read lengths. Altogether, these advances have significantly increased the number of genomic data generated for human pathogens in recent years (Table 3.1).

TIGR's state-of-the-art sequencing facility currently consists of 25 ABI Prism 3700 DNA analysers, 24 Applied Biosystems 3730xl DNA analysers, and six ABI Prism 3100 DNA Analysers. Approximately 50 projects are processed concurrently. More than 3.4 million sequencing reactions were performed in 2001, and over 6 million reactions were completed by the end of 2002. TIGR's current sequencing capacity is 14 000 000 sequence reads per year.

Sequencing protocols are based on the dideoxy sequencing method (Sanger *et al.*, 1977). To obtain paired sequence reads from opposite ends of each clone insert, two 384-well cycle sequencing reaction plates are prepared from each plate of plasmid template DNA. Sequencing reactions are carried out using Big Dye Terminator chemistry version 3.1 (Applied Biosystems) and standard M13 or custom forward and reverse primers. Sequencing reactions are set up by the Biomek FX (Beckman) pipetting workstations. The robots are used to aliquot templates and to combine them with the reaction mixes consisting of deoxy-nucleotides and fluorescently labelled dideoxy-nucleotides, the Taq thermo-stable DNA polymerase, sequencing primers, and reaction buffer. The template and reaction plates are bar coded and tracked by the bar code readers on the Biomek FX work-stations to assure error-free template and reaction mix transfer. Thirty to forty consecutive cycles of linear amplification steps are performed on MJ Research Tetrads or 9700 thermal cyclers (Applied Biosystems). Reaction products are efficiently precipitated by isopropanol, dried at

Table 3.1 Human bacterial pathogens whose complete genome sequence has been published

Human bacterial pathogen	Mb	Sequencing institution	Funding agency	Publication
Borrelia burgdorferi B31	1.5	TIGR/Brookhaven National Laboratory	Mathers Foundation/NIH	(Fraser *et al.*, 1997, Casjens *et al.*, 2000)
Brucella melitensis 16M	3.3	Univ. of Scranton/Integrated Genomics	DARPA	(DelVecchio *et al.*, 2002)
Brucella suis 1330	3.3	TIGR	DARPA/NIAID	(Paulsen *et al.*, 2002)
Campylobacter jejuni NCTC 11168	1.6	Sanger	Beowulf Genomics	(Parkhill *et al.*, 2000b)
Chlamydia muridarum Nigg	1.1	TIGR	NIAID	(Read *et al.*, 2000)
Chlamydia pneumoniae AR39	1.2	TIGR	NIAID	(Read *et al.*, 2000)
Chlamydia pneumoniae CWL029	1.2	Univ. of California Berkeley/Stanford	Incyte Genomics	(Kalman *et al.*, 1999)
Chlamydia pneumoniae J138	1.2	Japanese consortium	JSPS	(Shirai *et al.*, 2000)
Chlamydia trachomatis serovar D	1.0	Univ. of California Berkeley/Stanford	NIAID	(Stephens *et al.*, 1998)
Clostridium perfringens 13	3.1	Univ. of Tsukuba/Kyushu Univ. Kitasato Univ.	JSPS	(Shimizu *et al.*, 2002)
Enterococcus faecalis V583	3.4	TIGR	NIAID	(Paulsen *et al.*, 2003)
Escherichia coli K12-MG1655	4.6	Univ. of Wisconsin	NHGRI	(Blattner *et al.*, 1997)
Escherichia coli O157:H7 EDL933	5.5	Univ. of Wisconsin	NHGRI/NIAID	(Perna *et al.*, 2001)
Escherichia coli O157:H7 VT2-Sakai	5.5	Japanese consortium	JSPS	(Hayashi *et al.*, 2001)
Fusobacterium nucleatum ATCC 25586	2.2	Integrated Genomics	NIH	(Kapatral *et al.*, 2002)
Haemophilus influenzae KW20	1.8	TIGR	TIGR	
Helicobacter pylori 26695	1.7	TIGR	TIGR	(Tomb *et al.*, 1997)

(continued overleaf)

Table 3.1 (*continued*)

Human bacterial pathogen	Mb	Sequencing institution	Funding agency	Publication
Helicobacter pylori J99	1.6	Astra/Genome Therapeutics	Astra & Genome Therapeutics	(Alm et al., 1999)
Listeria innocua CLIP 11262	3.0	GMP	Institut Pasteur	(Glaser et al., 2001)
Listeria monocytogenes EGD-e	2.9	European consortium	European Union	(Glaser et al., 2001)
Mycobacterium leprae TN	3.3	Sanger Centre	Amici di Raoul Follereau/ The New York Community Trust	(Cole et al., 2001)
Mycobacterium tuberculosis CDC1551	4.4	TIGR	NIAID	(Fleischmann et al., 2002)
Mycobacterium tuberculosis H37Rv	4.4	Sanger Centre	Wellcome Trust	(Cole et al., 1998)
Mycoplasma genitalium G-37	0.6	TIGR	DOE	(Fraser et al., 1995)
Mycoplasma penetrans HF-2	1.4	NIID/Kitasango Univ.	JSPS	(Sasaki et al., 2002)
Mycoplasma pneumoniae M129	0.8	Univ. of Heidelberg	DFG	(Himmelreich et al., 1996)
Neisseria meningitidis MC58	2.3	TIGR	Chiron Corporation	(Tettelin et al., 2000)
Neisseria meningitidis Z2491	2.2	Sanger Centre	Wellcome Trust	(Parkhill et al., 2000a)
Pseudomonas aeruginosa PAO1	6.3	Univ. of Washington Genome Center/PathoGenesis Corporation	Cystic Fibrosis Foundation/ PathoGenesis Corporation	(Stover et al., 2000)
Rickettsia conorii Malish 7	1.3	Genoscope	Aventis	(Ogata et al., 2001)
Rickettsia prowazekii Madrid E	1.1	Univ. of Uppsala	SSF/NFR	(Andersson et al., 1998)
Salmonella enterica serovar Typhi CT18	4.8	Sanger Centre/Imperial College	Beowulf Genomics	(Parkhill et al., 2001a)
Salmonella typhimurium LT2 SGSC1412	5.0	Washington Univ.	NIH	(McClelland et al., 2001)

Organism	Size	Center	Funder	Reference
Shigella flexneri 2a 301	4.7	Microbial Genome Center	Chinese Ministry of Public Health	(Jin *et al.*, 2002)
Staphylococcus aureus MW2	2.8	NITE/Juntendo Univ.	NITE	(Baba *et al.*, 2002)
Staphylococcus aureus N315	2.8	NITE/Juntendo Univ.	NITE	(Kuroda *et al.*, 2001)
Streptococcus agalactiae 2603V/R	2.2	TIGR	Chiron Corporation	(Tettelin *et al.*, 2002)
Streptococcus agalactiae NEM316	2.2	GMP	MENRT	(Glaser *et al.*, 2002)
Streptococcus mutans UA159	2.0	Univ. of Oklahoma	NIDCR	(Ajdic *et al.*, 2002)
Streptococcus pneumoniae R6	2.0	Eli Lilly	Eli Lilly/Incyte Genomics	(Hoskins *et al.*, 2001)
Streptococcus pneumoniae TIGR4	2.2	TIGR	TIGR/NIAID/MGRI	(Tettelin *et al.*, 2001)
Streptococcus pyogenes MGAS315	1.9	Rocky Mountain Laboratories	NIAID	(Beres *et al.*, 2002)
Streptococcus pyogenes MGAS8232	1.9	Rocky Mountain Laboratories	NIAID	(Smoot *et al.*, 2002)
Streptococcus pyogenes SF370 serotype M1	1.9	Univ. of Oklahoma	NIAID	(Ferretti *et al.*, 2001)
Treponema pallidum Nichols	1.1	TIGR/Univ. of Texas	NIAID	(Fraser *et al.*, 1998)
Ureaplasma urealyticum serovar 3	0.8	Applied Biosystems/Univ. of Alabama/Eli Lilly	Applied Biosystems/NI/NIAID/Eli Lilly/UAB	(Glass *et al.*, 2000)
Vibrio cholerae El Tor N16961	4.0	TIGR	NIAID	(Heidelberg *et al.*, 2000)
Yersinia pestis CO92	4.8	Sanger Centre	Beowulf Genomics	(Parkhill *et al.*, 2001b)
Yersinia pestis KIM	4.6	Univ. Of Wisconsin	NIAID	(Deng *et al.*, 2002)

room temperature, and stored at 4 °C or resuspended in water and sealed. As sequencing machines become available, a sample sheet for each plate is automatically generated upon scanning the plate's barcode. The plates are then transferred to one of the ABI Prism 3700 or AB3730xl DNA analysers for electrophoresis. The current polymers and software allow for eight electrophoresis runs per day on an ABI 3700 and 12 runs per day on an AB 3730xl with a set-up time of less than one hour.

High-throughput sequencing facilities, with a large number of projects running in parallel, typically require automated management through a laboratory information management and sample tracking system (LIMS) (Kerlavage *et al.*, 1993). At TIGR, the system includes a suite of programs that track samples from the early stages of library construction through sequencing and closure. Throughout the process, data are maintained in Sybase relational database tables. The database stores and correlates all information collected during an entire genome sequencing project, allowing one to track the flow of data all the way back from an annotated gene to the original sequencing trace files underlying that gene. The system includes client/server applications for sample management, data entry, library management and sequence processing. This system is mature and robust and has been modified over the years to incorporate new laboratory methods, new types of equipment and new software, and for use in several other laboratories. These integrated applications include automated vector removal, identification and masking of repetitive elements, detection of contaminant clones and tracking of clone and template information.

The quality of the templates and sequences generated during the course of a sequencing project is monitored systematically on a daily basis through user-friendly interfaces (Figure 3.3). This allows rapid detection and correction of potential problems that can arise during a project. Typically, a quality control/quality assessment (QC/QA) group is put together to apply the quality criteria. The group is responsible for the testing and release of reagents to production teams, in-process testing of template quality, investigation of failures and deviations from standard performance matrices, monitoring of data quality, auditing, identifying areas for improvement, and creation of controlled documentation (standard operating procedures) to insure that the documentation maintains a consistent format and is technically accurate.

The theory for shotgun sequencing follows from the Lander–Waterman (Lander and Waterman, 1988) application of the equation for the Poisson distribution. Extrapolation of the Lander–Waterman equation for a typical bacterial genome (2–5 Mb) predicts that eight-fold sequence coverage of a genome should theoretically provide more than 99 per cent of the sequence of the genome (Table 3.2). The number of random clones that must be sequenced

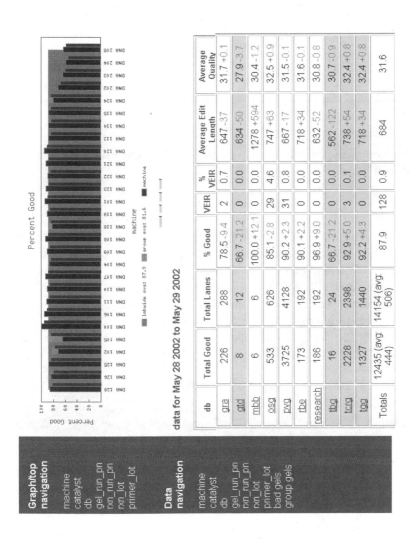

Figure 3.3 Quality control of sequences as they are generated in the high-throughput sequencing facility. Left-hand panel: the quality of the sequences can be interrogated by grouping them by sequencing machine, by person who initiated the reactions or the sequencing runs, by primer lot etc. The top panel represents a histogram of sequence success rates for each sequencing machine. The bottom panel represents a table of the daily number of sequences, success rates, average edited sequence read lengths and average quality for each sequencing project

Table 3.2 Lander–Waterman computer simulation of a whole-genome shotgun sequencing project where the genome size $L = 2\,000\,000$ and the average sequence length $w = 700$

Clones (n)	% unsequenced	bp unsequenced	No. of gaps	Ave. gap length (bp)	Fold coverage
2 860	36.75	735 024	1051	699	1.0
5 720	13.50	270 130	773	350	2.0
8 580	4.97	99 276	425	233	3.0
11 440	1.82	36 485	209	174	4.0
17 160	0.24	4 928	42	116	6.0
22 880	0.03	666	7	87	8.0

to obtain the complete genome sequence can be calculated using the Poisson distribution as follows:

$$p_x = m^x e^{-m}/x! \tag{3.1}$$

where p_x is the probability of x occurrences of an event and m is the mean number of occurrences. The probability that any given base is not yet sequenced after a certain number of clones have been sequenced can be determined as the probability that no clone originates at any of the w bases preceding the given base. The mean number of hits to this region is $m = nw/L$ where L is the genome length, w is the width of the sequenced region (the average sequence length) and n is the number of clones sequenced. The probability of no hits ($x = 0$) is

$$p_0 = e^{-m} = e^{-nw/L}. \tag{3.2}$$

From equation (3.2) we can obtain the total unsequenced bases (the total gap length)

$$\text{total gap length} = Le^{-nw/L} \tag{3.3}$$

and the total number of connected unsequenced regions (total number of gaps)

$$\text{number of gaps} = ne^{-nw/L}. \tag{3.4}$$

Equation (3.4) can be understood if one considers the sequence obtained from any given clone and calculates the probability that it terminates at an unsequenced base (p_0). The number of gaps is equal to the number of individual sequenced regions terminating at an unsequenced base (np_0). The number of contiguous sequenced regions (contigs) equals the number of gaps. This treatment is essentially that of Lander and Waterman, assuming that any overlap is detectable.

Since clones carry fragments in either of two orientations, each strand of the genome is sequenced approximately equally. Based on a simulation for $w = 700$, for a 2.0 Mb microbial genome, 23 000 sequence segments would result in approximately 8.0-fold coverage. The theoretical treatment assumes that even a single base overlap is detectable and further that the clones represent a perfectly uniform sample. In practice, assembly programs require at least a 40 bp overlap between fragments. In this example, an ideal library of random inserts and random clone picking would result in unsequenced DNA, i.e. not sequenced on either strand, of approximately 700 bases and only about eight gaps of totally unsequenced DNA (Table 3.2).

3.2.3 Genome sequence assembly

Once the shotgun sequencing phase is complete, all the sequences have to be assembled into contigs based on similarity between individual sequences. Repetitive regions of a genome represent a major challenge to assembly software. To help tackle this problem, TIGR Assembler (Sutton *et al.*, 1995; Pop, Salzberg and Shumway, 2002) identifies all repetitive regions in the set of sequences to assemble and makes use of forward and reverse sequence information generated from both ends of individual shotgun clone inserts. This information together with knowledge about the distribution of insert sizes helps generating correct assemblies of repeats if their size is shorter than that of clone inserts. In addition, TIGR Assembler provides a feature called 'jumpstart' designed to aid the finishing process: this allows an assembly to be re-started using a collection of previously assembled contigs and new sequences. This feature greatly facilitates the numerous re-assemblies necessary to close gaps during the genome finishing process.

3.2.4 Genome finishing

The assembly of shotgun sequence data into contigs typically results in more than 99 per cent of the genome being covered, with many gaps separating the contigs. Gaps are typically due to regions of the genome that were not represented in the genomic libraries (e.g. regions that are toxic when cloned in *E. coli*), to sequences of bad quality (e.g. from regions with secondary structures that hinder the sequencing reactions) and/or to complicated repeats. The first step of the finishing process consists in linking contigs into large scaffolds using forward and reverse sequence information. Software such as Bambus (Pop, Salzberg and Shumway, 2002) generate scaffolds that indicate the relative order and orientation of contigs linked by clone information. Contigs within scaffolds are separated by sequencing gaps that are easily sequenced by primer walking or transposon-associated sequencing (pGPS-1, New Englands Biolabs) of the linking clone inserts. The order of scaffolds with respect to one another is unknown because they are separated by physical gaps for which no cloned template is available. Before directed gap closure is initiated, it is often beneficial to automatically identify and re-sequence all clones for which only one end was successfully sequenced and is pointing towards a physical gap in order to obtain a pair of reads that could convert the gap into a sequence gap. In addition, all sequences of length shorter than average and pointing towards a gap are re-sequenced using long-run sequencing conditions in order to reduce the gap size and possibly close it.

To generate templates for physical gaps, PCR is performed using genomic DNA as template and PCR products are sequenced directly to fill the gaps or cloned and sequenced with transposons. Because the order of the scaffolds is unknown, it is necessary to test each contig extremity with all other extremities in a PCR experiment. While this can be achieved by combinatorial PCR that involves a large number of PCR reactions, the preferred method is multiplex PCR. POMP (pipette optimal multiplex PCR) is a method based on multiplex PCR combined with a pooling algorithm (Tettelin *et al.*, 1999). It allows rapid generation of PCR products to span most of the physical gaps remaining in a shotgun sequencing project using primers designed at the extremities of the gaps, while performing a minimal number of PCR reactions and pipetting operations. When gaps are flanked by repeats, the primers are designed before the repeat area in the contigs so that the PCR products generated are specific. Gaps for which no PCR product is obtained can be extended using the Genome Walker Kit (Clontech Laboratories). This method makes use of semi-directed PCR from adaptor-modified total genomic DNA restriction fragments in which only one end of the PCR product contains a genome-specific priming site and the other end contains an adaptor-specific primer. Alternatively, primer walking directly on genomic DNA can be attempted for non-repetitive regions with the Big Dye terminator chemistry.

The finishing process is usually the most time-consuming one but improvements in sequencing and gap-closure technologies, as well as automation of many steps, have greatly decreased the time and effort to complete microbial genomes. Once assembled into its final form, the genome sequence is edited for accuracy and correction of nucleotide ambiguities.

3.2.5 Complete and draft genome sequencing

The availability of complete genome sequences of human pathogens free of gaps and edited is of great utility as it gives access to complete gene sets, provides for studies of genome organization, genome comparisons and functional genomics, including the development of microarrays (Fraser *et al.*, 2002). Once one or two genomes of a given organism are sequenced to completion, the diversity of the species can be assessed through comparative genome hybridization (CGH) using microarrays (Tettelin *et al.*, 2001, Smoot *et al.*, 2002, Tettelin *et al.*, 2002, Chan *et al.*, 2003) or by generating draft sequences of other strains of interest and comparing them to the reference genome(s) (see section 3.3.2). These two approaches are rapid and cost effective while generating genome-scale information on species diversity.

3.2.6 Novel sequencing technologies

Several sequencing technologies that involve drastically different ways of identifying nucleotides present in DNA fragments are currently under development but have not yet been applied to high-throughput genome sequencing. Pyrosequencing (Ronaghi, Uhlen and Nyren, 1998) is a technique based on the detection of released pyrophosphate during DNA synthesis; visible light is generated that is proportional to the number of incorporated nucleotides in a polymerase reaction. Polony sequencing (Mitra and Church, 1999) relies on the generation of 'PCR colonies' or polonies in a thin polyacrylamide film poured on a glass microscope slide; single-base extension sequencing reactions are performed on polonies and the sequence is deduced from the fluorescence patterns. Nanopore/ion channel sequencing (Deamer and Branton, 2002) uses the blockade of ionic current produced when single-stranded DNA molecules are driven through a nanoscopic pore by an applied electric field; modulations of the current blockade indicate the length, composition, structure and dynamic motion of the molecules. The optical characterization of ordered labelled DNA templates method (http://www.usgenomics.com/technology/technologies.shtml) is based on the optical interrogation of fluorescent-dye-labelled DNA molecules that are linearized within the fluid stream of specialized microchannels and posts. Molecular motors (http://www.usgenomics.com/technology/technologies. shtml) make use of the fluorescence resonance energy transfer (FRET) that occurs when a fluorophore-linked DNA polymerase excites fluorophores linked to the DNA bases from the molecule it processes. Two-dye sequencing (http://www.lecb.ncifcrf.gov/~toms/patent/dnasequencing/) and molecular data stream reading (http://www.visigenbio.com/sys-tmpl/door/) are also based on

interactions such as FRET. Finally, molecular resonance sequencing (http://
www.mobious.com/) is based on measuring the changes in electromagnetic
radiation intensity that result from the conformational alteration of an immobilized polymerase that binds and processes DNA molecules.

It is currently difficult to judge which of these technologies will prove
amenable to large-scale sequencing projects and what impact they will have on
production, accuracy and cost. It seems clear, however, that the use of
nanotechnology and automation holds great promise of decreasing costs and
increasing sequencing speed. It is possible that in the near future one will be
able to obtain the complete genome sequence of any isolate of a human
pathogen within hours/days of isolation together with its comparison to related
strains/species, allowing for rapid typing and diagnostics.

3.3   Genome analysis

3.3.1   Annotation pipeline

All large-scale genome sequencing centres rely on strong bioinformatics
support to analyse the large numbers of data that they generate. Levels and
types of analysis vary but there is a common need for robust tools combining
many types of evidence for annotation and providing user-friendly interfaces
for both the annotators and the public.

At TIGR, the annotation tools applied to genomes have evolved over the past
years to combine a mixture of automated open reading frame (ORF) identification, ORF and non-ORF feature identification and assignment of database
matches and functional role categories to genes. Glimmer (Salzberg *et al.*,
1998; Delcher *et al.*, 1999a), an algorithm using interpolated Markov models, is
used for the identification of ORFs in prokaryotes. Predicted coding regions are
searched against public databases with BlastP (Altschul *et al.*, 1990). The
protein–protein matches are aligned with a modified Smith–Waterman (Waterman, 1988) algorithm that maximally extends regions of similarity across
frameshifts (Fleischmann *et al.*, 1995; Fraser *et al.*, 1995). Gene identification
is facilitated by searching against a database of non-redundant bacterial
proteins (nraa) developed at TIGR and curated from the public archives
GenBank, Genpept, PIR and SwissProt. ORFs matching entries in nraa are
automatically assigned the common name and functional role category (Riley,
1993) of the match. Regions of the genome without predicted coding regions
and Glimmer predictions with no database match are re-evaluated using BlastX
(Altschul *et al.*, 1990) as the initial search and new genes are extrapolated from

regions of alignment. Work is currently under way to incorporate the Gene Ontology (GO) system (Ashburner *et al.*, 2000) into the annotation pipeline

In order to enhance our ability to make potential gene identifications, approaches and tools based on multiple sequence alignment and family building are employed (Eddy, 1998). Paralogous gene families are created from multiple sequence alignments made with the predicted protein sequences and built with the MKDOM software (Gouzy *et al.*, 1997). The multiple sequence alignments are used to group similar proteins into families for verification of annotation and identification of family members perhaps not recognized by simple pairwise alignments. The proteins are also searched against a database of hidden Markov models (HMMs, PFam v3.1 (Bateman *et al.*, 2000), TIGRFam (Haft *et al.*, 2001)) built on protein family/superfamily multiple sequence alignments.

In addition to ORF analysis and gene discovery by similarity searches, a number of other features are routinely analysed and documented. TopPred is used to identify potential membrane-spanning domains in proteins (Claros and von Heijne, 1994). Signal peptides and their probable cleavage site in secreted proteins are predicted with SignalP (Nielsen *et al.*, 1997). Genes coding for untranslated RNAs are identified by database searches at the nucleotide level. Searches for tRNA genes are performed using tRNAScan-SE (Lowe and Eddy, 1997). Repeated sequences in a genome are identified with MUMmer, an algorithm based on suffix trees (Delcher *et al.*, 1999b, 2002), which can very rapidly identify all repeats in large genome sequences. The initial set of repeats is further processed to group repeats into classes, which are used to guide both assembly and annotation.

The automated annotation is curated manually by teams of annotators using the interface Manatee, which also allows outside collaborators to access the data during preparation of journal publications. Manatee is freely available (http://www.tigr.org) and has been installed successfully at several other institutions.

Finally, a suite of programs called Genome Control (GC) automatically handles the day-to-day processing of annotation data from microbial genomes. It runs a nightly process control system that updates information on any ORFs whose characteristics have been modified either manually or automatically. This ensures that the annotation is kept up to date and synchronized.

3.3.2   Comparative genomics

The Comprehensive Microbial Resource (CMR) is a database describing all sequenced microbial genomes (Peterson *et al.*, 2001). It provides the scientific community with information pertaining to inter- and intra-genomic relationships for comparative genomics, genome diversity and evolutionary studies. It displays whole-genome analyses like genome-wide alignment dot plots or circular representations, and tables of genes shared across several species. The interface allows for a wide variety of data retrievals based on gene properties including molecular weight, hydrophobicity, GC content, functional role assignments and taxonomy. Users examining an individual gene can also compare it to genes from other organisms that have a similar function based on TIGRFam (Haft *et al.*, 2001) or PFam (Bateman *et al.*, 2000) protein family membership, clusters of orthologous groups (COGs) (Tatusov, Koonin and Lipman, 1997), sequence similarity, common EC numbers or common functional role categories (Riley, 1993).

In particular, the CMR is extremely useful to compare the genomes of various human pathogens. For instance, genes shared between a group of species/strains, aside from those coding for housekeeping functions like replication and translation, are usually indicative of metabolic pathways or cell structures that are characteristic of the species compared. On the other hand, genes identified as unique to certain species usually comprise mobile elements like phages and pathogenicity islands, as well as features including virulence determinants, resistance genes or host–pathogen-interaction-mediating proteins. The analysis of both shared and unique genes provides insights into the species biology, lifestyle and virulence, especially when correlated to previously known information like disease causing phenotypes, tissue tropism, epidemiological data etc.

As discussed previously, there are advantages in comparing complete genome sequence data to draft (incomplete) genome sequences from additional related species or strains. To achieve this goal, contigs from unfinished genomes are automatically ordered as best as possible along the reference complete genome using MUMmer and derivatives (Delcher *et al.*, 2002), and pasted together into a pseudo-genome, incorporating stop codons in all six frames between all contigs. ORFs are then automatically predicted, annotated and assigned functional roles as described above. Pseudo-genome sequences and genes sets thereof can then be analysed with tools from the CMR. Because these processes are completely automated, many species/strains can be analysed rapidly in this manner with minimal human intervention. In addition to global tools from the CMR designed mostly for cross-species analyses, an extension of Manatee currently under development at TIGR enables the analysis of several closely related species, their whole-genome nucleotide and protein-level alignments, SNPs and larger-sized polymorphisms, as well as synonymous and non-synonymous substitutions between protein pairs. Finally, the system also provides a display of the contigs of an unfinished genome aligned to a complete genome with the scaffolding information superimposed and distances between contigs calculated. This accelerates the finishing process of additional genomes for which having a completed genome is judged relevant. Indeed, physical gaps are then converted into gaps of known size where a simple PCR can be attempted. While most of the gaps between highly related species/strains will close this way, some will not due to inversions or insertions in the new genome.

3.3.3 Impact of whole-genome analyses

The study of whole-genome sequences from numerous microbial species has revealed a tremendous amount of information on the physiology and evolution of microbial species, and provided novel approaches to the diagnosis and treatment of infectious diseases. One of the major observations is that almost half of the ORFs in each species are of unknown function (Fraser *et al.*, 2000). Elucidating the function of these new genes through functional assays is likely to lead to new biochemical pathways, new virulence determinants etc. From the pool of genes whose biological function could be determined by *in silico* analyses, it is possible to reconstruct all the major metabolic pathways that a given organism relies on to live. The correlation of these pathways to the ability of the organism to transport substances across its membrane(s) through detailed analysis of its various transporters provides for hypotheses on modes of survival and niche adaptation (see an example in Tettelin *et al.*, (2001)). Microbial evolutionary biology has also greatly benefited from whole-genome sequences. Analyses suggested for instance that horizontal gene transfer is more frequent than previously expected, and revealed how genomes evolve on short timescales (Eisen, 2000). Potential virulence factors and pathogenicity islands can be identified through analysis of protein similarities and secondary structure (amino acid motifs), nucleotide composition and organization of chromosomal regions. Finally, genome-scale prediction of proteins likely to be exposed on the surface of an organism leads to the identification of novel potential vaccine candidates as described below.

3.3.4 Identification of vaccine candidates

Conventional approaches to subunit vaccine development involve the identification of individual components of the pathogen by biochemical, serological or genetic methods. A new high-throughput method, reverse vaccinology, based on the identification of vaccine candidates from whole-genome sequence information, was developed recently (Rappuoli, 2000; Grandi, 2001; Rappuoli,

2001a,b; Masignani, Rappuoli and Pizza, 2002; Adu-Bobie *et al.*, 2003). It was applied to *Neisseria meningitidis* (Pizza *et al.*, 2000; Tettelin *et al.*, 2000), *Streptococcus pneumoniae* (Wizemann *et al.*, 2001), *Streptococcus agalactiae* (Tettelin *et al.*, 2002) and *Bacillus anthracis* (Ariel *et al.*, 2002). Another genome-based method aiming at the identification of immunologically active epitopes was applied to *Staphylococcus aureus* (Etz *et al.*, 2002). Some of these applications are discussed in separate chapters of this book. Here we will focus on the first step of reverse vaccinology after generation of the genome sequence. It involves the use of specific bioinformatics tools for *in silico* identification of potential vaccine candidates that will be subsequently characterized experimentally.

For most bacterial pathogens, the proteins that are likely to induce a protective immune response are localized on the cell surface because of their availability to come into contact with the host immune system. Proteins of less than 100 amino acid residues are unlikely to code for good protective antigens and are not considered.

All the predicted proteins from a pathogen's genome are screened systematically and automatically for the presence of amino acid motifs characteristic of cell surface localization. First, surface-exposed proteins or virulence factors known based on previous experimental characterizations are selected (e.g. members of the functional category 'Cell envelope' or 'Pathogenesis'), while those with known cytoplasmic functions are rejected. Then basic motifs are searched with SignalP (Nielsen *et al.*, 1997) for signal peptides and TopPred (Claros and von Heijne, 1994) for membrane-spanning domains. Proteins carrying more than four predicted membrane-spanning domains are usually discarded. They are likely to be completely embedded in the cell membrane and therefore inaccessible to antibodies, and they are difficult to express in *E. coli*.

Subsequently, proteins matching a suite of HMMs from PFam v3.1 (Bateman *et al.*, 2000) and TIGRFam (Haft *et al.*, 2001) characteristic of surface proteins are also selected. The motifs include lipoproteins, extra-cellular proteases, Gram-positive cell-wall anchors, Gram-negative outer membrane anchors, hemolysins, signal recognition proteins and pili. Shorter motifs for which no HMM is available such as the host integrin-binding domain (RGD) and choline-binding domain (WYY) are also searched by classical pattern matching. The program PSORT (Nakai and Horton, 1999) can also be used to estimate the probability that a protein is surface exposed based on some of the motifs described above.

In addition, proteins containing two or more repeated amino acid motifs can be identified with Wordmatch (http://www.hgmp.mrc.ac.uk/Software/EMBOSS/Apps/#current). Repeated motifs are usually indicative of surface-exposed proteins interacting with the host.

The list of potential vaccine candidates obtained from this combination of analyses typically contains 500–800 proteins for an average-sized prokaryotic genome of 2–4 Mb.

When more than one genome sequence of a given pathogen is available, the application of comparative genomics analyses allows the identification of candidates that are conserved across strains, increasing the likelihood that a vaccine based on such antigens will be efficacious against various outbreaks of the disease. Another approach to assess conservation across strains or closely related species is to PCR amplify the selected candidates from the genomic DNA of many strains, ideally sampling the whole panel of isolates for the pathogen considered, and to sequence and compare the PCR products (Pizza *et al.*, 2000; Tettelin *et al.*, 2002). Finally, the integration of CGH results (Tettelin *et al.*, 2002) or expression studies of host-pathogen interactions (Grifantini *et al.*, 2002) provide for a more informed choice of candidates that are likely to lead to effective vaccines.

3.4   Conclusion

The ability to identify hundreds of novel potential vaccine candidates through the use of whole-genome sequence data has revolutionized vaccinology. Classical approaches identified only a few candidates that often proved inappropriate after investment of years of vaccine research. Advances in sequencing technologies provided the capacity to rapidly generate whole genome sequences for many human pathogens (Table 3.1). This breakthrough combined with large-scale highly automated bioinformatics approaches aimed at characterizing candidates *in silico* provided for major advances in vaccine development of several important disease-causing organisms for which no effective vaccine is available so far. It is anticipated that reverse vaccinology will become increasingly streamlined and will be applied more and more broadly, rapidly improving human health.

References

Adu-Bobie, J., Capecchi, B., Serruto, D., Rappuoli, R., and Pizza, M. (2003) *Vaccine* **21**, 605–610.

Ajdic, D., McShan, W. M., McLaughlin, R. E., Savic, G., Chang, J., Carson, M. B., Primeaux, C., Tian, R., Kenton, S., Jia, H., Lin, S., Qian, Y., Li, S., Zhu, H., Najar, F., Lai, H., White, J., Roe, B.A., and Ferretti, J. J. (2002) *Proc. Natl. Acad. Sci. USA* **99**, 14434–14439.

Alm, R. A., Ling, L. S., Moir, D. T., King, B. L., Brown, E. D., Doig, P. C., Smith, D. R., Noonan, B., Guild, B. C., deJonge, B. L., Carmel, G., Tummino, P. J., Caruso, A., Uria-Nickelsen, M., Mills, D. M., Ives, C., Gibson, R., Merberg, D., Mills, S. D., Jiang, Q., Taylor, D. E., Vovis, G. F., and Trust, T. J. (1999) *Nature* **397**, 176–180.

Altschul, S. F., Gish, W., Miller, W., Myers, E. W., and Lipman, D. J. (1990) *J. Mol. Biol.* **215**, 403–410.

Andersson, S. G., Zomorodipour, A., Andersson, J. O., Sicheritz-Ponten, T., Alsmark, U. C., Podowski, R. M., Naslund, A. K., Eriksson, A. S., Winkler, H. H., and Kurland, C. G. (1998) *Nature* **396**, 133–140.

Ariel, N., Zvi, A., Grosfeld, H., Gat, O., Inbar, Y., Velan, B., Cohen, S., and Shafferman, A. (2002) *Infect. Immun.* **70**, 6817–6827.

Ashburner, M., Ball, C. A., Blake, J. A., Botstein, D., Butler, H., Cherry, J. M., Davis, A. P., Dolinski, K., Dwight, S. S., Eppig, J. T., Harris, M. A., Hill, D. P., Issel-Tarver, L., Kasarskis, A., Lewis, S., Matese, J. C., Richardson, J. E., Ringwald, M., Rubin, G. M., and Sherlock, G. (2000) *Nat. Genet.* **25**, 25–29.

Baba, T., Takeuchi, F., Kuroda, M., Yuzawa, H., Aoki, K., Oguchi, A., Nagai, Y., Iwama, N., Asano, K., Naimi, T., Kuroda, H., Cui, L., Yamamoto, K., and Hiramatsu, K. (2002) *Lancet* **359**, 1819–1827.

Bateman, A., Birney, E., Durbin, R., Eddy, S. R., Howe, K. L., and Sonnhammer, E. L. (2000) *Nucleic Acids Res.* **28**, 263–266.

Beres, S. B., Sylva, G. L., Barbian, K. D., Lei, B., Hoff, J. S., Mammarella, N. D., Liu, M. Y., Smoot, J. C., Porcella, S. F., Parkins, L. D., Campbell, D. S., Smith, T. M., McCormick, J. K., Leung, D. Y., Schlievert, P. M., and Musser, J. M. (2002) *Proc. Natl. Acad. Sci. USA* **99**, 10 078–10 083.

Blattner, F. R., Plunkett, G. III, Bloch, C. A., Perna, N. T., Burland, V., Riley, M., Collado-Vides, J., Glasner, J. D., Rode, C. K., Mayhew, G. F., Gregor, J., Davis, N. W., Kirkpatrick, H. A., Goeden, M. A., Rose, D. J., Mau, B., and Shao, Y. (1997) *Science* **277**, 1453–1474.

Bloom, B. R. (1995) *Nature* **378**, 236.

Casjens, S., Palmer, N., van Vugt, R., Huang, W. M., Stevenson, B., Rosa, P., Lathigra, R., Sutton, G., Peterson, J., Dodson, R. J., Haft, D., Hickey, E., Gwinn, M., White, O., and Fraser, C. M. (2000) *Mol. Microbiol.* **35**, 490–516.

Chan, K., Baker, S., Kim, C. C., Detweiler, C. S., Dougan, G., and Falkow, S. (2003) *J. Bacteriol.* **185**, 553–563.

Claros, M. G., and von Heijne, G. (1994) *Comput. Appl. Biosci.* **10**, 685–686.

Cole, S. T., Brosch, R., Parkhill, J., Garnier, T., Churcher, C., Harris, D., Gordon, S. V., Eiglmeier, K., Gas, S., Barry, C. E. III, Tekaia, F., Badcock, K., Basham, D., Brown, D., Chillingworth, T., Connor, R., Davies, R., Devlin, K., Feltwell, T., Gentles, S., Hamlin, N., Holroyd, S., Hornsby, T., Jagels, K., Barrell, B. G. *et al.* (1998) *Nature* **393**, 537–544.

Cole, S. T., Eiglmeier, K., Parkhill, J., James, K. D., Thomson, N. R., Wheeler, P. R., Honore, N., Garnier, T., Churcher, C., Harris, D., Mungall, K., Basham, D., Brown, D., Chillingworth, T., Connor, R., Davies, R. M., Devlin, K., Duthoy, S., Feltwell, T., Fraser, A., Hamlin, N., Holroyd, S., Hornsby, T., Jagels, K., Lacroix, C., Maclean, J.,

Moule, S., Murphy, L., Oliver, K., Quail, M. A., Rajandream, M. A., Rutherford, K. M., Rutter, S., Seeger, K., Simon, S., Simmonds, M., Skelton, J., Squares, R., Squares, S., Stevens, K., Taylor, K., Whitehead, S., Woodward, J. R., and Barrell, B. G. (2001) *Nature* **409**, 1007–1011.

Deamer, D. W., and Branton, D. (2002) *Acc. Chem. Res.* **35**, 817–825.

Delcher, A. L., Harmon, D., Kasif, S., White, O., and Salzberg, S. L. (1999a) *Nucleic Acids Res.* **27**, 4636–4641.

Delcher, A. L., Kasif, S., Fleischmann, R. D., Peterson, J., White, O., and Salzberg, S. L. (1999b) *Nucleic Acids Res.* **27**, 2369–2376.

Delcher, A. L., Phillippy, A., Carlton, J., and Salzberg, S. L. (2002) *Nucleic Acids Res.* **30**, 2478–2483.

DelVecchio, V. G., Kapatral, V., Redkar, R. J., Patra, G., Mujer, C., Los, T., Ivanova, N., Anderson, I., Bhattacharyya, A., Lykidis, A., Reznik, G., Jablonski, L., Larsen, N., D'Souza, M., Bernal, A., Mazur, M., Goltsman, E., Selkov, E., Elzer, P. H., Hagius, S., O'Callaghan, D., Letesson, J. J., Haselkorn, R., Kyrpides, N., and Overbeek, R. (2002) *Proc. Natl. Acad. Sci. USA* **99**, 443–448.

Deng, W., Burland, V., Plunkett, G., 3rd, Boutin, A., Mayhew, G. F., Liss, P., Perna, N. T., Rose, D. J., Mau, B., Zhou, S., Schwartz, D. C., Fetherston, J. D., Lindler, L. E., Brubaker, R. R., Plano, G. V., Straley, S. C., McDonough, K. A., Nilles, M. L., Matson, J. S., Blattner, F. R., and Perry, R. D. (2002) *J. Bacteriol.* **184**, 4601–4611.

Eddy, S. R. (1998) *Bioinformatics* **14**, 755–763.

Eisen, J. A. (2000) *Curr. Opin. Microbiol.* **3**, 475–480.

Etz, H., Minh, D. B., Henics, T., Dryla, A., Winkler, B., Triska, C., Boyd, A. P., Sollner, J., Schmidt, W., von Ahsen, U., Buschle, M., Gill, S. R., Kolonay, J., Khalak, H., Fraser, C. M., von Gabain, A., Nagy, E., and Meinke, A. (2002) *Proc. Natl. Acad. Sci. USA* **99**, 6573–6578.

Ferretti, J. J., McShan, W. M., Ajdic, D., Savic, D. J., Savic, G., Lyon, K., Primeaux, C., Sezate, S., Suvorov, A. N., Kenton, S., Lai, H. S., Lin, S. P., Qian, Y., Jia, H. G., Najar, F. Z., Ren, Q., Zhu, H., Song, L., White, J., Yuan, X., Clifton, S. W., Roe, B. A., and McLaughlin, R. (2001) *Proc. Natl. Acad. Sci. USA* **98**, 4658–4663.

Fleischmann, R. D., Adams, M. D., White, O., Clayton, R. A., Kirkness, E. F., Kerlavage, A. R., Bult, C. J., Tomb, J. F., Dougherty, B. A., Merrick, J. M., McKenney, K., Sutton, G. *et al.* (1995) *Science* **269**, 496–512.

Fleischmann, R. D., Alland, D., Eisen, J. A., Carpenter, L., White, O., Peterson, J., DeBoy, R., Dodson, R., Gwinn, M., Haft, D., Hickey, E., Kolonay, J. F., Nelson, W. C., Umayam, L. A., Ermolaeva, M., Salzberg, S. L., Delcher, A., Utterback, T., Weidman, J., Khouri, H., Gill, J., Mikula, A., Bishai, W., Jacobs, W.R. Jr., Venter, J. C., and Fraser, C. M. (2002) *J. Bacteriol.* **184**, 5479–5490.

Frangeul, L., Nelson, K. E., Buchrieser, C., Danchin, A., Glaser, P., and Kunst, F. (1999) *Microbiology* **145** (Pt 10), 2625–2634.

Fraser, C. M., Casjens, S., Huang, W. M., Sutton, G. G., Clayton, R., Lathigra, R., White, O., Ketchum, K. A., Dodson, R., Hickey, E. K., Gwinn, M., Dougherty, B., Tomb, J. F., Fleischmann, R. D., Richardson, D., Peterson, J., Kerlavage, A. R.,

Quackenbush, J., Salzberg, S., Hanson, M., van Vugt, R., Palmer, N., Adams, M. D., Gocayne, J., Venter, J. C. *et al.* (1997) *Nature* **390**, 580–586.

Fraser, C. M., Eisen, J., Fleischmann, R. D., Ketchum, K. A., and Peterson, S. (2000) *Emerg. Infect. Dis.* **6**, 505–512.

Fraser, C. M., Eisen, J. A., Nelson, K. E., Paulsen, I. T., and Salzberg, S. L. (2002) *J. Bacteriol.* **184**, 6403–6405; discusion 6405.

Fraser, C. M., and Fleischmann, R. D. (1997) *Electrophoresis* **18**, 1207–1216.

Fraser, C., Gocayne, J. D., White, O., Adams, M. D., Clayton, R. A., Fleischmann, R. D., Bult, C. J., Kerlavage, A. R., Sutton, G., Kelley, J. M., Fritchman, J. L., Weidman, J. F., Small, K. V., Sandusky, M., Fuhrmann, J., Nguyen, D., Utterback, T. R., Saudek, D. M., Phillips, C. A., Merrick, J. M., Tomb, J. F., Dougherty, B. A., Bott, K. F., Hu, P. C., Lucier, T. S., Peterson, S. N., Smith, H. O., Hutchinson, C. A., and Venter, J. C. (1995) *Science* **270**, 397–403.

Fraser, C. M., Norris, S. J., Weinstock, G. M., White, O., Sutton, G. G., Dodson, R., Gwinn, M., Hickey, E. K., Clayton, R., Ketchum, K. A., Sodergren, E., Hardham, J. M., McLeod, M. P., Salzberg, S., Peterson, J., Khalak, H., Richardson, D., Howell, J. K., Chidambaram, M., Utterback, T., McDonald, L., Artiach, P., Bowman, C., Cotton, M. D., Venter, J. C. *et al.* (1998) *Science* **281**, 375–388.

Glaser, P., Frangeul, L., Buchrieser, C., Rusniok, C., Amend, A., Baquero, F., Berche, P., Bloecker, H., Brandt, P., Chakraborty, T., Charbit, A., Chetouani, F., Couve, E., de Daruvar, A., Dehoux, P., Domann, E., Dominguez-Bernal, G., Duchaud, E., Durant, L., Dussurget, O., Entian, K. D., Fsihi, H., Portillo, F. G., Garrido, P., Gautier, L., Goebel, W., Gomez-Lopez, N., Hain, T., Hauf, J., Jackson, D., Jones, L. M., Kaerst, U., Kreft, J., Kuhn, M., Kunst, F., Kurapkat, G., Madueno, E., Maitournam, A., Vicente, J. M., Ng, E., Nedjari, H., Nordsiek, G., Novella, S., de Pablos, B., Perez-Diaz, J. C., Purcell, R., Remmel, B., Rose, M., Schlueter, T., Simoes, N., Tierrez, A., Vazquez-Boland, J. A., Voss, H., Wehland, J., and Cossart, P. (2001) *Science* **294**, 849–852.

Glaser, P., Rusniok, C., Buchrieser, C., Chevalier, F., Frangeul, L., Msadek, T., Zouine, M., Couve, E., Lalioui, L., Poyart, C., Trieu-Cuot, P., and Kunst, F. (2002) *Mol. Microbiol.* **45**, 1499–1513.

Glass, J. I., Lefkowitz, E. J., Glass, J. S., Heiner, C. R., Chen, E. Y., and Cassell, G. H. (2000) *Nature* **407**, 757–762.

Gouzy, J., Eugene, P., Greene, E. A., Kahn, D., and Corpet, F. (1997) *Comput. Appl. Biosci.* **13**, 601–608.

Grandi, G. (2001) *Trends Biotechnol.* **19**, 181–188.

Grifantini, R., Bartolini, E., Muzzi, A., Draghi, M., Frigimelica, E., Berger, J., Ratti, G., Petracca, R., Galli, G., Agnusdei, M., Giuliani, M. M., Santini, L., Brunelli, B., Tettelin, H., Rappuoli, R., Randazzo, F., and Grandi, G. (2002) *Nat. Biotechnol.* **20**, 914–921.

Haft, D. H., Loftus, B. J., Richardson, D. L., Yang, F., Eisen, J. A., Paulsen, I. T., and White, O. (2001) *Nucleic Acids Res.* **29**, 41–43.

Hayashi, T., Makino, K., Ohnishi, M., Kurokawa, K., Ishii, K., Yokoyama, K., Han, C. G., Ohtsubo, E., Nakayama, K., Murata, T., Tanaka, M., Tobe, T., Iida, T., Takami,

H., Honda, T., Sasakawa, C., Ogasawara, N., Yasunaga, T., Kuhara, S., Shiba, T., Hattori, M., and Shinagawa, H. (2001) *DNA Res.* **8**, 11–22.

Heidelberg, J. F., Eisen, J. A., Nelson, W. C., Clayton, R. A., Gwinn, M. L., Dodson, R. J., Haft, D. H., Hickey, E. K., Peterson, J. D., Umayam, L., Gill, S. R., Nelson, K. E., Read, T. D., Tettelin, H., Richardson, D., Ermolaeva, M. D., Vamathevan, J., Bass, S., Qin, H., Dragoi, I., Sellers, P., McDonald, L., Utterback, T., Fleishmann, R. D., Nierman, W. C., and White, O. (2000) *Nature* **406**, 477–483.

Himmelreich, R., Hilbert, H., Plagens, H., Pirkl, E., Li, B. C., and Herrmann, R. (1996) *Nucleic Acids Res.* **24**, 4420–4449.

Hoskins, J., Alborn, W. E. Jr., Arnold, J., Blaszczak, L. C., Burgett, S., DeHoff, B. S., Estrem, S. T., Fritz, L., Fu, D. J., Fuller, W., Geringer, C., Gilmour, R., Glass, J. S., Khoja, H., Kraft, A. R., Lagace, R. E., LeBlanc, D. J., Lee, L. N., Lefkowitz, E. J., Lu, J., Matsushima, P., McAhren, S. M., McHenney, M., McLeaster, K., Mundy, C. W., Nicas, T. I., Norris, F. H., O'Gara, M., Peery, R. B., Robertson, G. T., Rockey, P., Sun, P. M., Winkler, M. E., Yang, Y., Young-Bellido, M., Zhao, G., Zook, C. A., Baltz, R. H., Jaskunas, S. R., Rosteck, P. R. Jr., Skatrud, P. L., and Glass, J. I. (2001) *J. Bacteriol.* **183**, 5709–5717.

Jin, Q., Yuan, Z., Xu, J., Wang, Y., Shen, Y., Lu, W., Wang, J., Liu, H., Yang, J., Yang, F., Zhang, X., Zhang, J., Yang, G., Wu, H., Qu, D., Dong, J., Sun, L., Xue, Y., Zhao, A., Gao, Y., Zhu, J., Kan, B., Ding, K., Chen, S., Cheng, H., Yao, Z., He, B., Chen, R., Ma, D., Qiang, B., Wen, Y., Hou, Y., and Yu, J. (2002) *Nucleic Acids Res.* **30**, 4432–4441.

Kalman, S., Mitchell, W., Marathe, R., Lammel, C., Fan, J., Hyman, R. W., Olinger, L., Grimwood, J., Davis, R. W., and Stephens, R. S. (1999) *Nat. Genet.* **21**, 385–389.

Kapatral, V., Anderson, I., Ivanova, N., Reznik, G., Los, T., Lykidis, A., Bhattacharyya, A., Bartman, A., Gardner, W., Grechkin, G., Zhu, L., Vasieva, O., Chu, L., Kogan, Y., Chaga, O., Goltsman, E., Bernal, A., Larsen, N., D'Souza, M., Walunas, T., Pusch, G., Haselkorn, R., Fonstein, M., Kyrpides, N., and Overbeek, R. (2002) *J. Bacteriol.* **184**, 2005–2018.

Kerlavage, A. R., Adams, M., Kelley, J. C., Dubnick, M., Powell, J., Shanmugam, P., Venter, J. C., and Fields, C. (1993) In *Proceedings of the 26th Annual Hawaii International Conference on System Sciences*. IEEE Computer Society Press, Hawaii.

Kuroda, M., Ohta, T., Uchiyama, I., Baba, T., Yuzawa, H., Kobayashi, I., Cui, L., Oguchi, A., Aoki, K., Nagai, Y., Lian, J., Ito, T., Kanamori, M., Matsumaru, H., Maruyama, A., Murakami, H., Hosoyama, A., Mizutani-Ui, Y., Takahashi, N. K., Sawano, T., Inoue, R., Kaito, C., Sekimizu, K., Hirakawa, H., Kuhara, S., Goto, S., Yabuzaki, J., Kanehisa, M., Yamashita, A., Oshima, K., Furuya, K., Yoshino, C., Shiba, T., Hattori, M., Ogasawara, N., Hayashi, H., and Hiramatsu, K. (2001) *Lancet* **357**, 1225–1240.

Lander, E. S., and Waterman, M. S. (1988) *Genomics* **2**, 231–239.

Lowe, T. M., and Eddy, S. R. (1997) *Nucleic Acids Res.* **25**, 955–964.

Masignani, V., Rappuoli, R., and Pizza, M. (2002) *Expert Opin. Biol. Ther.* **2**, 895–905.

McClelland, M., Sanderson, K. E., Spieth, J., Clifton, S. W., Latreille, P., Courtney, L.,

Porwollik, S., Ali, J., Dante, M., Du, F., Hou, S., Layman, D., Leonard, S., Nguyen, C., Scott, K., Holmes, A., Grewal, N., Mulvaney, E., Ryan, E., Sun, H., Florea, L., Miller, W., Stoneking, T., Nhan, M., Waterston, R., and Wilson, R. K. (2001) *Nature* **413**, 852–856.

Meldrum, D. (2000a) *Genome Res.* **10**, 1081–1092.

Meldrum, D. (2000b) *Genome Res.* **10**, 1288–1303.

Mitra, R. D., and Church, G. M. (1999) *Nucleic Acids Res.* **27**, e34.

Nakai, K., and Horton, P. (1999) *Trends Biochem. Sci.* **24**, 34–36.

Nielsen, H., Engelbrecht, J., Brunak, S., and von Heijne, G. (1997) *Protein Eng.* **10**, 1–6.

Ogata, H., Audic, S., Renesto-Audiffren, P., Fournier, P. E., Barbe, V., Samson, D., Roux, V., Cossart, P., Weissenbach, J., Claverie, J. M., and Raoult, D. (2001) *Science* **293**, 2093–2098.

Parkhill, J., Achtman, M., James, K. D., Bentley, S. D., Churcher, C., Klee, S. R., Morelli, G., Basham, D., Brown, D., Chillingworth, T., Davies, R. M., Davis, P., Devlin, K., Feltwell, T., Hamlin, N., Holroyd, S., Jagels, K., Leather, S., Moule, S., Mungall, K., Quail, M. A., Rajandream, M. A., Rutherford, K. M., Simmonds, M., Skelton, J., Whitehead, S., Spratt, B. G., and Barrell, B. G. (2000a) *Nature* **404**, 502–506.

Parkhill, J., Dougan, G., James, K. D., Thomson, N. R., Pickard, D., Wain, J., Churcher, C., Mungall, K. L., Bentley, S. D., Holden, M. T., Sebaihia, M., Baker, S., Basham, D., Brooks, K., Chillingworth, T., Connerton, P., Cronin, A., Davis, P., Davies, R. M., Dowd, L., White, N., Farrar, J., Feltwell, T., Hamlin, N., Haque, A., Hien, T. T., Holroyd, S., Jagels, K., Krogh, A., Larsen, T. S., Leather, S., Moule, S., O'Gaora, P., Parry, C., Quail, M., Rutherford, K., Simmonds, M., Skelton, J., Stevens, K., Whitehead, S., and Barrell, B. G. (2001a) *Nature* **413**, 848–852.

Parkhill, J., Wren, B. W., Mungall, K., Ketley, J. M., Churcher, C., Basham, D., Chillingworth, T., Davies, R. M., Feltwell, T., Holroyd, S., Jagels, K., Karlyshev, A. V., Moule, S., Pallen, M. J., Penn, C. W., Quail, M. A., Rajandream, M. A., Rutherford, K. M., van Vliet, A. H., Whitehead, S., and Barrell, B. G. (2000b) *Nature* **403**, 665–668.

Parkhill, J., Wren, B. W., Thomson, N. R., Titball, R. W., Holden, M. T., Prentice, M. B., Sebaihia, M., James, K. D., Churcher, C., Mungall, K. L., Baker, S., Basham, D., Bentley, S. D., Brooks, K., Cerdeno-Tarraga, A. M., Chillingworth, T., Cronin, A., Davies, R. M., Davis, P., Dougan, G., Feltwell, T., Hamlin, N., Holroyd, S., Jagels, K., Karlyshev, A. V., Leather, S., Moule, S., Oyston, P. C., Quail, M., Rutherford, K., Simmonds, M., Skelton, J., Stevens, K., Whitehead, S., and Barrell, B. G. (2001b) *Nature* **413**, 523–527.

Paulsen, I. T., Banerjei, L., Myers, G. S., Nelson, K. E., Seshadri, R., Read, T. D., Fouts, D. E., Eisen, J. A., Gill, S. R., Heidelberg, J. F., Tettelin, H., Dodson, R. J., Umayam, L., Brinkac, L., Beanan, M., Daugherty, S., DeBoy, R. T., Durkin, S., Kolonay, J., Madupu, R., Nelson, W., Vamathevan, J., Tran, B., Upton, J., Hansen, T., Shetty, J., Khouri, H., Utterback, T., Radune, D., Ketchum, K. A., Dougherty, B. A., and Fraser, C. M. (2003) *Science* **299**, 2071–2074.

Paulsen, I. T., Seshadri, R., Nelson, K. E., Eisen, J. A., Heidelberg, J. F., Read, T. D., Dodson, R. J., Umayam, L., Brinkac, L. M., Beanan, M. J., Daugherty, S. C., Deboy, R. T., Durkin, A. S., Kolonay, J. F., Madupu, R., Nelson, W. C., Ayodeji, B., Kraul, M., Shetty, J., Malek, J., Van Aken, S. E., Riedmuller, S., Tettelin, H., Gill, S. R., White, O., Salzberg, S. L., Hoover, D. L., Lindler, L. E., Halling, S. M., Boyle, S. M., and Fraser, C. M. (2002) *Proc. Natl. Acad. Sci. USA* **99**, 13148–13153.

Perna, N. T., Plunkett, G. III, Burland, V., Mau, B., Glasner, J. D., Rose, D. J., Mayhew, G. F., Evans, P. S., Gregor, J., Kirkpatrick, H. A., Posfai, G., Hackett, J., Klink, S., Boutin, A., Shao, Y., Miller, L., Grotbeck, E. J., Davis, N. W., Lim, A., Dimalanta, E. T., Potamousis, K. D., Apodaca, J., Anantharaman, T. S., Lin, J., Yen, G., Schwartz, D. C., Welch, R. A., and Blattner, F. R. (2001) *Nature* **409**, 529–533.

Peterson, J. D., Umayam, L. A., Dickinson, T., Hickey, E. K., and White, O. (2001) *Nucleic Acids Res.* **29**, 123–125.

Pizza, M., Scarlato, V., Masignani, V., Giuliani, M. M., Arico, B., Comanducci, M., Jennings, G. T., Baldi, L., Bartolini, E., Capecchi, B., Galeotti, C. L., Luzzi, E., Manetti, R., Marchetti, E., Mora, M., Nuti, S., Ratti, G., Santini, L., Savino, S., Scarselli, M., Storni, E., Zuo, P., Broeker, M., Hundt, E., Knapp, B., Blair, E., Mason, T., Tettelin, H., Hood, D. W., Jeffries, A. C., Saunders, N. J., Granoff, D. M., Venter, J. C., Moxon, E. R., Grandi, G., and Rappuoli, R. (2000) *Science* **287**, 1816–1820.

Pop, M., Salzberg, S., and Shumway, M. (2002) *IEEE Computer* **35**, 47–54.

Rappuoli, R. (2000) *Curr. Opin. Microbiol.* **3**, 445–450.

Rappuoli, R. (2001a) *Vaccine* **19**, 2319–2322.

Rappuoli, R. (2001b) *Vaccine* **19**, 2688–2691.

Read, T. D., Brunham, R. C., Shen, C., Gill, S. R., Heidelberg, J. F., White, O., Hickey, E. K., Peterson, J., Utterback, T., Berry, K., Bass, S., Linher, K., Weidman, J., Khouri, H., Craven, B., Bowman, C., Dodson, R., Gwinn, M., Nelson, W., DeBoy, R., Kolonay, J., McClarty, G., Salzberg, S. L., Eisen, J., and Fraser, C. M. (2000) *Nucleic Acids Res.* **28**, 1397–1406.

Riley, M. (1993) *Microbiol. Rev.* **57**, 862–952.

Ronaghi, M., Uhlen, M., and Nyren, P. (1998) *Science* **281**, 363, 365.

Salzberg, S. L., Delcher, A. L., Kasif, S., and White, O. (1998) *Nucleic Acids Res.* **26**, 544–548.

Sambrook, J., Fritsch, E. F., and Maniatis, T. (1989) *Molecular Cloning: a Laboratory Manual*. Cold Spring Harbor Laboratory Press, Cold Spring Harbor.

Sanger, F., Nicklen, S., and Coulson, A. R. (1977) *Proc. Natl. Acad. Sci. USA* **74**, 5463–5467.

Sasaki, Y., Ishikawa, J., Yamashita, A., Oshima, K., Kenri, T., Furuya, K., Yoshino, C., Horino, A., Shiba, T., Sasaki, T., and Hattori, M. (2002) *Nucleic Acids Res.* **30**, 5293–5300.

Shimizu, T., Ohtani, K., Hirakawa, H., Ohshima, K., Yamashita, A., Shiba, T., Ogasawara, N., Hattori, M., Kuhara, S., and Hayashi, H. (2002) *Proc. Natl. Acad. Sci. USA* **99**, 996–1001.

Shirai, M., Hirakawa, H., Kimoto, M., Tabuchi, M., Kishi, F., Ouchi, K., Shiba, T., Ishii,

K., Hattori, M., Kuhara, S., and Nakazawa, T. (2000) *Nucleic Acids Res.* **28**, 2311–2314.

Smoot, J. C., Barbian, K. D., Van Gompel, J. J., Smoot, L. M., Chaussee, M. S., Sylva, G. L., Sturdevant, D. E., Ricklefs, S. M., Porcella, S. F., Parkins, L. D., Beres, S. B., Campbell, D. S., Smith, T. M., Zhang, Q., Kapur, V., Daly, J. A., Veasy, L. G., and Musser, J. M. (2002) *Proc. Natl. Acad. Sci. USA* **99**, 4668–4673.

Stephens, R. S., Kalman, S., Lammel, C., Fan, J., Marathe, R., Aravind, L., Mitchell, W., Olinger, L., Tatusov, R. L., Zhao, Q., Koonin, E. V., and Davis, R. W. (1998) *Science* **282**, 754–759.

Stover, C. K., Pham, X. Q., Erwin, A. L., Mizoguchi, S. D., Warrener, P., Hickey, M. J., Brinkman, F. S., Hufnagle, W. O., Kowalik, D. J., Lagrou, M., Garber, R. L., Goltry, L., Tolentino, E., Westbrock-Wadman, S., Yuan, Y., Brody, L. L., Coulter, S. N., Folger, K. R., Kas, A., Larbig, K., Lim, R., Smith, K., Spencer, D., Wong, G. K., Wu, Z., Paulsen, I. T., Reizer, J., Saier, M. H., Hancock, R. E., Lory, S., and Olson, M. V. (2000) *Nature* **406**, 959–964.

Sutton, G., White, O., Adams, M., and Kerlavage, A. R. (1995) *Genome Sci. Technol.* **1**, 9–19.

Tatusov, R. L., Koonin, E. V., and Lipman, D. J. (1997) *Science* **278**, 631–637.

Tettelin, H., Masignani, V., Cieslewicz, M. J., Eisen, J. A., Peterson, S., Wessels, M. R., Paulsen, I. T., Nelson, K. E., Margarit, I., Read, T. D., Madoff, L. C., Wolf, A. M., Beanan, M. J., Brinkac, L. M., Daugherty, S. C., DeBoy, R. T., Durkin, A. S., Kolonay, J. F., Madupu, R., Lewis, M. R., Radune, D., Fedorova, N. B., Scanlan, D., Khouri, H., Mulligan, S., Carty, H. A., Cline, R. T., Van Aken, S. E., Gill, J., Scarselli, M., Mora, M., Iacobini, E. T., Brettoni, C., Galli, G., Mariani, M., Vegni, F., Maione, D., Rinaudo, D., Rappuoli, R., Telford, J. L., Kasper, D. L., Grandi, G., and Fraser, C. M. (2002) *Proc. Natl. Acad. Sci. USA* **99**, 12391–12396.

Tettelin, H., Nelson, K. E., Paulsen, I. T., Eisen, J. A., Read, T. D., Peterson, S., Heidelberg, J., DeBoy, R. T., Haft, D. H., Dodson, R. J., Durkin, A. S., Gwinn, M., Kolonay, J. F., Nelson, W. C., Peterson, J. D., Umayam, L. A., White, O., Salzberg, S. L., Lewis, M. R., Radune, D., Holtzapple, E., Khouri, H., Wolf, A. M., Utterback, T. R., Hansen, C. L., McDonald, L. A., Feldblyum, T. V., Angiuoli, S., Dickinson, T., Hickey, E. K., Holt, I. E., Loftus, B. J., Yang, F., Smith, H. O., Venter, J. C., Dougherty, B. A., Morrison, D. A., Hollingshead, S. K., and Fraser, C. M. (2001) *Science* **293**, 498–506.

Tettelin, H., Radune, D., Kasif, S., Khouri, H., and Salzberg, S. L. (1999) *Genomics* **62**, 500–507.

Tettelin, H., Saunders, N. J., Heidelberg, J., Jeffries, A. C., Nelson, K. E., Eisen, J. A., Ketchum, K. A., Hood, D. W., Peden, J. F., Dodson, R. J., Nelson, W. C., Gwinn, M. L., DeBoy, R., Peterson, J. D., Hickey, E. K., Haft, D. H., Salzberg, S. L., White, O., Fleischmann, R. D., Dougherty, B. A., Mason, T., Ciecko, A., Parksey, D. S., Blair, E., Cittone, H., Clark, E. B., Cotton, M. D., Utterback, T. R., Khouri, H., Qin, H., Vamathevan, J., Gill, J., Scarlato, V., Masignani, V., Pizza, M., Grandi, G., Sun, L., Smith, H. O., Fraser, C. M., Moxon, E. R., Rappuoli, R., and Venter, J. C. (2000) *Science* **287**, 1809–1815.

Tomb, J. F., White, O., Kerlavage, A. R., Clayton, R. A., Sutton, G. G., Fleischmann, R. D., Ketchum, K. A., Klenk, H. P., Gill, S., Dougherty, B. A., Nelson, K., Quackenbush, J., Zhou, L., Kirkness, E. F., Peterson, S., Loftus, B., Richardson, D., Dodson, R., Khalak, H. G., Glodek, A., McKenney, K., Fitzegerald, L. M., Lee, N., Adams, M. D., Venter, J. C., *et al.* (1997) *Nature* **388**, 539–547.

Waterman, M. S. (1988) *Methods Enzymol.* **164**, 765–793.

Wizemann, T. M., Heinrichs, J. H., Adamou, J. E., Erwin, A. L., Kunsch, C., Choi, G. H., Barash, S. C., Rosen, C. A., Masure, H. R., Tuomanen, E., Gayle, A., Brewah, Y. A., Walsh, W., Barren, P., Lathigra, R., Hanson, M., Langermann, S., Johnson, S., and Koenig, S. (2001) *Infect. Immun.* **69**, 1593–1598.

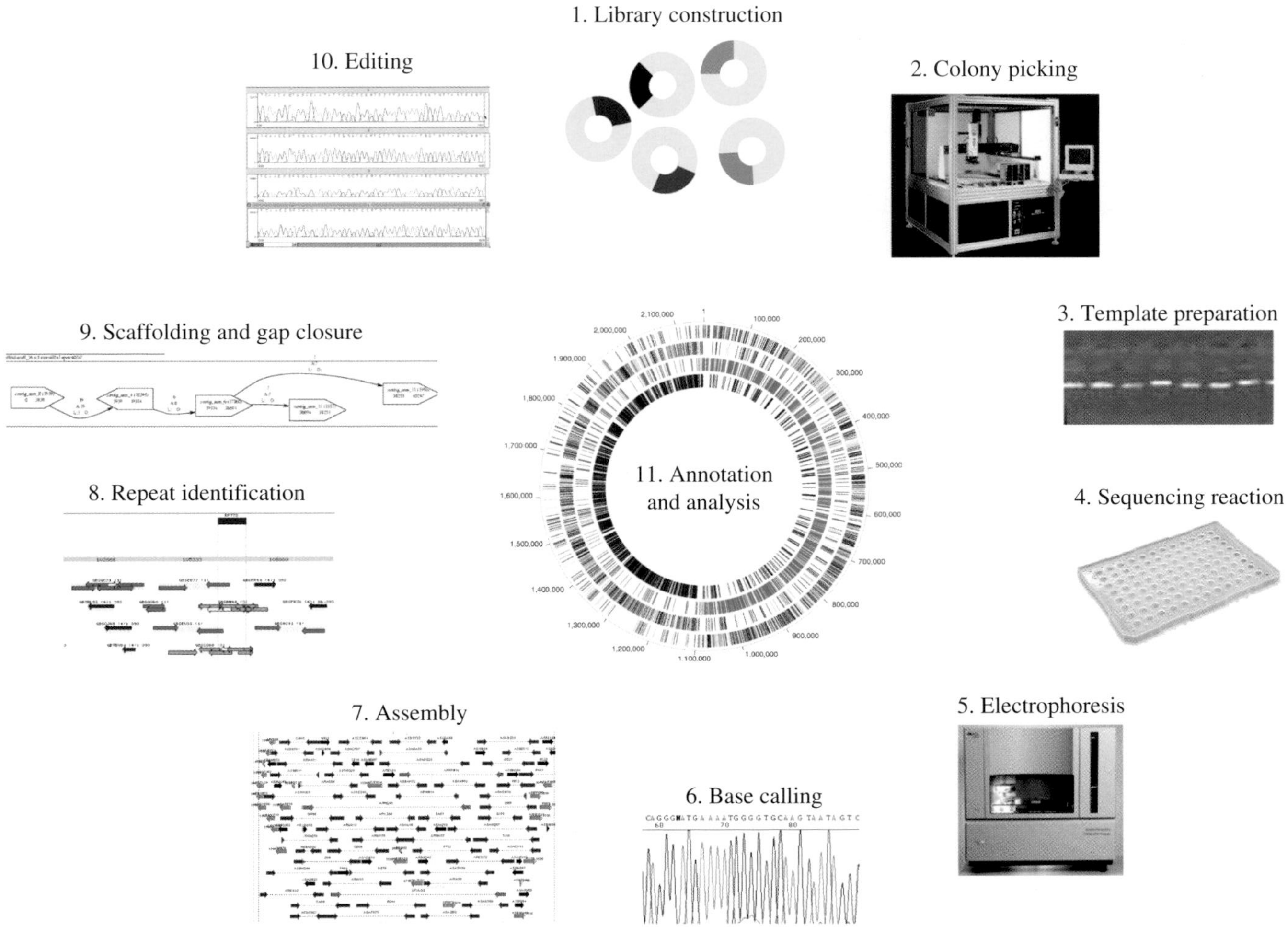
1. Library construction
2. Colony picking
3. Template preparation
4. Sequencing reaction
5. Electrophoresis
6. Base calling
CAGGGNATGAAAATGGGGTGCAAGTAATAGTC
60
70
80
7. Assembly
8. Repeat identification
9. Scaffolding and gap closure
10. Editing
11. Annotation and analysis
1
100,000
200,000
300,000
400,000
500,000
600,000
700,000
800,000
900,000
1,000,000
1,100,000
1,200,000
1,300,000
1,400,000
1,500,000
1,600,000
1,700,000
1,800,000
1,900,000
2,000,000
2,100,000

Figure 3.1

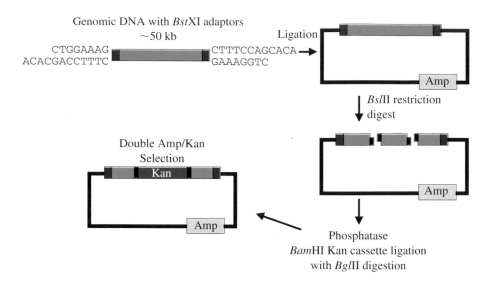

Figure 3.2

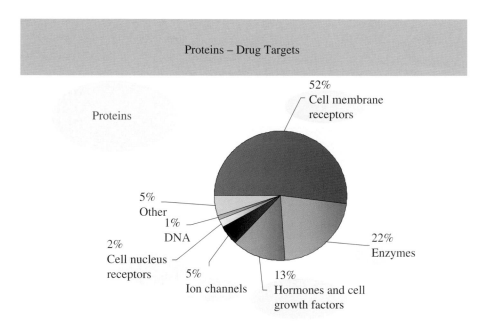

Figure 5.1

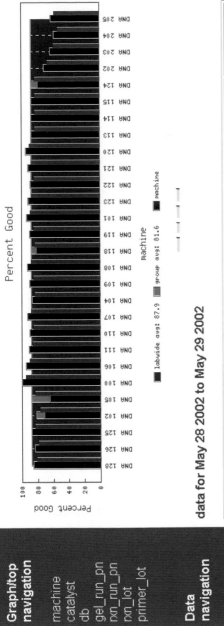

Graph/top navigation

machine
catalyst
db
gel_run_pn
rxn_run_pn
rxn_lot
primer_lot

Data navigation

machine
catalyst
db
gel_run_pn
rxn_run_pn
rxn_lot
primer_lot
bad gels
group gels

data for May 28 2002 to May 29 2002

db	Total Good	Total Lanes	% Good	VEIR	% VEIR	Average Edit Length	Average Quality
gra	226	288	78.5 -9.4	2	0.7	647 -37	31.7 +0.1
gtd	8	12	66.7 -21.2	0	0.0	634 -50	27.9 -3.7
mbb	6	6	100.0 +12.1	0	0.0	1278 +594	30.4 -1.2
osg	533	626	85.1 -2.8	29	4.6	747 +63	32.5 +0.9
pvg	3725	4128	90.2 +2.3	31	0.8	667 -17	31.5 -0.1
rbe	173	192	90.1 +2.2	0	0.0	718 +34	31.6 -0.1
research	186	192	96.9 +9.0	0	0.0	632 -52	30.8 -0.8
tbg	16	24	66.7 -21.2	0	0.0	562 -122	30.7 -0.9
tcrg	2228	2398	92.9 +5.0	3	0.1	738 +54	32.4 +0.8
tgg	1327	1440	92.2 +4.3	0	0.0	718 +34	32.4 +0.8
Totals	12435 (avg: 444)	14154 (avg: 506)	87.9	128	0.9	684	31.6

Figure 3.3

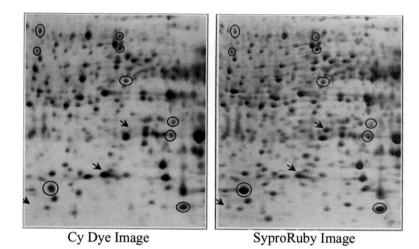

Cy Dye Image   SyproRuby Image

Figure 5.4

Multicompartment electrolyser fractionation

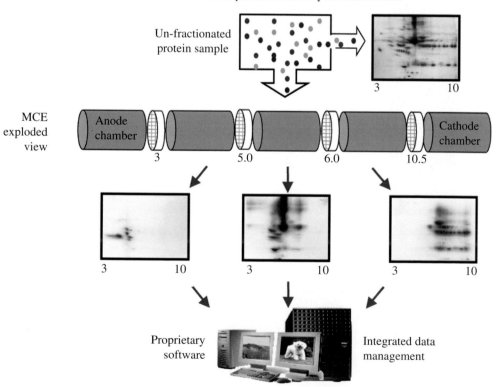

Un-fractionated
protein sample

MCE
exploded
view

Anode
chamber

Cathode
chamber

3   5.0   6.0   10.5

3   10

3   10   3   10   3   10

Proprietary
software

Integrated data
management

Figure 5.5

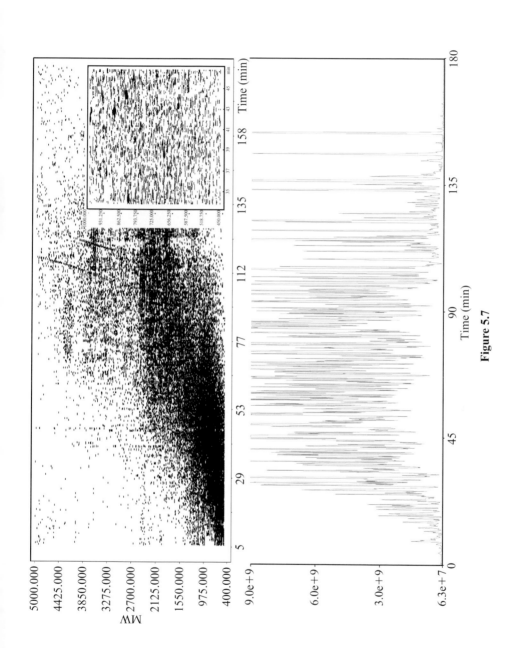

Figure 5.7

ProteinChip® Arrays

Chemical surfaces

(Reverse Phase) (Cation Exchange) (Anion Exchange) (Metal Ion) (Normal Phase)

Bioaffinity surfaces

(Reactive surface) (Antibody–Antigen) (Receptor–Ligand) (DNA–Protein)

Figure 5.8

Principle of ProteinChip® Process

- Sample  goes *directly* onto the ProteinChip Array
- Proteins ● are captured and washed directly on the chip (chromatographic capture ◯)
- Surface is 'read' by surface-enhanced laser desorption/ionization (SELDI)-MS

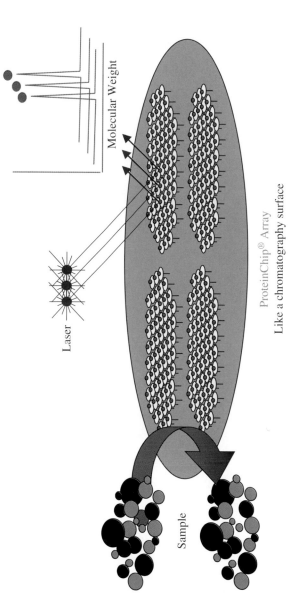

Laser

Molecular Weight

Sample

ProteinChip® Array
Like a chromatography surface

Figure 5.9

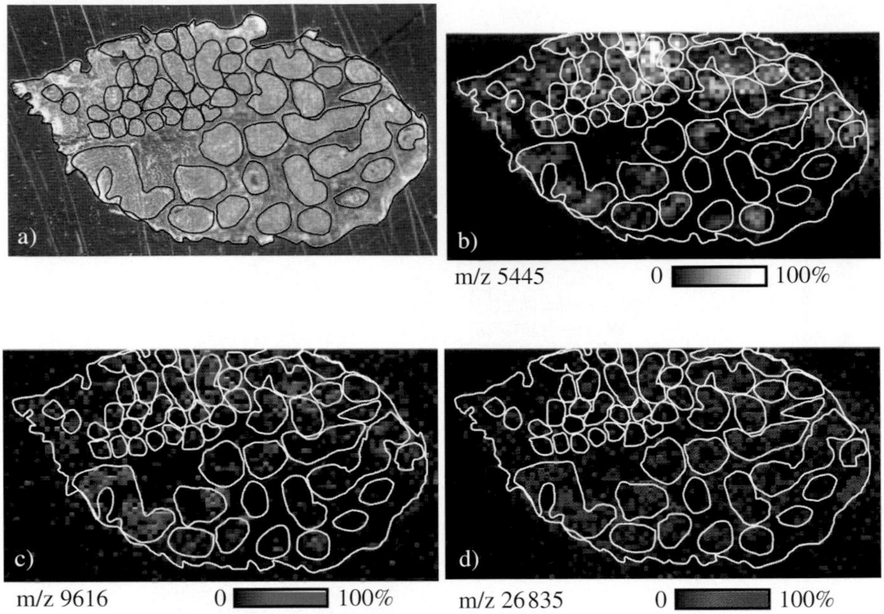
a)
b)
m/z 5445
0
100%
c)
d)
m/z 9616
0
100%
m/z 26835
0
100%

Figure 5.10

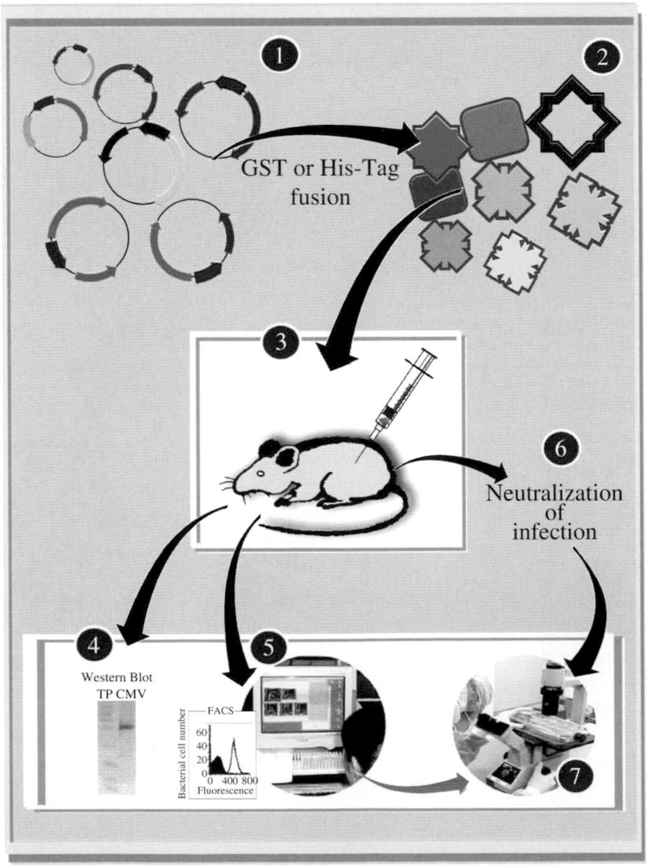
1
2
GST or His-Tag fusion
3
6
Neutralization of infection
4
Western Blot
TP CMV
5
FACS
Bacterial cell number
Fluorescence
7

Figure 11.1

4

Understanding DNA Microarrays: Sources and Magnitudes of Variances in DNA Microarray Data Sets

She-pin Hung, Suman Sundaresh, Pierre F. Baldi and **G. Wesley Hatfield**

4.1 Introduction

From time to time, scientific breakthroughs or new technologies are born that forever change the way we do science. Advances in high-throughput DNA sequencing represent such a breakthrough that has led us into the genomics era. Just in the past few years, we have obtained genomic sequences for scores of organisms from man to bacteria, and the accumulation of new genome sequences is continuing at a rapid pace. Based on these accomplishments, many investigators are turning their attention to the study of complex biological systems at molecular levels never before possible. This systems biology approach is particularly evident in the gene expression field, where the availability of microarrays containing DNA probes for all of the genes of an organism makes whole-cell gene expression profiling possible. However, with these new genomic methods come new challenges. Foremost among these challenges is the development of computational methods to analyse and interpret the masses of data generated by genomic experiments. To meet these needs, biologists have developed dialogues with scientists in other fields, predominantly computer scientists, statisticians, and mathematicians, and great

Genomics, Proteomics and Vaccines edited by Guido Grandi
© 2004 John Wiley & Sons, Ltd   ISBN 0 470 85616 5

progress is being made (Baldi and Hatfield, 2002). In this chapter, we briefly review some of the most popular current DNA microarray platforms with an emphasis on experimental design and computational methods to deal with experimental and biological variances. Finally, we present analyses of highly replicated DNA microarray data sets that highlight the sources of this variance that complicate the analysis of DNA array data.

4.2   DNA array formats

Array technologies monitor the combinatorial interaction of a set of molecules, such as DNA fragments or proteins, with a library of molecular probes. The most advanced of these technologies is DNA arrays, also called DNA chips or biochips, for simultaneously measuring the level of all the mRNA gene products expressed in a living cell. DNA arrays in common used today contain an orderly arrangement of hundreds to thousands of unique DNA molecules (probes) of known sequence. There are two basic sources for the DNA probes on an array. Either each unique probe is individually synthesized on a rigid surface (usually glass), or pre-synthesized probes (oligonucleotides or PCR products) are attached directly to the array platform (usually glass slides or nylon membranes).

4.2.1   *In situ* synthesized oligonucleotide arrays

The first *in situ* probe synthesis method for manufacturing DNA arrays was the photolithographic method developed by Fodor *et al*. (1993) and commercialized by Affymetrix Inc. (Santa Clara, CA). First, a set of oligonucleotide DNA probes (each 25 or so nucleotides in length) is defined based on its ability to hybridize to complementary sequences in target genomic loci or genes of interest. With this information, computer algorithms are used to design photo-lithographic masks for use in manufacturing the probe arrays. Selected addresses on a photo-protected glass surface are illuminated through holes in the photolithographic mask, the glass surface is flooded with the first nucleotide of the probes to be synthesized at the selected addresses, and photo-chemical coupling occurs at these sites. For example, the addresses on the glass surface for all probes beginning with guanosine are photo-activated and chemically coupled to guanine bases. This step is repeated three more times with masks for all addresses with probes beginning with adenosine, thymine or cytosine. The cycle is repeated with masks designed for adding the appropriate second nucleotide of each probe. During the second cycle, modified phosphoramidite moieties on each of the nucleosides attached to the glass surface in the first step are light activated through appropriate masks for the addition of the second base to each growing oligonucleotide probe. This process is continued until unique probe oligonucleotides of a defined length and sequence have been synthesized at each of thousands of addresses on the glass surface (Figure 4.1).

While Affymetrix GeneChips[TM] are the most widely used *in situ* synthesized DNA microarrays, non-lithographic chip manufacturing technologies including electronic methods are also being developed.

4.2.2 Pre-synthesized DNA arrays

The method of attaching pre-synthesized DNA probes (usually 50–5000 bases long) to a solid surface such as glass (or nylon filter) supports was conceived 25 years ago by Ed Southern. Modern glass slide DNA array manufacturing methods pioneered in the laboratory of Patrick O. Brown at Stanford University has made DNA arrays affordable for academic research laboratories. As early as 1996 the Brown laboratory published step-by-step plans for the construction of a robotic DNA arrayer on the internet. Since that time, many commercial DNA arrayers have become available. Besides the commercially produced Affymetric GeneChips[TM], these Brown-type glass slide DNA arrays are currently the most popular format for gene expression profiling experiments.

The Brown method for printing glass slide DNA arrays involves the robotic spotting of small volumes (in the nanoliter to picoliter range) of a DNA probe sample onto a $25 \times 76 \times 1$ mm^3 glass slide surface previously coated with poly-lysine or poly-amine for elecrostatic adsorption of the DNA probes onto the slide. Depending upon the pin type and the exact printing technology employed, as many as 20 000 or more spots ranging in size from 50 to 75 μm can be spotted in a 1 cm^2 area. Many public and private research institutions in the US and abroad have developed core facilities for the in-house manufacture of custom glass slide DNA arrays.

4.2.3 Filter-based DNA arrays

Although solid support matrices such as glass offer many advantages for high-throughput processing, nylon-filter-based arrays continue to be a popular format. This is because gene expression profiling with nylon filters is based on standard Southern blotting protocols familiar to molecular biologists, and because equipment to perform filter hybridizations with ^{33}P-labelled cDNA

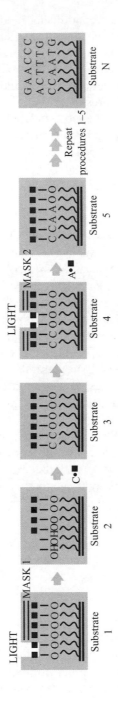

Affymetrix GeneChip probe array synthesis process

Figure 4.1 The Affymetrix method for the manufacture of *in situ* synthesized DNA microarrays. 1. A photo-protected glass substrate is selectively illuminated by light passing through a photolithographic mask. 2. Deprotected areas are activated. 3. The surface is flooded with a nucleoside solution and chemical coupling occurs at photo-activated positions. 4. A new photolithographic mask pattern is applied. 5. The coupling step is repeated. *N*. This process is repeated until the desired set of probes is obtained

targets and for data acquisition, such as a phosphorimager, are available at most research institutions.

Although non-radioactive labelling methods are preferred, radioactive labelling of the targets for hybridization to nylon filter arrays offers the advantage of greater sensitivity compared with fluorescently labelled targets, and intensity measurements linear over a four to five log range are achieved with radio-labelled targets, whereas linear ranges of only three logs are typically observed with fluorescently labelled targets. Additional advantages of nylon filter arrays are that many types are available on the commercial market, and the cost advantage inherent in the fact that nylon filters can be stripped and reused several times without significant deterioration (Arfin et al., 2000). However, because of the porous nature of nylon filters, they are not amenable to miniaturization. Nevertheless, they are suited for gene expression profiling studies in organisms with small genome sizes such as bacteria, or for the production of custom arrays containing DNA probes for a functional subset of an organism's genes or for disease diagnosis. For example, a popular nylon filter DNA array for the model organism *Escherichia coli* manufactured by Sigma-Genosys (Woodland, TX) measures 11×21 cm^2, contains over 18 000 addresses spotted, in duplicate, with full-length PCR products of each of the 4290 *E. coli* open reading frames (ORFs). The remaining empty addresses dispersed throughout the array are used for background measurements.

Detailed discussions of the instrumentation and methods for manufacturing glass slide DNA arrays can be found in a book edited by Mark Schena titled *DNA Microarrays: a Practical Approach* (1999). Experimental methods and protocols for preparing and labelling microarray targets, as well as hybridization protocols and data acquisition and normalization methods for filters, glass slides and Affymetrix GeneChips, are described in a recent book by Baldi and Hatfield titled *DNA Microarrays and Gene Expression: from Experiments to Data Analysis and Modeling* (2002).

4.3   Data analysis methods

Many experimental designs and applications of gene expression profiling experiments are possible. However, no matter what the purpose of the experiment, a sufficient number of measurements must be obtained for statistical analysis of the data, either through multiple measurements of homogeneous samples (replication) or multiple sample measurements (e.g. across time or subjects). This is basically because each gene expression profiling experiment results in the simultaneous measurement of the expression levels of thousands of genes. In such a high-dimensional experiment, many genes will show large changes in expression levels between two experimental conditions without being significant. In the same manner, many truly differentially expressed genes will show small changes. These false positive and false negative observations arise from chance occurrences exacerbated by uncontrolled biological variance as well as experimental and measurement errors.

For example, if we measure the gene expression patterns of cells simply grown under two different treatment conditions or between two genotypes, experimental replication will be required for the assignment of statistical significance to differential gene measurements. However, if, as recently reported by the NCI (Scheif *et al.*, 2000), we measure the effects of 188 drugs on the expression patterns of 1376 genes in 60 cancer cell line samples, statistical significance is achieved by the averaging of measurements across samples instead of across individually replicated experiments. In this latter case, clusters of gene expression patterns can be measured with statistical significance in the absence of individual experimental replications. However, most DNA microarray experiments are designed to examine gene expression patterns between a limited numbers of treatment conditions. While these types of experiment demand replication, such replications quickly become labour intensive and prohibitively expensive. This leads to the question of how many replicates are required.

4.3.1 Robust estimation of standard deviation with a small number of replicates

The short answer to the above question is – enough to provide a robust estimate of the standard deviation of each gene measurement. Obviously, if we could repeat our microarray experiments 1000 times for each treatment condition we could accurately determine the difference of gene expression levels between two treatment conditions simply by comparing the means of the two measurements, for example using a simple fold approach or a *t*-test. The problem is that neither approach performs well with a small number of replicates typical of DNA array experiments because of poor estimates of the standard deviations. Thus, the problem reduces to finding a method that will produce more robust estimates of the standard deviation of a small set of individual gene measurements.

To address this problem, we have shown that the confidence in the interpretation of DNA microarray data with a low number of replicates can be improved by using a Bayesian statistical approach that incorporates information of within treatment measurements (Long *et al.*, 2001; Baldi and Long, 2001). The

Bayesian approach is based on the observation that genes of similar expression levels exhibit similar variance. Thus, more robust estimates of the variance of a gene can be derived by pooling neighboring genes with comparable expression levels. For this analysis, gene expression data can be ranked according to the mean expression levels of the replicate experiments. For example, we can use a sliding window of 101 genes, and assign the average variance of the 50 genes ranked below and above each gene as the background variance for that gene. The variance of any gene within any given treatment then can be estimated by a weighted average of the treatment-specific background variance and the treatment-specific empirical variance across experimental replicates. In turn, it is this regularized variance that is used for the t-test. Using this Bayesian approach, the weight given to the within experiment gene variance estimate is a function of the number of experimental replicates. This leads to the desirable property that the Bayesian approach employing such a regularized t-test converges on the same set of differentially expressed genes as the simple t-test but with fewer replicates (Long $et\ al.$, 2001). This Bayesian approach has been implemented in a DNA microarray data analysis program, Cyber-T, available for online use at www.igb.uci.edu. Several examples of its utility and complete mathematical details of its algorithms have been published (Baldi and Hatfield, 2002; Arfin $et\ al.$, 2000; Long $et\ al.$, 2001; Hatfield, Hung and Baldi, 2003; Hung, Baldi and Hatfield, 2002).

4.3.2   Estimation of global false positive levels

Although the Bayesian approach allows us to identify differentially expressed genes at a higher confidence level with fewer replicates, we still expect many false positive and false negative observations based on chance occurrences inherent in a high-dimensional experiment involving thousands of simultaneous measurements. Therefore, to more accurately interpret the results of a high-dimensional DNA array experiment it is necessary to determine the global false positive and false negative levels at any given statistical threshold inherent in the data set being analysed. For example, a simple way to infer the false positive rate is to compare controls to controls. With this information, a global level of confidence can be calculated for differentially expressed genes measured at any given statistical significance level. For example, consider an experiment comparing the gene expression profiles of two genotypes where an average of 10 genes are observed to be differentially expressed with a p-value less than 0.0001 when gene expression profiles from one genotype are compared to data of the same genotype. Since no differential expression is expected, this comparison identifies 10 false positives generated by chance occurrences driven by experimental errors and biological variance. Now if 100 genes are differentially expressed with a p-value less than 0.0001 when the data from one genotype is compared to the data from the other genotype, it is reasonable to infer that we can be only 90 per cent confident that the differential expression of any one of these 100 genes is biologically meaningful since 10 false positives are expected from this data set. This example demonstrates that, although the local confidence level based on the measurement for an individual gene may exceed 99.99 per cent for two treatment conditions (local confidence 0.0001), the global confidence that this gene is differentially expressed might be only 90 per cent (global confidence 0.9).

A computational method for the estimation of false positive and false negative levels based on this reasoning has recently been described (Allison *et al.*, 2002) and incorporated into the Cyber-T software. The basic idea, briefly described here, is to consider the p-values as a new data set and to build a probabilistic model for this new data. When control data sets are compared with one another (i.e. no differential gene expression), it is easy to see that the p-values ought to have a uniform distribution between zero and one. In contrast, when data sets from different genotypes or treatment conditions are compared with one another, the distribution of p-values will tend to cluster more closely to zero than one; that is, there will be a subset of differentially expressed genes with 'significant' p-values. The goal becomes to model the observed mixed distribution of p-values in a way that estimates the probability that any given gene measurement is associated with the uniform distribution (unchanged) or a member of the non-uniform distribution (changed). The result is a posterior probability of differential expression (PPDE) value ranging from zero (unchanged) to one (changed). A mathematical description of this method and an experimental example of its utility has been published by Hung, Baldi and Hatfield (2002).

A plot of the p-value distributions from an experiment comparing two genotypes of otherwise isogenic *E. coli* strains (lrp^+ versus lrp^-) is shown in Figure 4.2. The curved dashed line along the top of the binned p-values describes the mixed (uniform plus non-uniform) distribution. The straight horizontal line describes the uniform distribution. The arrows identify the true positive, true negative, false positive and false negative regions at a confidence threshold level of 95 per cent ($p = 0.05$). It is clear from this plot that as the confidence threshold is lowered more true positives will be identified but at the expense of more false positives. This trade-off is further illustrated in the receiver operating characteristic (ROC) curve shown in Figure 4.3. Here it can again be seen that as the confidence threshold is lowered more false positives are observed. For example, if we accept a 5.0 per cent false hit rate we will identify 45 per cent of our differentially expressed genes. However, if we are

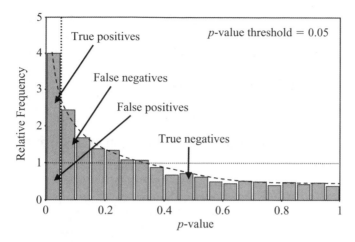

Figure 4.2 Distribution of the p-values from the lrp^+ versus lrp^- data. The curved dashed line along the top of the binned p-values describes the mixed (uniform plus non-uniform) distribution. The straight horizontal line describes the uniform distribution. The arrows identify the true positive, true negative, false positive and false negative regions at a confidence threshold level of 95 per cent ($p = 0.05$)

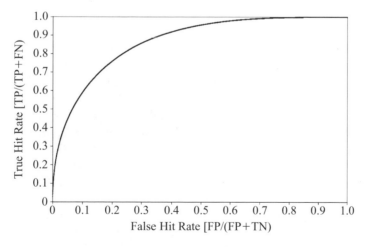

Figure 4.3 Receiver operating characteristic (ROC) curve. This plot relates the fraction of correctly identified differentially expressed genes (y-axis) to the fraction of falsely identified differentially expressed genes (x-axis).

willing to accept a false hit rate of 25 per cent we will identify more than 80 per cent of our differentially expressed genes. This clearly illustrates that the only way to identify all of the differentially expressed genes in a high-dimensional DNA microarray experiment with experimental and biological noise is to accept a false hit rate of 100 per cent.

4.4   Sources and magnitudes of noise in DNA microarray experiments

The mathematical approaches we describe above are designed to handle the large amount of variance inherent in DNA microarray experiments. However, they cannot perform miracles. The only way to improve the quality of a DNA microarray data set is to reduce the variances driven by differences in experimental conditions and scores of manipulations, as well as instrument noise and biological differences (Baldi and Hatfield, 2002). Thus, in order to maximize the quality of information obtained from such experiments these sources of variance should be suppressed as much as possible. However, to rationally approach this problem we need to know the sources and magnitudes of these experimental, instrument and biological variables, which we refer to here as experimental and biological variance.

4.4.1   Data sets

To approach this problem we have analysed the sources and magnitudes of variance inherent in two highly replicated data sets from different DNA microarray formats. The first set contains 32 measurements from 16 nylon filter DNA microarrays containing duplicate probe sites for each of 4290 open reading frames (ORFs) hybridized with ^{33}P-labelled cDNA targets from wild-type *Escherichia coli* cells cultured at 37 °C under balanced growth conditions in glucose minimal salt medium (Arfin *et al.*, 2000; Hung, Baldi and Hatfield, 2002). The experimental design and methods for these experiments are illustrated in Figure 4.4 and described in detail elsewhere (Baldi and Hatfield, 2002; Arfin *et al.*, 2000; Hung, Baldi and Hatfield, 2002).

To address another DNA microarray technology, a second set contains data from four Affymetrix GeneChip experiments that measured the expression levels of the genes of the same *E. coli* cells cultured under the same growth conditions. The experimental design and methods for these experiments are illustrated in Figure 4.5 and described in detail by Hung, Baldi and Hatfield

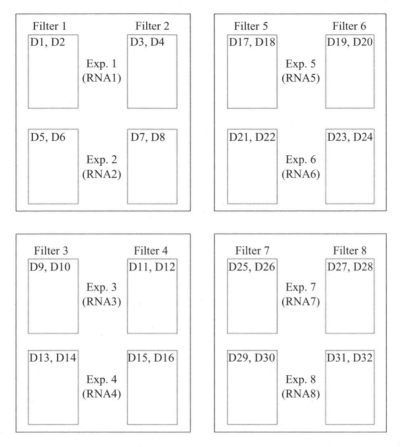

Figure 4.4 Experimental design for nylon filter DNA array experiments. In Experiment 1, filters 1 and 2 were hybridized with ^{33}P-labelled, random-hexamer-generated, cDNA targets complementary to each of three independently prepared RNA preparations (RNA1) obtained from the cells of three individual cultures of a wild-type (*wt*) *E. coli* strain. These three ^{33}P-labelled cDNA target preparations were pooled prior to hybridization to the full-length ORF probes on the filters (experiment 1). Following phosphorimager analysis, these filters were stripped and again hybridized with pooled, ^{33}P-labelled cDNA targets complementary to each of another three independently prepared RNA preparations (RNA2) from the *wt* strain (experiment 2). This procedure was repeated twice more with filters 3 and 4 using two more independently prepared pools of cDNA targets (experiment 3, RNA3; experiment 4, RNA4). Another set of filters, filters 3 and 4, was used for experiments 3 and 4 as described for experiments 1 and 2. This protocol results in duplicate filter data for four experiments performed with cDNA targets complementary to four independent prepared sets of pooled RNA. Thus, since each filter contains duplicate spots for each ORF and duplicate filters were used for each experiment, 16 measurements (D1–D16) for each ORF from four experiments were obtained. These procedures were performed with another two pairs of filters 5–8, for experiments 5–8 to obtain another 16 measurements (D17–D32) for each ORF

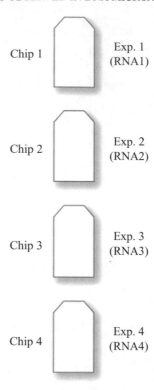

Figure 4.5 Experimental design of the Affymetrix GeneChip experiments. The same 12 total RNA preparations as used for the four pooled RNA sets (RNA1–RNA4) used for filter experiments 1–4 were used for the preparation of biotin-labelled RNA targets for hybridization to four Affymetrix GeneChips. The *.cel file generated by the scanner was used as the raw data source for all subsequent analyses

(2002). The GeneChip measurements were processed by two separate methods, either by the MAS 4.0 or MAS 5.0 software of Affymetrix or by the dChip software of Li and Wong (2001).

Theoretically, all of these experiments should produce the same results since they were performed – as far as possible – under the same conditions with the same cell type. However, as demonstrated below, even these data sets exhibit differences that can be correlated with differences among RNA preparations, differences among filters, differences among GeneChips, differences between filters and GeneChips and even the different times when the experiments were performed.

4.4.2 Data analysis

The gene expression measurements for each experiment of both data sets from the filters and the GeneChips were globally normalized by dividing each expression measurement of each filter or GeneChip by the sum of all the gene expression measurements of that filter or GeneChip. Thus, the signal for each measurement can be expressed as a fraction of the total signal for each filter of GeneChip, or by implication the fraction of total mRNA.

Of the 32 filter measurements, the data from one set of experiments, 1–16, were obtained in October 2001 and the data from the second set, 17–32, were obtained six months earlier in April 2001. During this period of time the efficiency of our ^{33}P labelling was improved. Consequently, more signals marginally above background were detected on the filters for experiments 1–16. In fact, when we edit out all of the genes that contain one or more measurements at or below background in at least one experiment we observe the expression of 2607 genes for all experiments 1–16 and 1194 genes for experiments 17–32. If we consider the whole data set for all 32 experiments, we find that only 959 genes have all 32 expression measurements above background. The log-transformed values of the 959 gene expression measurements above background for all 32 filter experiments and the log-transformed values of 2370 gene expression measurements above background for all GeneChip measurements were used for subsequent analyses.

To investigate correlations between conditions such as different filters or different cDNA target preparations we computed and analysed a correlation matrix. For any two variables x and y that are each vectors of length n, the sample correlation coefficient is defined as

$$\text{cor}(x, y) = \frac{\text{cov}(x, y)}{\sigma_x \sigma_y}$$

where σ_x and σ_y are the sample standard deviations of x and y, respectively. $\text{cov}(x, y)$ is the sample covariance between x and y and is defined as

$$\text{cov}(x, y) = \frac{1}{n} \sum_{i=1}^{n} (x(i) - \bar{x})(y(i) - \bar{y})$$

where $\bar{x}$ and $\bar{y}$ are the sample means of x and y, respectively

The correlation coefficient indicates the strength of the similarity between two variables (from +1 to −1). Values of +1 or −1 indicate strong correlations in the same or opposite directions, respectively. Values close to zero indicate weak correlations. Given m variables, we can construct an $m \times m$ correlation matrix by computing the sample correlation coefficient for each pair of variables.

4.4.3 Correlations between measurements from filter experiments

A 32×32 correlation matrix of the duplicate measurements of all above background target signals present on the 16 filters illustrated in Figure 4.4 is shown in Table 4.1. These results clearly demonstrate strong correlations among the measurements of the experiments performed in October 2001 (D1– D16 versus D1–D16) as well as strong correlations among the measurements of the experiments performed in April 2001 (D17–D32 versus D17–D32). However, lesser correlations are observed among measurements of experiments performed at different times (D1–D16 versus D17–D32). These results demonstrate that significant variance can be introduced into a DNA microarray experiment when experimental parameters such as personnel, reagents, protocols and experimental methods vary.

Another informative observation comes from the results of one of the duplicate measurements on filter 8 of experiment 7 performed in April 2001 (Figure 4.4, filter 8; Table 4.1, D28). In this case, poor correlation is observed between this measurement and other measurements performed at the same time. These results clearly identify this measurement as an outlier that should be further examined for its accuracy.

4.4.4 Correlations by pairs of experiments

Because two filters were hybridized with the same cDNA targets and two cDNA target preparations were hybridized to the same filter for each of eight cDNA target preparations, we can examine the correlations both between filters and cDNA target preparations (Figure 4.4). A 16×16 correlation matrix comparing the average of duplicate measurements of each of the 16 filters is shown in Table 4.2. These data show a consistently high correlation when the same cDNA targets were hybridized to different filters (Table 4.2, black highlights). On the other hand, lesser correlations are observed when different cDNA targets are hybridized to the same filters (Table 4.2, light gray highlights).

In addition to confirming earlier suggestions that the experimental and biological variables of a DNA microarray experiment contribute more variance than differences among the microarrays themselves (Arfin *et al.*, 2000), these data demonstrate both the subtle differences among replicated gene measure-

Table 4.1 Single data point correlation values. A 32 × 32 correlation matrix of the duplicate measurements of all above background target signals present on the 16 filters illustrated in Figure 4.1

	D1	D2	D3	D4	D5	D6	D7	D8	D9	D10	D11	D12	D13	D14	D15	D16	D17	D18	D19	D20	D21	D22	D23	D24	D25	D26	D27	D28	D29	D30	D31	D32
D1	1.00																															
D2	0.97	1.00																														
D3	0.96	0.96	1.00																													
D4	0.94	0.96	0.97	1.00																												
D5	0.95	0.95	0.94	0.93	1.00																											
D6	0.95	0.96	0.94	0.93	0.98	1.00																										
D7	0.92	0.92	0.92	0.92	0.95	0.95	1.00																									
D8	0.90	0.92	0.91	0.91	0.94	0.94	0.97	1.00																								
D9	0.69	0.69	0.70	0.69	0.78	0.78	0.81	0.81	1.00																							
D10	0.66	0.67	0.67	0.67	0.76	0.75	0.78	0.78	0.96	1.00																						
D11	0.68	0.68	0.68	0.68	0.76	0.76	0.80	0.79	0.96	0.93	1.00																					
D12	0.68	0.70	0.69	0.69	0.78	0.78	0.80	0.81	0.95	0.94	0.95	1.00																				
D13	0.83	0.83	0.83	0.82	0.90	0.89	0.90	0.89	0.93	0.90	0.92	0.92	1.00																			
D14	0.83	0.84	0.83	0.83	0.89	0.90	0.90	0.90	0.91	0.92	0.90	0.92	0.99	1.00																		
D15	0.81	0.82	0.81	0.81	0.88	0.88	0.89	0.89	0.92	0.90	0.92	0.92	0.98	0.97	1.00																	
D16	0.80	0.82	0.81	0.80	0.88	0.88	0.88	0.89	0.91	0.89	0.92	0.92	0.97	0.97	0.98	1.00																
D17	0.50	0.51	0.51	0.51	0.48	0.49	0.44	0.43	0.21	0.19	0.20	0.20	0.37	0.37	0.35	0.36	1.00															
D18	0.51	0.52	0.52	0.52	0.48	0.50	0.45	0.44	0.21	0.20	0.20	0.21	0.38	0.38	0.36	0.37	0.98	1.00														
D19	0.50	0.51	0.51	0.51	0.48	0.49	0.45	0.44	0.21	0.19	0.21	0.20	0.38	0.38	0.36	0.36	0.96	0.96	1.00													
D20	0.51	0.52	0.52	0.52	0.49	0.51	0.46	0.45	0.22	0.20	0.21	0.21	0.38	0.39	0.37	0.38	0.96	0.96	0.98	1.00												
D21	0.48	0.49	0.50	0.50	0.45	0.46	0.42	0.41	0.18	0.16	0.17	0.17	0.34	0.34	0.32	0.32	0.95	0.94	0.94	0.94	1.00											
D22	0.48	0.49	0.49	0.50	0.44	0.46	0.41	0.41	0.16	0.15	0.16	0.15	0.33	0.33	0.31	0.31	0.95	0.95	0.94	0.94	0.99	1.00										
D23	0.48	0.48	0.50	0.49	0.44	0.45	0.41	0.41	0.17	0.15	0.16	0.16	0.34	0.34	0.32	0.32	0.94	0.94	0.93	0.94	0.97	0.97	1.00									
D24	0.48	0.49	0.50	0.50	0.44	0.46	0.42	0.41	0.17	0.16	0.16	0.16	0.33	0.34	0.32	0.33	0.94	0.94	0.93	0.94	0.97	0.97	0.98	1.00								
D25	0.49	0.50	0.50	0.50	0.47	0.48	0.43	0.43	0.22	0.20	0.22	0.22	0.37	0.37	0.35	0.35	0.91	0.90	0.91	0.91	0.94	0.93	0.92	0.92	1.00							
D26	0.49	0.50	0.50	0.50	0.47	0.48	0.43	0.43	0.22	0.20	0.21	0.21	0.36	0.36	0.35	0.35	0.92	0.92	0.91	0.92	0.94	0.94	0.93	0.93	0.98	1.00						
D27	0.48	0.49	0.49	0.49	0.46	0.47	0.43	0.43	0.21	0.18	0.21	0.21	0.36	0.36	0.34	0.34	0.91	0.91	0.91	0.91	0.93	0.93	0.92	0.92	0.96	0.97	1.00					
D28	0.18	0.21	0.20	0.20	0.18	0.19	0.17	0.18	0.07	0.06	0.06	0.06	0.13	0.13	0.12	0.11	0.41	0.41	0.41	0.42	0.42	0.42	0.43	0.44	0.42	0.43	0.47	1.00				
D29	0.46	0.45	0.46	0.46	0.41	0.42	0.38	0.37	0.11	0.10	0.11	0.10	0.27	0.27	0.26	0.26	0.89	0.88	0.88	0.88	0.93	0.93	0.92	0.92	0.90	0.90	0.92	0.42	1.00			
D30	0.44	0.45	0.46	0.46	0.41	0.42	0.37	0.37	0.11	0.10	0.10	0.11	0.27	0.27	0.25	0.25	0.89	0.88	0.88	0.88	0.93	0.92	0.90	0.92	0.90	0.90	0.92	0.42	0.96	1.00		
D31	0.42	0.43	0.43	0.43	0.38	0.39	0.35	0.35	0.10	0.09	0.09	0.09	0.25	0.25	0.23	0.24	0.86	0.85	0.85	0.85	0.90	0.89	0.88	0.88	0.86	0.86	0.85	0.40	0.90	0.90	1.00	
D32	0.43	0.44	0.44	0.45	0.39	0.40	0.36	0.36	0.11	0.10	0.10	0.11	0.26	0.26	0.24	0.25	0.85	0.84	0.85	0.85	0.89	0.89	0.87	0.88	0.86	0.86	0.87	0.41	0.90	0.90	0.94	1.00

Table 4.2 Pair-wise correlation of average expression values. A 16×16 correlation matrix comparing the average of all above background duplicate target signals present on the 16 filters illustrated in Figure 4.1. Cells highlighted in black indicate cases where the same cDNA targets were hybridized to different filters. Cells highlighted in light grey indicate cases where different cDNA targets were hybridized to the same filters

	D1,D2	D3,D4	D5,D6	D7,D8	D9,D10	D11,D12	D13,D14	D15,D16	D17,D18	D19,D20	D21,D22	D23,D24	D25,D26	D27,D28	D29,D30	D31,D32
D1,D2	1															
D3,D4	0.971	1														
D5,D6	0.960	0.944	1													
D7,D8	0.929	0.929	0.959	1												
D9,D10	0.690	0.695	0.778	0.807	1											
D11,D12	0.699	0.698	0.786	0.813	0.965	1										
D13,D14	0.841	0.836	0.902	0.905	0.923	0.929	1									
D15,D16	0.823	0.819	0.890	0.898	0.920	0.929	0.980	1								
D17,D18	0.515	0.520	0.491	0.445	0.202	0.205	0.377	0.363	1							
D19,D20	0.518	0.523	0.498	0.454	0.211	0.211	0.384	0.370	0.973	1						
D21,D22	0.490	0.502	0.454	0.415	0.164	0.163	0.335	0.320	0.957	0.950	1					
D23,D24	0.488	0.502	0.453	0.416	0.166	0.162	0.340	0.326	0.946	0.947	0.979	1				
D25,D26	0.500	0.503	0.476	0.434	0.213	0.215	0.367	0.352	0.919	0.920	0.946	0.936	1			
D27,D28	0.395	0.400	0.376	0.352	0.157	0.158	0.285	0.266	0.761	0.767	0.783	0.782	0.811	1		
D29,D30	0.458	0.468	0.419	0.378	0.108	0.109	0.275	0.258	0.901	0.891	0.939	0.929	0.923	0.764	1	
D31,D32	0.439	0.447	0.399	0.363	0.102	0.100	0.261	0.245	0.866	0.866	0.906	0.897	0.878	0.740	0.925	1

ments obtained from DNA microarray experiments as well as more dramatic differences that are observed when basic changes in experimental protocols are adopted. However, for any given set of experiments performed under the same experimental conditions the most important source of variance to consider is those experimental and biological variables for which we have, at least, limited control. Here we have analysed these variables as correlations among target preparations and microarrays. These data are summarized in Table 4.3.

Table 4.3 The comparison of average correlation values obtained from the correlation matrix of Table 4.1

Comparison	Average correlation	
	All measurements	All measurements except one set of duplicate measurements on filter 8 of experiment 7 (D28)
Duplicate measurements from each filter[1]	0.941	0.973
Same targets hybridized to different filters[2]	0.920	0.952
Different targets hybridized to the same filters[3]	0.890	0.922
Different targets hybridized to different filters[4]	0.830	0.858

[1] The average correlation values from the correlation matrix illustrated in Table 4.1 of D1vD2, D3vD4, D5vD6, D7vD8, D9vD10, D11vD12, D13vD14, D15vD16, D17vD18, D19vD20, D21vD22, D23vD24, D25vD26, D27vD28, D29vD30 and D31vD32.

[2] The average correlation values from the correlation matrix illustrated in Table 4.1 of D1vD3, D1vD4, D2vD3, D2vD4, D5vD7, D5vD8, D6vD7, D6vD8, D9vD11, D9vD12, D10vD11, D10vD12, D13vD15, D13vD16, D14vD15, D14vD16, D17vD19, D17vD20, D18vD19, D18vD20, D21vD23, D21vD24, D22vD23, D22vD24, D25vD27, D25vD28, D26vD27, D26vD28, D29vD31, D29vD32, D30vD31 and D30vD32.

[3] The average correlation values from the correlation matrix illustrated in Table 4.1 of D1vD5, D1vD6, D2vD5, D2vD6, D3vD7, D3vD8, D4vD7, D4vD8, D9vD13, D9vD14, D10vD13, D10vD14, D11vD15, D11vD16, D12vD15, D12vD16, D17vD21, D17vD22, D18vD21, D18vD22, D19vD23, D19vD24, D20vD23, D20vD24, D25vD29, D25vD30, D26vD29, D26vD30, D27vD31, D27vD32, D28vD31 and D28vD32.

[4] The average correlation values from the correlation matrix illustrated in Table 4.1 of all cells in quadrants 1 and 4 except those belonging to the other three categories above. We do not consider cells in quadrants 2 and 3 because they measure correlations of experiments across the time gap, the effects of which we do not want to include in this comparison.

In Table 4.3 we see that when all of the duplicate measurements of each filter are compared to one another, a high correlation (0.94) is observed. However, when the D28 outlier described above is removed, the correlation is even better (0.97). This illustrates that this single outlier of one experiment can significantly compromise the entire data set. A reasonably high correlation is also observed when we compare the measurements among different filters hybridized with the same targets. However, less correlation is observed when different targets are hybridized to the same filters. This demonstrates greater variance among target preparations (biological variance) than among filters (experimental variance). Thus, it stands to reason that the variance is even greater when different target preparations are hybridized to different filters. It should be noted that it has been demonstrated that the variability among target preparations can be significantly reduced by pooling independently prepared target samples prior to hybridization (Baldi and Hatfield, 2002; Arfin *et al.*, 2000; Hung, Baldi and Hatfield, 2002).

4.4.5   Correlations by groups of four experiments

The data in Table 4.4 examine the variance contributed by differences among both filters and cDNA target preparations. Each group (row or column) in the table corresponds to the average of the expression levels of groups of four experiments. For example, group 1 is an average of gene expression levels of experiments (1–4), group 2 is an average of experiments (5–8) and so on. As shown in Figure 4.4, group 1 is an average of four experiments, which are measurements from two filters (1 and 2 from filter 1 and 3 and 4 from filter 2). We now compare four measurements from a pair of different filters with four measurements of either the same pair of filters hybridized with different cDNA targets, or a different pair of filters hybridized with different cDNA targets. Here we see that the variance among common filters hybridized with different

Table 4.4   Correlation of average expression values among experiments

	Exp. 1	Exp. 2	Exp. 3	Exp. 4	Exp. 5	Exp. 6	Exp. 7	Exp. 8
Exp. 1	1.000							
Exp. 2	0.957	1.000						
Exp. 3	0.707	0.811	1.000					
Exp. 4	0.840	0.913	0.938	1.000				
Exp. 5	0.527	0.480	0.210	0.378	1.000			
Exp. 6	0.502	0.442	0.166	0.334	0.962	1.000		
Exp. 7	0.480	0.438	0.199	0.338	0.896	0.916	1.000	
Exp. 8	0.465	0.401	0.108	0.266	0.904	0.941	0.890	1.000

cDNA targets again is less than the variance observed when different filters hybridized with different cDNA targets are compared. For example, the correlation between the four measurements of filters 1 and 2 hybridized with RNA1 compared with filters 1 and 2 hybridized with RNA2 is better than the correlation between the four measurements of filters 1 and 2 hybridized with RNA1 and the four measurements of filters 3 and 4 hybridized with RNA3 (Figure 4.4, Table 4.4). Again it is apparent that less variance is observed when common filters hybridized with different cDNA targets are compared than when different filters hybridized with different cDNA targets are compared. The data in Table 4.4 also illuminate another interesting observation which is that the correlations of the comparisons in quadrant 3 of Table 4.4 are much weaker than those of quadrants 1 and 4. This observation is correlated with the fact that the measurements in quadrant 3 compare experiments performed at different times. Experiments 1–4 were performed in October 2001 and experiments 5–8 were performed in April 2001. While these two sets of experiments show high correlations among themselves (quadrants 1 and 4, Table 4.4), poor correlations are observed when they are compared with one another (quadrant 3, Table 4.4).

4.4.6 Correlations between GeneChip data processed with Affymetrix MAS 4.0, MAS 5.0 or dChip software

When different array formats are used that require different target preparation methods, the magnitudes and sources of these experimental errors are surely different. Thus, we wished to examine the correlations among data sets of another DNA microarray format. For this purpose we used data sets obtained with Affymetrix GeneChips. This microarray format differs in that Affymetrix GeneChips are manufactured by the *in situ* synthesis of short single-stranded oligonucleotide probes complementary to sequences within each ORF directly synthesized on a glass surface, whereas nylon filter arrays are manufactured by the attachment of full-length, double-stranded, DNA probes of each *E. coli* ORF directly onto the filter as described earlier.

For comparison of the GeneChip and filter data, the exact same four pooled total RNA preparations as used for the nylon filter (experiments 1–4, Figure 4.4) were used for hybridization to four *E. coli* Affymetrix GeneChips (Figure 4.5) as described by Hung, Baldi and Hatfield (2002). In this case, however, instead of having four measurements for each gene expression level, as for each filter experiment (Figure 4.4), only one measurement was obtained from each GeneChip (Figure 4.5). On the other hand, this single measurement is the average of the difference between hybridization signals from approximately 15 perfect match (PM) and mismatch (MM) probe pairs for each ORF.

While these are not equivalent to duplicate measurements because different probes are used, these data can increase the reliability of each gene expression level measurement (Baldi and Hatfield, 2002). Nevertheless, large differences in the average difference of individual probe pairs are often observed. Thus, individual outliers can have a large effect on the average difference for probe pair sets of individual GeneChips. In fact, it has been reported that this variance can be five times greater than the variance observed among GeneChips. Thus, it is clear that proper treatment of probe effects is essential to the analysis of Affymetrix GeneChip array data. This problem was first addressed by Li and Wong (2001). They developed a model-based analysis method to detect and handle cross-hybridizing probes and image and/or GeneChip defects, and to identify outliers across GeneChip sets. These methods are implemented in a software program, dChip, that can be obtained from the authors (Li and Wong, 2001).

In the standard analysis implemented in the Affymetrix MAS 4.0 software, the mean and standard deviation of the PM−MM differences of a probe set in one array are computed after excluding the maximum and the minimum values obtained for that probe set. If, among the remaining probe pairs, a difference deviates by more than 3SD from the mean, that probe pair is declared an outlier and not used for the average difference calculation of both the control and the experimental array. A flaw of this approach is that a probe with a large response might well be the most informative but may be consistently discarded. Furthermore, if multiple arrays are compared at the same time, this method tends to exclude too many probes and/or discard probes inconsistently measured among GeneChips.

In the Li−Wong approach, a probe set from multiple chips is modelled and the standard deviation between a fitted curve and the actual curve for each probe set for each GeneChip is calculated. Probe pair sets containing (an) anomalous probe pair measurement(s) are declared outliers and discarded. The remaining probe pair sets are remodelled and the fitted curve data is used for average difference calculations.

A different empirical approach to improve the consistency of average difference measurements has been implemented in the more recent Affymetrix MAS 5.0 software. In this implementation, if the MM value is less than the PM value MAS 5.0 uses the MM value directly. However, if the MM value is larger than the PM value, MAS 5.0 creates an adjusted MM value based on the average difference intensity between the ln PM and ln MM, or if the measurement is too small some fraction of PM. The adjusted MM values are used to calculate the ln PM − ln-adjusted MM for each probe pair. The signal for a probe set is calculated as a one-step biweight estimate of the combined differences of all of the probe pairs of the probe set.

The data in Table 4.5 compare the correlations of GeneChip expression data processed with either Affymetrix MAS 4.0, MAS 5.0 or dChip software. The first four rows and columns of Table 4.5 compare the consistency of measurements obtained from four GeneChips processed with the MAS 4.0 software, the data in rows 5–8 and columns 5–8 of Table 4.5 compare the consistency of measurements obtained from four GeneChips processed with the dChip software, and the data in rows 9–12 and columns 9–12 of Table 4.5 compare the consistency of measurements obtained from four GeneChips processed with the MAS 5.0 software. These correlations reveal that the most consistent set of measurements are obtained with the dChip analysis (average correlation = 0.85), and that the MAS 5.0 software performs marginally better (average correlation = 0.83) than the previous MAS 4.0 version (average correlation = 0.81). It is also apparent that close correlations are observed when MAS 4.0 and MAS 5.0 processed data are compared to one another (average correlation = 0.82, Table 4.5 rows 9–12 and columns 1–4), and that this correlation is much better than between dChip and MAS 4.0 or MAS 5.0 processed data (average correlation 0.78, rows 5–8 and 9–12 and columns 1–4 and 5–8, respectively, Table 4.5). The data in Table 4.6 further demonstrate that the correlations between the GeneChip and the filter data are improved when the GeneChip data is processed with dChip. The data in Table 4.7 containing within group averages further shows that dChip modelled values are more self-consistent than values obtained with either MAS 4.0 or MAS 5.0.

Table 4.5   Correlation values among data processed with dChip, MAS 4.0 or MAS 5.0

	MAS 4.0				dChip				MAS 5.0			
	Chip 1	Chip 2	Chip 3	Chip 4	Chip 1	Chip 2	Chip 3	Chip 4	Chip 1	Chip 2	Chip 3	Chip 4
MAS 4.0 Chip 1	1.000											
Chip 2	0.846	1.000										
Chip 3	0.821	0.816	1.000									
Chip 4	0.783	0.817	0.777	1.000								
dChip Chip 1	0.789	0.827	0.766	0.740	1.000							
Chip 2	0.747	0.762	0.736	0.818	0.847	1.000						
Chip 3	0.850	0.795	0.777	0.733	0.896	0.802	1.000					
Chip 4	0.793	0.780	0.832	0.739	0.866	0.813	0.868	1.000				
MAS 5.0 Chip 1	0.912	0.835	0.809	0.780	0.777	0.745	0.835	0.782	1.000			
Chip 2	0.841	0.906	0.811	0.813	0.816	0.770	0.799	0.785	0.887	1.000		
Chip 3	0.831	0.824	0.915	0.788	0.780	0.757	0.789	0.829	0.854	0.857	1.000	
Chip 4	0.764	0.782	0.749	0.836	0.731	0.782	0.727	0.734	0.796	0.822	0.789	1.000

Table 4.6 Correlation values between filter and Affymetrix GeneChip experiments processed with dChip, MAS 4.0 or MAS 5.0 software[1]

		Filter				MAS 4.0				dChip				MAS 5.0			
		Exp. 1	Exp. 2	Exp. 3	Exp. 4	Chip 1	Chip 2	Chip 3	Chip 4	Chip 1	Chip 2	Chip 3	Chip 4	Chip 1	Chip 2	Chip 3	Chip 4
Filter	Exp. 1	1.000															
	Exp. 2	0.902	1.000														
	Exp. 3	0.597	0.753	1.000													
	Exp. 4	0.733	0.863	0.917	1.000												
MAS 4.0	Chip 1	0.342	0.293	−0.010	0.129	1.000											
	Chip 2	0.347	0.296	−0.027	0.120	0.865	1.000										
	Chip 3	0.343	0.300	−0.006	0.143	0.831	0.842	1.000									
	Chip 4	0.347	0.309	0.008	0.152	0.796	0.840	0.794	1.000								
dChip	Chip 1	0.347	0.303	0.004	0.145	0.804	0.844	0.787	0.765	1.000							
	Chip 2	0.342	0.294	0.008	0.146	0.770	0.795	0.757	0.836	0.868	1.000						
	Chip 3	0.335	0.298	0.019	0.148	0.860	0.818	0.794	0.756	0.911	0.829	1.000					
	Chip 4	0.338	0.304	0.008	0.153	0.803	0.806	0.838	0.761	0.882	0.836	0.882	1.000				
MAS 5.0	Chip 1	0.357	0.302	−0.002	0.139	0.917	0.862	0.826	0.800	0.807	0.778	0.860	0.807	1.000			
	Chip 2	0.365	0.314	−0.018	0.135	0.849	0.922	0.826	0.829	0.838	0.797	0.820	0.860	0.898	1.000		
	Chip 3	0.365	0.325	0.009	0.162	0.842	0.849	0.917	0.805	0.806	0.784	0.813	0.846	0.870	0.874	1.000	
	Chip 4	0.347	0.306	0.020	0.156	0.780	0.809	0.772	0.855	0.762	0.755	0.762	0.761	0.817	0.838	0.814	1.000

[1] When the filter data was added to the comparison, we had to use common genes in all the four technologies being compared and the number of genes used for the analysis for Table 4.6 (1794) is *less* than for Table 4.5 (2370) (hence the minor variations between Tables 5 and 6 in the numbers comparing the MAS4, MAS5 and dChip with each other).

Table 4.7 Average correlation values for Affymetrix GeneChip data processed with dChip, MAS 4.0 or MAS 5.0 software

	MAS 4.0	dChip	MAS 5.0
MAS 4.0	0.873		
dChip	0.845	0.893	
MAS 5.0	0.863	0.848	0.889

4.5 Conclusions

The results presented here demonstrate that the variability inherent in DNA microarray data can result from a large number of disparate factors operating at different times and levels in the course of a typical experiment. These numerous factors are often interrelated in complex ways, but for the purpose of simplicity we have broken them down into two major categories: biological variability and experimental variability. Other sources of variability involve DNA microarray fabrication methods as well as differences in imaging technology, signal extraction and data processing.

In this study we used simple statistical methods to quantitatively assess the experimental and biological variance observed among 32 replicated DNA microarray data sets obtained under 'identical' experimental conditions. This study highlights the magnitudes of these sources of variance that necessitate the use of appropriate statistical methods for the analysis and interpretation of DNA microarray experimental results (Baldi and Hatfield, 2002).

This study confirms earlier assertions that, even with carefully controlled experiments with isogenic model organisms, the major source of variance comes from uncontrolled biological factors (Hatfield, Hung and Baldi, 2003). Our ability to control biological variation in a model organism such as *E. coli* with an easily manipulated genetic system is an obvious advantage for gene expression profiling experiments. However, most systems are not as easily controlled. For example, human samples obtained from biopsy materials not only differ in genotype but also in cell types. Thus, care should be taken to reduce this source of biological variability as much as possible, for example, with the use of laser-capture techniques for the isolation of single cells from animal and human tissues.

An additional source of biological variation, even when comparing the gene profiles of isogenic cell types, comes from the conditions under which the cells are cultured. In this regard, it has been recommended that standard cell-specific media should be adopted for the growth of cells queried by DNA array experiments (Baldi and Hatfield, 2002). While this is not possible in every case, many experimental conditions for the comparison of two different genotypes of common cell lines can be standardized. The adoption of such medium standards would greatly reduce experimental variations and facilitate the cross-comparison of experimental data obtained from different experiments, different microarray formats, and/or different investigators. However, even employing these precautions, non-trivial and sometimes substantial variance in gene expression levels, even between genetically identical cells cultured in the same environment such as those revealed in this study, are observed. This simple fact can result from a variety of influences including environmental differences, phase differences between the cells in the culture, periods of rapid change in gene expression and multiple additional stochastic effects. To emphasize the importance of microenvironments encountered during cell growth, Piper *et al.* (2002) have recently demonstrated that variance among replicated gene measurements is dramatically decreased when isogenic yeast cells are grown in chemostats rather than batch cultures.

Biological variance can be even further exacerbated by experimental errors. For example, if extreme care in the treatment and handling of the RNA is not taken during the extraction of the RNA from the cell and its subsequent processing. It is often reported that the cells to be analysed are harvested by centrifugation and frozen for RNA extraction at a later time. It is important to consider the effects of these experimental manipulations on gene expression and mRNA stability. If the cells encounter a temperature shift during the centrifugation step, even for a short time, this could cause a change in the gene expression profiles due to the consequences of temperature stress. If the cells are centrifuged in a buffer with even small differences in osmolarity from the growth medium, this could cause a change in the gene expression profiles due to the consequences of osmotic stress. Also, removal of essential nutrients during the centrifugation period could cause significant metabolic perturbations that would result in changes in gene expression profiles. Each of these and other experimentally caused gene expression changes will confound the interpretation of the experiment. These are not easy variables to control. Therefore, the best strategy is to harvest the RNA as quickly as possible under conditions that 'freeze' it at the same levels at which it occurs in the cell population at the time of sampling. Several methods are available that address this issue (Baldi and Hatfield, 2002). There are numerous other sources of experimental variability such as differences among protocols different techniques employed by different personnel, differences between reagents and differences among instruments and their calibrations, as well as others. While these sources of variance are usually less than those that come from biological sources, they can dominate the results of a DNA microarray experiment. This is illustrated by the poor correlation between the two replicated data sets reported here, one obtained eight months after the other. Although there is good correlation among the replicated measurements of each set, there is much less correlation among the measurements between these sets. In this case, the major difference can be attributed to improvements in the cDNA target labelling protocol.

With regard to the second dataset obtained from Affymetrix GeneChip experiments, one possibility for poor correlation between signal intensities between filter and Affymetrix GeneChip experiments can be attributed to probe effects. Probe effects are due to differences in the hybridization efficiencies of different probes, even when target sequences are physically linked in the same nucleic acid polymer and present at the same concentration. However, these differences are less for filters containing full-length ORF probes hybridized to targets generated with random hexamers than for Affymetrix GeneChips that query only a limited number of target sequences. Thus, it is to be expected that the signal intensities obtained from Affymetrix GeneChips are less correlated to *in vivo* transcript levels than signal intensities obtained from filters. This provides a rationale for why signal intensities obtained from different micro-array platforms do not correlate well with one another. Nevertheless, signal ratios obtained from the same probe on two different arrays can ameliorate these probe effects. Thus, the overall differential expression profiles obtained from different microarray platforms should be comparable. In support of this conclusion, Hung, Baldi and Hatfield (2002) have demonstrated that, with appropriate statistical analysis, similar results can be obtained when the same experiments are performed with pre-synthesized filters containing full-length ORF probes and Affymetrix GeneChips.

In summary, we have used simple statistical tests to analyse repeated experiments using DNA microarrays. These techniques have illuminated patterns in the data and have shown that even minor differences while conducting an experiment generate variability in the data values. We also show how we can discover potential anomalies or possible errors in the experimental set-up by performing such an analysis. These pre-processing steps may be used to produce a better awareness of what the data contain before more sophisticated data analysis techniques are employed.

Finally, it is reassuring to observe that carefully executed and replicated DNA microarray experiments produce data with high global correlations (in the 0.9 range). This high correlation, however, should not be interpreted as a sign that replication is not necessary. Replication as well as proper statistical analysis remain important in order to monitor experimental variability and because the variability of *individual* genes can be high. It is also reassuring to know that, while correlations of expression measurements across technologies remain low, overall differential expression profiles obtained from different microarray platforms can be compared (Hung, Baldi and Hatfield, 2002).

Acknowledgements

This work was supported in part by the University of California, Irvine, Institute for Genomics and Bioinformatics, and by NIH grants GM55073 and GM68903 to GWH. S.-P. H. is the recipient of NIH postdoctoral fellowship LM07443.

References

Allison, D. B., Gadbury, G. L., Heo, M., Fernndez, J. R., Lee, C. K., Prolla, T. A., and Weindruch, R. (2002) A mixture model approach for the analysis of microarray gene expression data. *Comput. Statist. Data Anal.* **39**, 1–20.

Arfin, S. M., Long, A. D., Ito, E. T., Tolleri, L., Riehle, M. M., Paegle, E. S., and Hatfield, G. W. (2000) Global gene expression profiling in *Escherichia coli* K12. The effects of integration host factor. *J. Biol. Chem.* **275**, 29 672–29 684.

Baldi, P., and Hatfield, G. W. (2002) *DNA Microarrays and Gene Expression: from Experiments to Data Analysis and Modeling.* Cambridge University Press, Cambridge, UK.

Baldi, P., and Long, A. D. (2001) A Bayesian framework for the analysis of microarray expression data: regularized *t*-test and statistical inferences of gene changes. *Bioinformatics* **17**, 509–519.

Fodor, S. P., Rava, R. P., Huang, X. C., Pease, A. C., Holmes, C. P., and Adams, C. L. (1993) Multiplexed biochemical assays with biological chips. *Nature* **364**, 555–556.

Hatfield, G. W., Hung, S. P., and Baldi, P. (2003) Differential analysis of DNA microarray gene expression data. *Mol. Microbiol.* **47**, 871–877.

Hung, S.-P., Baldi, P., and Hatfield, G. W. (2002) Global gene expression profiling in *Escherichia coli* K12: the effects of leucine-responsive regulatory protein. *J. Biol. Chem.* **277**, 40 309–40 323.

Li, C., and Wong, W. H. (2001) Model-based analysis of oligonucleotide arrays: expression index computation and outlier detection. *Proc. Natl. Acad. Sci. USA* **98**, 31–36.

Long, A. D., Mangalam, H. J., Chan, B. Y., Tolleri, L., Hatfield, G. W., and Baldi, P. (2001) Improved statistical inference from DNA microarray data using analysis of variance and a Bayesian statistical framework. Analysis of global gene expression in *Escherichia coli* K12. *J. Biol. Chem.* **276**, 19937–19944.

Piper, M. D., Daran-Lapujade, P., Bro, C., Regenberg, B., Knudsen, S., Nielsen, J., and Pronk, J. T. (2002) Reproducibility of oligonucleotide microarray transcriptome analyses. An interlaboratory comparison using chemostat cultures of *Saccharomyces cerevisiae*. *J. Biol. Chem.* **277**, 37001–37008.

Schena, M. (Ed.). (1999) *DNA Microarrays: a Practical Approach*, Oxford University Press, Oxford, England.

Scherf, U., Ross, D. T., Waltham, M., Smith, L. H., Lee, J. K., Tanabe, L., Kohn, K. W., Reinhold, W. C., Myers, T. G., Andrews, D. T., Scudiero, D. A., Eisen, M. B., Sausville, E. A., Pommier, Y., Botstein, D., Brown, P. O., and Weinstein, J. N. (2000) A gene expression database for the molecular pharmacology of cancer. *Nat. Genet.* **24**, 236–244.

5

The Proteome, *Anno Domini* Two Zero Zero Three

Pier Giorgio Righetti[*], **Mahmoud Hamdan, Frederic Reymond** and **Joël S. Rossier**

By the turn of the millennium, if not much sooner, we will see a dramatic shift of emphasis from DNA sequencing and mRNA profiling to proteomics. Considered objectively, there is every reason to expect that proteomics will ultimately exceed genomics in total effort, though this growth will sorely be limited by the availability of scientists able to deal with protein's non-ideal properties...

5.1   Introduction

Was it Tiresias, the blind soothsayer of Thebes, who prophesied the above prediction? Well, poor Tiresias, smart as he was, could not predict his assassination by order of Creon. In fact, the above predicament came from Anderson and Anderson (1998), the celebrated couple who fathered, among others, the explosion of modern proteome science. The fact is that the world of functional genomics had become so entrenched in its credo that a sound lesson was required to chase them out in the open fields. For years they had been propagating the notion that one did not need to study protein levels in order to obtain data on up- and down-regulation in any biologically relevant process (from cancer to ontogenesis): all one needed was to obtain the profiles of

[*] E-mail of the corresponding author: righetti@sci.univr.it

Genomics, Proteomics and Vaccines edited by Guido Grandi
© 2004 John Wiley & Sons, Ltd   ISBN 0 470 85616 5

mRNA levels (or expression). A big blow to this credo came in 1997, when Anderson presented (in the proceedings of the third Siena meeting) a paper offering the first multi-gene comparison plot of mRNA versus protein abundance for cellular gene products, and found a correlation coefficient of 0.43 between them (Anderson and Seilhamer, 1997). This resulted in notable comments, ranging from consternation that it was so low to amazement that it was so high!

Today there are more than a few altar-fires alight in the proteome field: in fact the entire arena is ablaze with fires, with diggers from all corners of science excavating deeper and deeper in search of the mother lode. Because this is today's scenario in science: nobody does it for science's sake; most people do it for the sake of the green dollar, the polychromatic euro, the yen, you name it (given that the rouble is now defunct). Figure 5.1 gives an explanation for this furious digging: proteins are drug targets, so not only the pharmaceutical industry, but just about any research fellow in the field, looks for proteins with

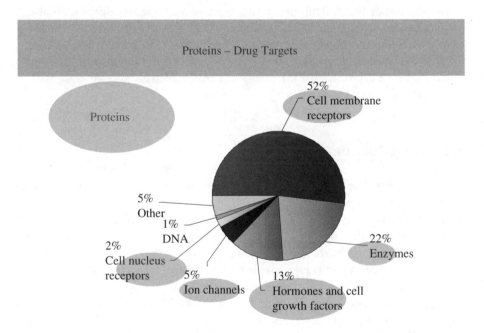

Figure 5.1   Pie chart of the distribution of important proteins acting as drug targets. It can be appreciated that about 52 per cent of them are cell membrane receptors, the second most abundant category being enzymes

key activities, whose modulation could be brought about by active-site-directed drugs. Among these, up to 52 per cent could be cell membrane receptors and this gives a feeling of how complex are the investigations in the field: in a total cell lysate, these receptors would be among the least abundant species; thus, their detection requires special skills and techniques, as illustrated in this review. The damnation of proteome research is that we keep seeing only the tip of the iceberg, the vast majority of precious and rare proteins being still hidden to sight. This can be easily appreciated in the vast body of papers published weekly, if not daily: when the spots are analysed by mass spectrometry and their identifications given, it is quite clear that the various laboratories are 're-discovering', over and over, the very same set of proteins, i.e. the most abundant ones, in general house-keeping and structural proteins. The rare (and precious) ones are missing the roll-call.

It is not quite possible here to thoroughly review methodologies and the enormous literature in the field. We will discuss only some hot, recent topics and developments. The smart reader can now find quite a bit of information in the following new journals: *Proteomics* (Wiley), which began publication in the year 2000, and *Journal of Proteome Research* (American Chemical Society) and *Molecular and Cellular Proteomics* (American Biochemical Society), both founded in 2001. Of course, plenty of papers will still appear in *Electrophoresis* and *Analytical Chemistry*, the two leading journals in separation science. Among recent books dedicated to this topic one could recommend those of Wilkins *et al.* (1997), Kellner, Lottspeich and Meyer (1999), Rabilloud (2000), James (2001), Righetti, Stoyanov and Zhukov (2001) and Figeys and Ross (2003), and among book chapters Westermeier (1997) and Hanash (1998), to name just a few.

5.2   Some definitions

Before entering the proteome arena, some definitions are due, given the fact that today's science jargon is becoming more and more unintelligible. We will start first with the obscene ones, which have surreptitiously penetrated the fortress of modern science, as depicted in Table 5.1: were it not enough to have to digest such odd terms as genomics, transcriptomics, proteomics and metabo-lomics, we have to gobble down, appearing on a camel's back from the desert dunes, like Rommel's troops, such a ridiculous term as 'systeomics'. Fear not, though, a whole host of them is appearing on the horizon: there are rumours of 'glycomics', 'complexomics' and the like. Moreover, just to increase the entropy in the tower of Babel, here come more hair-splitting definitions: Jeremy

Table 5.1   The sloppy vernacular of biology, or how too many 'omics' could lead us to 'comics'

GOD ALMIGHTY, GIVE US THIS DAY OUR DAILY 'OMICS':
Genomics: all the genes you could find
Transcriptomics: all the messengers on the loose (perhaps 100 000 strong)
Proteomics: all the king's horses and all the king's men. . . (an army: perhaps 1 000 000!)
Metabolomics: the few survivors of the wrath of all previous macromolecules (a handful, perhaps only 2000)
Systeomics: a cocktail of all the above!
Sed libera nos a. . .
Comics!

Nicholson (personal communication) insists on a distinction between 'metabolomics' (cellular metabolism) and 'metabonomics' (intact systems metabolism). They might all be right, but who will save us from 'comics'? Some more serious definitions can be found in Tables 5.2 (Celis *et al.*, 1998) and 5.3, which refer to such clear terms as functional and structural proteomics. Table 5.4 offers a glimpse of the tremendous complexity of the proteome: assuming a total asset of 45 000 genes, the human race could display as many as 10^8 polypeptides, a very large number indeed (Landen and Weinberg, 2000)! Table 5.5 throws down some intriguing questions waiting for answers. Whereas some of them might be too far-fetched, questions 2, 3 and 4 definitely fall within the domain of proteome research.

Table 5.2   The concept of functional proteomics

The description of changes in protein expression during differentiation, proliferation and signalling of cells, in both qualitative and quantitative terms, falls under the field of functional proteomics. It also includes studies of co-ordinate expression of genes, as well as elucidation of the sequence of regulatory events during all stages which a cell or an organism undergoes during its entire life span.

Table 5.3   The definition of structural proteomics

The aim of structural proteomics is to identify the molecular structure, i.e. the amino acid sequence of the protein entities involved in a given process, and to relate this information to the data base of identified genes. The most powerful method, one that has revolutionized proteomics, is mass spectrometry analysis.

Table 5.4   A guess estimate of the repertoire of the proteome in humans

Proteome of a cell: 5000 polypeptides
Proteome of an individual as a snap-shot: 10^6 polypeptides
Proteome of an individual during the entire life span: 10^7 polypeptides
Proteome of a species: 10^8 polypeptides
Based on an estimate of 45 000 genes in the human genome

Table 5.5   Some questions waiting for answers in modern genomic/proteomic analysis. We do not need to know everything about everything, but we might want to be able to answer at least some of the following questions

- How many and which genes are activated in a cell?
- How many polypeptides, and in what amounts, are synthesized?
- What kinds of post-translational modification occur on a synthesized protein?
- What is the half-life of protein molecules and how does this affect their function?
- What are the rules for moving proteins into and out from any cellular compartment?

5.3   What methods exist to tackle the proteome complexity?

Given the outstanding complexity of the proteome field, as illustrated in Table 5.4, it is evident that, to pursue such a problem, we will need a panoply of tools, as listed in Table 5.6. It is here evident that, in addition to the classical electrophoretic approach, two-dimensional (2D) map analysis, today we have quite a number of chromatographic techniques competing and trying to replace

Table 5.6   A (partial) list of methods available today for tackling the proteome complexity

- Classical 2-D map analysis (IEF followed by SDS-PAGE)
- Chromatographic and electrophoretic pre-fractionation
- Multidimensional separations, such as
   (a) coupled size exclusion and RP-HPLC;
   (b) coupled ion exchange and RP-HPLC;
   (c) coupled RP-HPLC and RP-HPLC at 25/60 °C;
   (d) coupled RP-HPLC and capillary electrophoresis (CE);
   (e) metal affinity chromatography coupled with CE
- Protein chips

the good old O'Farrell (1975) approach, incriminated on the grounds of its complexity and labour-intensive manipulations. Curiously, chromatographers do not seem to be content with just a 2D approach, but they boast a 'multi-dimensional' procedure, although all the papers we screened appear to offer, at best, a 2D approach. By our reckoning, a multidimensional scheme was only concocted once, in Jorgenson's laboratory (Hooker, Jeffrey and Jorgenson, 1998). It was in fact a three-dimensional protocol, size exclusion chromatography, coupled with reversed-phase fluid chromatography, followed by high-speed capillary zone electrophoresis. The results were quite disastrous and led these authors to the following conclusions: 'the increased peak capacity of this system may not be worth the extra effort and added complexity that is entailed'. This was the understatement of the year: the complexity of such an approach produced such terrible hardship that most post-docs in Jorgenson's laboratory escaped and enrolled in the Foreign Legion, in search of relief.

5.3.1 Standard 2D map analysis

Although the power of two-dimensional (2D) electrophoresis as a biochemical separation technique has been clearly recognized since its introduction, its application, nevertheless, has become particularly significant in the past few years, as a result of a number of developments, outlined below.

- The 2D technique has been tremendously improved to generate 2D maps that are superior in terms of resolution and reproducibility. This new technique utilizes a unique first dimension, that replaces the carrier ampholyte-generated pH gradients with immobilized pH gradients (IPGs) and replaces the tube gels with gel strips supported by a plastic film backing (Bjellqvist *et al.*, 1982).

- Methods for the rapid analysis of proteins have been improved to the point that single spots eluted or transferred from single 2D gels can be rapidly identified. Mass spectroscopic techniques have been developed that allow analysis of very small quantities of proteins and peptides (Aebersold and Leavitt, 1990; Patterson and Aebersold, 1995; Lahm and Langen, 2000). Chemical microsequencing and amino acid analysis can be performed on increasingly smaller samples (Lottspeich, Houthaeve and Kellner, 1999). Immunochemical identification is now possible with a wide assortment of available antibodies.

- More powerful, less expensive computers and software are now available, allowing routine computerized evaluations of the highly complex 2D patterns.

- Data on entire genomes (or substantial fractions thereof) for a number of organisms are now available, allowing rapid identification of the genes encoding a protein separated by 2D electrophoresis.

- The World Wide Web (WWW) provides simple, direct access to spot pattern databases for the comparison of electrophoretic results and to genome sequence databases for assignment of sequence information.

In 2D PAGE, one of the most critical steps is the initial sample solubilization. For decades, the most popular lysis solutions has been the O'Farrell cocktail (9 M urea, 2 per cent Nonidet P-40, 2 per cent β-mercaptoethanol and 2 per cent carrier ampholytes, in any desired pH interval). Although much in vogue also in present times, over the years new, even more powerful, solubilizing mixtures have been devised, in view of the fact that many authors have noted that hydrophobic proteins were largely absent from 2D maps (Wilkins *et al.*, 1998). For example, in three different species analysed (*Escherichia coli, Bacillus subtilis* and *Saccharomyces cerevisiae*), all proteins above a given hydrophobicity value were completely missing, independent from the mode of IEF (soluble CAs or IPG). This suggested that the initial sample solubilization was the primary cause for loss of such hydrophobic proteins. The progress made in solubilizing cocktails can be summarized as follows (see also Molloy (2000) and Rabilloud and Chevallet (2000)).

- *Chaotropes*. Although urea (up to 9.5 M) has been for decades the only chaotrope used in isoelectric focusing (IEF), recently thiourea has been found to further improve solubilization, especially of membrane proteins (Musdante, Candiano and Ghiggeri, 1998). The inclusion of thiourea is recommended for use with IPGs, which are prone to adsorptive losses of hydrophobic and isoelectrically neutral proteins. Typically, thiourea is added at concentrations of 2 M in conjunction with 5–7 M urea. The high concentration of urea is essential for solvating thiourea, which is poorly water soluble (Rabilloud *et al.*, 1997). It should also be remembered that urea in water exists in equilibrium with ammonium cyanate, whose level increases with increasing pH and temperature. Since cyanate can react with amino groups in proteins, such as the N-terminus α-amino or the ε-amino groups of Lys, these reactions should be avoided, since they will produce a false sample heterogeneity and give wrong M_r values upon peptide analysis by MALDI-TOF MS. Thus, fresh solutions of pure grade urea should be used, in concomitance with low temperatures and with the use of scavengers of cyanate (such as the primary amines of carrier ampholytes or other suitable amines). In addition, protein mixtures solubilized in high levels of urea should be subjected to separation in the electric field as soon as possible: in the presence of the high voltage gradients typical of the IEF protocol, cyanate ions are quickly removed and no carbamylation can possibly take place (McCarthy *et al.*, 2003), whereas it will if the protein/ urea solution is left standing on the bench.

- *Surfactants*. Detergents are important in preventing hydrophobic interactions due to exposure of protein hydrophobic domains induced by chaotropes. Both the hydrophobic tails and the polar head groups of detergents play an important role in protein solubilization. The surfactant tail binds to hydrophobic residues, allowing dispersal of these domains in an aqueous medium, while the polar head groups of detergents can disrupt ionic and hydrogen bonds, aiding in dispersion. Detergents typically used in the past included Nonidet P-40 or Triton X-100, in concentrations ranging from 0.5 to 4 per cent. More and more, zwitterionic surfactants, such as CHAPS, are replacing these neutral detergents (Hochstrasser *et al.*, 1988), often in combination with low levels (0.5 per cent) of Triton X-100 and of CAs (<1 per cent), which appear to reduce protein–matrix hydrophobic interactions and overcome detrimental effects caused by salt boundaries (Rimpilainen and Righetti, 1985). Linear sulphobetaines are now emerging as perhaps the most powerful surfactants, especially ASB 14, an amidosulphobetaine containing a 14-C linear alkyl tail (Chevallet *et al.*, 1998). This reagent has since been used successfully in combination with urea and thiourea for solubilizing integral membrane proteins of both *E. coli* (Molloy *et al.*, 2000) and *Arabidopsis thaliana* (Santoni *et al.*, 2000).

- *Reducing agents*. Thiol agents are typically used to break intramolecular and intermolecular disulphide bridges. Cyclic reducing agents, such as dithiothreitol (DTT) or dithioerythritol (DTE), are admixed to solubilizing cocktails, in large excess (e.g. 20–40 mM), so as to shift the equilibrium toward oxidation of the reducing agent with concomitant reduction of the protein disulphides. Because this is an equilibrium reaction, loss of the reducing agent through migration of proteins away from the sample application zone can permit reoxidation of free Cys to disulphides in proteins, which would result not only in horizontal streaking, but also, possibly, in formation of spurious extra bands due to scrambled $-S-S-$ bridges and their cross-linking different polypeptide chains. Even if the sample is directly reswollen in the dried IPG strip, as customary today, the excess DTT or DTE will not remain in the gel at a constant concentration, since, due to their weakly acidic character, both compounds will migrate towards pH 7 and be depleted from the alkaline gel region. Thus, this will aggravate focusing of alkaline proteins and be one of the multiple factors responsible for poor focusing in the alkaline pH scale (Bordini, Hamdan and Righetti, 1999). The situation would be aggravated when using β-mercaptoethanol, since the latter compound has an even lower pK value, thus it is more depleted in the alkaline region and will form a concentration gradient towards pH 7, with a distribution in the gel following its degree of ionization at any given pH value along the IEF strip (Righetti, Tudor and Gianazza, 1982). This is probably the reason for the dramatic loss of any pH gradient above pH 7.5, lamented by most users of conventional IEF in CAs, when generating 2D maps. The most modern solution to all the above problems appears to be the use of phosphines as alternative reducing agents. Phosphines operate in a stoichiometric reaction, thus allowing the use of low concentrations (barely 2 mM). The use of tributyl phosphine (TBP) was recently proposed by Herbert et al. (1998), who reported much improved protein solubility for intractable, highly disulphide cross-linked wool keratins. TBP thus offers two main advantages: it can be used at much reduced levels as compared with other thiolic reagents (at least one order of magnitude lower concentration) and, additionally, it can be uniformly distributed in the IPG gel strip (when rehydrated with the entire sample solution) since, being uncharged, it will not migrate in the electric field. A major drawback of TBP, though, is that it is volatile, toxic and rather flammable in concentrated stocks. No matter which strategy is used, it is imperative that all samples, prior to the IEF/IPG step, are reduced and alkylated; failure to implement this procedure will result in an impressive number of spurious spots in the final 2D map, due to mixed disulphide bridges among like and unlike polypeptide chains (Herbert et al., 2001; Galvani et al., 2001a,b). In addition, protection of the $-SH$ groups seems to strongly quench a noxious and unexpected artefact in proteome analysis: β-elimination (or desulphuration), which results in the loss of an H_2S group (34 Da) from Cys residues for proteins focusing in the alkaline pH region. Upon such an elimination event, a dehydro-alanine residue is generated at the Cys site. In turn, the presence of a double bond in this position elicits lysis of the peptide bond, generating a number of peptides of fairly large size from an intact protein (Herbert et al., 2003).

5.4   Quantitative proteomics

When analysing normal/pathological tissues undergoing changes in protein expression profiles (e.g. during growth, differentiation, drug exposure or onset of tumours) it is important to be able to quantify such changes, so as to detect univocally those polypeptides undergoing up- or down-regulation. Such proteins might be key targets for drug action, thus their detection would be most important in the pharmaceutical industry and medical treatment. One of the earliest ways for achieving that has been the proper use of computer programs (e.g. Melanie or PD Quest) able to compare standard maps generated from, e.g., normal and pathological tissues. The minimum statistical requirements for obtaining such standard maps would be to collect from three to five samples in each state, combine them, and generate, simultaneously, a minimum of five 2D maps runs side by side under the same experimental conditions, so as to minimize errors. Sophisticated computer programs would then combine the five different maps and generate one reference and one 'pathological' map, which could then be overlapped with confidence. The synthetic comparative map thus obtained would detect proteins being up- or down-regulated via variations in uptake of the Coomassie blue stain. An example of such 'differential maps' is given in Figure 5.2(a) and (b): these maps refer to a peculiar category of tumours, called mantle cell lymphomas (non-Hodgkins). Figure 5.2(a) displays nine spots exhibiting increments of spot density from 500 to 1000 per cent, whereas Figure 5.2(b) discloses a set of seven spots with stain density decrements of 500 to 1300 per cent (Antonucci *et al.*, 2003). This method is reliable, although extremely labour intensive and lengthy. Thus, in recent years, other approaches have emerged, based on the principle of differential isotope labelling of certain residues in proteins. The first one to appear, termed isotope-coded affinity tags (ICATs), is used for modifying the −SH groups of Cys residues (Gygi *et al.*, 1999; Han *et al.*, 2001). This chemical is composed of three regions, a terminal iodine tail for adding onto −SH groups; an intermediate or linker region, coded with either d_0 or d_8 (d = deuterium, thus imparting a difference of 8 Da to the two differentially labelled protein populations) and an affinity-capture extremity, containing a biotin bait. The two samples to be compared are labelled with either the light or heavy reagent, mixed in a 1:1 ratio and digested with trypsin. This would result in an extremely heterogeneous concoction of peptides, of the order of a few hundred thousand in a total cell lysate, not amenable to direct MS analysis. In order to strongly reduce this enormous sample complexity, the peptide mixture is subjected to affinity purification on an avidin column. Only the peptide population containing Cys residues (about 8–9 per cent of the total) will be captured and analysed by MS. Each peptide will be divided into two peaks, separated by 8 Da, representing

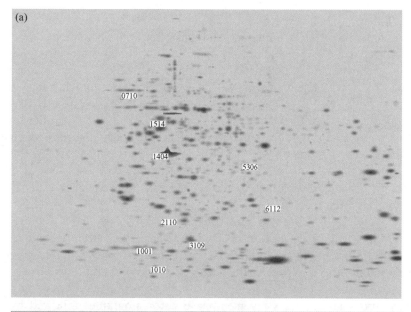

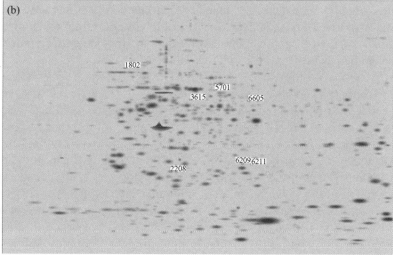

Figure 5.2   2D pattern of mantle cell lymphoma (MCL) tissue (a) and reactive lymph node (b). The 2D electrophoresis was performed on a linear Immobiline gradient pH 3–10, followed by 8–18 per cent SDS-PAGE. Proteins were detected by colloidal Coomassie staining. Image analysis of scanned gels was carried out using the PDQUEST software. The 2D gels were matched to find out quantitative differences of spot pattern. In panel (a) are highlighted nine proteins that are over-expressed on the MCL tissue map. In panel (b) are marked in red seven proteins that are over-expressed on a reactive lymph-node map (reprinted, by permission, from Antonucci *et al.*, 2003)

the light/heavy label, respectively. The ratio of these two peaks will give their ratio in the original sample mixture, and thus offer an immediate insight on up- or down-regulated polypeptides. Since the ICAT chemical commands a princely price, recently Sechi (2002) and Gehanne $et\ al.$ (2002) have proposed an inexpensive set of chemicals, namely d_0/d_3 acrylamide, which add with the same efficiency to $-SH$ groups and perform the same task as the ICAT (with the added benefit that they can be used in standard 2D maps, i.e. in the analysis of intact proteins, not of their peptide digests). Although the mass difference between the two isotope-coded, Cys-bearing peptides is only 3 Da, this does not pose any problem with modern mass spectrometers, whose mass accuracy allows for baseline resolution for adjacent peaks spaced apart even by this minute mass difference.

There are other disturbing drawbacks on the use of the ICAT protocol, reported by Zhang and Regnier (2002), which apply to capture techniques that are different from those reported by the inventors of the ICAT, based on avidin–biotin affinity. When dealing with such complex mixtures arising from a tryptic digest of a total cell lysate, one might want to use different chromato- graphic protocols, such as reversed-phase chromatography followed by ion- mobility separation. It turns out that, if ICAT-labelled peptides are separated through a reversed-phase column, a disturbing isotope effect takes place, by which the deuterated peptide elutes earlier than its non-deuterated counterpart; this separation causes an enormous variation in isotope ratio across the two different elution profiles of the isoforms. The effect is more pronounced the smaller the tagged peptide. For example, in the case of a simple, Cys-bearing octapeptide, the chromatographic resolution between the light/heavy species was as high as $R_s = 0.74$ (it is here recalled that an $R_s = 1.2$ means just baseline resolved peaks). Because the column eluate is continuously analysed by ESI- MS, obtaining the correct quantitative peak ratio according to the isotopic ratio becomes an extremely difficult task. Zhang and Regnier (2002) reported that the resolution of the isoforms, in a C_{18} column, exceeded 0.5 with 20 per cent of the peptides in the digest. Conversely, the same authors reported the complete absence of such an isotope effect in the case of peptides differentially labelled with ^{13}C- and ^{12}C-succinate, so they strongly recommended this type of peptide coding when attempting separations in C_{18} columns. More recently, Liu and Regnier (2002) have proposed a global coding strategy, which exploits differential derivatization of amine and carboxyl groups generated during proteolysis. Carboxyl groups produced during tryptic digestion would incorpo- rate ^{18}O from $H_2^{18}O$. Primary amines from control and experimental samples would be differentially acylated after proteolysis with either $^1H_3$- or $^2H_3$-N- acetoxysuccinimide. This global coding procedure is illustrated in Figure 5.3. A number of other approaches have also been reported. For example, Goodlett

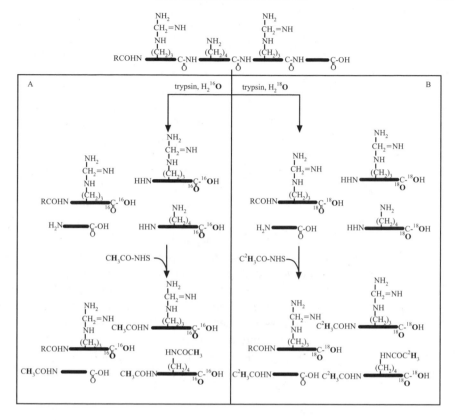

Figure 5.3 Differential isotope coding of peptides from control and experimental samples with both ^{18}O and ^{2}H labelling (reprinted, by permission, from Liu and Regnier, 2002)

et al. (2001) described methyl esterification (using d_0- or d_3-methanol) of peptides, a procedure that converts carboxylic acids on the side chains of aspartic and glutamic acids, as well as the carboxyl terminus, to their corresponding methyl esters. Derivatization of lysine to homoarginine was proposed by several groups (Hale *et al.*, 2000; Keough, Lacey and Youngquist, 2000; Beardsley, Karty and Reilly, 2000; Brancia, Oliver and Gaskell, 2000; Peters *et al.*, 2001) to enhance their intensity in the investigation of proteins/peptides by MALDI-TOF-MS. The same derivatization procedure was later adapted by Cagney and Emili (2002), who termed the approach 'mass-coded abundance tagging' (MCAT), where C-terminal Lys residues of tryptic peptides are modified through differential guanidination (resulting in a mass difference of 42 Da, as opposed to 8 Da in the case of ICAT).

All of the above methods are indeed similar, in that they exploit mass differences among polypeptides for inferring their quantitative expression ratio. In a totally different approach to quantitative proteomics, fluorescent labelling has been recently described. The first report came from Unlu, Morgan and Minden (1997), and the technique was aptly termed DIGE: differential in-gel electrophoresis. In fluorescence 2D DIGE, each sample is covalently labelled with a different mass and charge-matched fluorophor: Cy3 (1-(5-carboxypentyl)-1′-propylindocarbocyanine halide N-hydroxysuccinimidyl ester) and Cy5 (1-(5-carboxypentyl)-1′-methylindodicarbocyanine halide N-hydroxysuccinimidyl ester), before mixing the samples and analysing them on the same 2D gel. These structurally similar, but spectrally different (Cy3, $\lambda_{em} = 569$, orange colour; Cy5, $\lambda_{em} = 645$, colour far red), fluorophors undergo nucleophilic substitution reaction with the lysine ε-amino groups on proteins to form an amide. The dyes have very similar molecular masses and are positively charged, so as to match the charge on the lysine group. Charge matching should ensure that there should be little shift over the unlabelled protein on the pI axis, although it would appear that more than 95 per cent of the protein can be significantly shifted away from the unlabelled protein in the mass dimension, particularly at lower protein masses. When the two Cy3/Cy5 mixed samples are analysed on the same 2D gel, and the latter is monitored under different fluorescent gel imaging, the following scenario will occur: all matched spots, expressed in a 1:1 ratio in control and experimental tissues, will appear in a

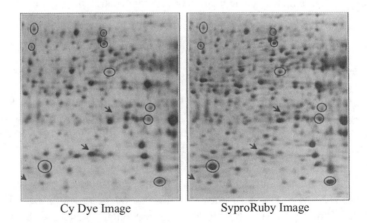

Cy Dye Image   SyproRuby Image

Figure 5.4 Differential labelling with N-hydroxysuccinimidyl esters of fluorescent Cy3 and Cy5 dyes (DIGE technique) of breast cancer cells ErbB-2-transformed (reprinted, by permission, from Gharbi *et al.*, 2002)

violet colour; those in which the Cy5 predominates will have progressively higher red hues, according to the quantitative Cy5/Cy3 ratio, and those in which the Cy3 marker is more abundant will tend to progressively bluish colours (special filters transform the orange fluorescence of Cy3 into blue; this is why all articles deal in terms of blue/red fluorescence). Special image analysis software matches the images, quantifies the spots, normalizes the signals and provides the difference of expression of any set of two proteins by comparison (Tonge *et al.*, 2001; Zhou *et al.*, 2002; Gharbi *et al.*, 2002). An example of DIGE analysis can be found in Figure 5.4. More on this topic can be found in an ample review by Hamdan and Righetti (2002).

5.5 Pre-fractionation in proteome analysis

When analysing protein spots from 2D maps by mass spectrometry, it appears that only generally abundant proteins (codon bias > 0.2) can be properly identified. Thus, the number of spots on a 2D gel is not representative of the overall number or classes of expressed genes that can be analysed. Gygi *et al.* (2000) have calculated that, when loading only 40 μg total yeast lysate, as done in the early days of 2D mapping, only polypeptides with an abundance of at least 51 000 copies/cell could be detected. With 0.5 mg of starting protein, proteins present at 1000 copies/cell could now be visualized by silvering, but those present at 100 and 10 copies/cell could not be revealed. These authors thus concluded that the large range of protein expression levels limits the ability of the 2D-MS approach to analyse proteins of medium to low abundance, and thus the potential of this technique for proteome analysis is likewise limited. This is a severe limitation, since it is quite likely that the portion of proteome we are currently missing is the most interesting one from the point of view of understanding cellular and regulatory proteins, since such low-abundance polypeptide chains will typically be regulatory proteins.

A way out of this impasse would be pre-fractionation. At present, two major approaches have been described: chromatographic and electrophoretic. Fountoulakis's group has extensively developed this approach, as presented below. In a first procedure, Fountoulakis *et al.* (1997) and Fountoulakis and Takàcs (1998) adopted affinity chromatography on heparin gels as a pre-fractionation step for enriching certain protein fractions in the bacterium *Haemophilus influenzae*. In a second approach (Fontoulakis *et al.*, 1998), the same lysate of *H. influenzae* was pre-fractionated by chromatofocusing on Polybuffer Exchanger. In another procedure, the cytosolic soluble proteins of *H. influenzae* were pre-fractionated by Fountoulakis, Takàcs and Takàcs (1999) by hydrophobic interaction chromatography (HIC) on a TSK Phenyl column. In yet another variant, Fountoulakis *et al.* (1999) reported enrichment of low-abundance proteins of *E. coli* by hydroxyapatite chromatography. All these different chromatographic steps allowed the discovery and characterization of several hundred new polypeptide chains, which were present, in the original, unfractionated lysate, at too low levels to be detected.

In terms of electrophoretic pre-fractionation, a most efficient procedure is that based on multicompartment electrolysers (MCEs), as devised by Righetti *et al.* (1990), whose scheme is depicted in Figure 5.5. This method relies on isoelectric membranes, fabricated with the same Immobiline chemicals as adopted in IPG fractionations (Righetti, 1990). The advantages of such a procedure are immediately apparent:

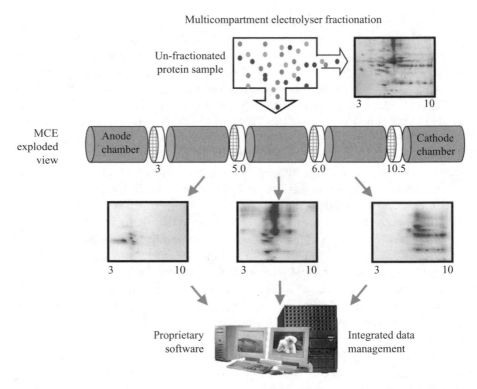

Figure 5.5 Scheme for sample pre-fractionation based on multicompartment electrolyser. The top right panel shows a 2D map of unfractionated human serum, versus three different 2D maps (bottom panels) of three isoelectric fractions, captured in traps having membranes with p*I* 3.0–5.0, 5.0–6.0 and 6.0–10.5 (Herbert and Righetti, unpublished)

- it offers a method that is fully compatible with the subsequent first-dimension separation, a focusing step based on the Immobiline technology;

- it permits harvesting a population of proteins having pI values precisely matching the pH gradient of any narrow (or wider) IPG strip;

- as a corollary of the above point, much reduced chances of protein precipitation will occur (in fact, when an entire cell lysate is analysed in a wide gradient, there are fewer risks of protein precipitation; in contrast, when the same mixture is analysed in a narrow gradient, massive precipitation of all non-isoelectric proteins could occur, with a strong risk of co-precipitation of proteins that would otherwise focus in the narrow pH interval);

- due to the fact that only the proteins co-focusing in the same IPG interval will be present, much higher sample loads can be operative, permitting detection of low-abundance proteins;

- in samples containing extreme ranges in protein concentrations (such as human serum, where a single protein, albumin, represents more than 60 per cent of the total species), one could assemble an isoelectric trap narrow enough to just eliminate the unwanted protein from the entire complex, this too permitting much higher sample loads without interference from the most abundant species;

- the apparatus depicted in Figure 5.5 has been miniaturized by Herbert and Righetti (2000; Righetti, Castagna and Herbert, 2001). In this particular case, a set-up made of five chambers is divided by four membranes, having pI values of 3.0, 5.0, 6.0 and 10.5, selected so as to trap in a central, narrow pI chamber albumin, the most abundant protein in human serum. This permits concentration of other fractions and detection of many more, dilute species in other regions of the pH scale. By properly exploiting this pre-fractionation device, we have been able to capture and detect much more of the 'unseen' yeast membrane proteome (Pedersen *et al.*, 2003).

The importance of some of these findings is now highlighted: integral membrane proteins are rarely seen in 2D maps; proteins with CBI (codon bias index) < 0.2 represent low-abundance proteins and are scarcely detected in 2D maps unless enriched by some pre-fractionation protocol. The power of our methodology, which not only relies on pre-fractionation steps but also deals with analysis of intact proteins, rather than of proteolytic digests, as customary in most protocols exploiting coupled chromatographic processes (see below), can also be appreciated in Figure 5.6. In yeast, there exist two forms of NADH-cytochrome b5 reductase, one called p34 (pI 8.7, M_r 34 kDa) and the other p32 (pI 7.8, M_r 30 kDa). The first species is an integral outer membrane protein, which mediates the reduction of cytochrome b5; the second one is derived from the first one by an *in vivo* proteolytic cleavage, is soluble and resides in the intermembrane space. Our enrichment protocol, coupled to 2D analysis of all intact polypeptide chains, is able to detect both species. In addition, we can easily observe two isoforms of the p34 chain, one with pI 8.7 and the other one more alkaline, pI 9.1. Chromatographic techniques, which rely on the presence of just one (or a few) peptides in a total cell digest, would not have been able to detect any of these, biologically relevant, forms!

p32/p34 NADH-cytochrome b5 reductase

Membrane prep, 7-10 MCE fraction

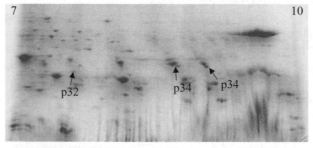

• p34 isoform (8.7/34 kDa): integral outer membrane protein mediates reduction of cyt B5

• p32 isoform (7.8/30kDa): intermembrane space, soluble

Figure 5.6 Example of pre-fractionation/enrichment of yeast membrane proteins and detection of truncated and isoforms of the p34 NADH-cytochrome b5 reductase (reprinted, by permission, from Pedersen *et al.*, 2002)

5.6   Multi-dimensional chromatography

In chromatographic approaches the proteins, in most cases, are digested into peptides prior to separation. The advantage is that peptides (especially from membrane proteins) are more soluble in a wide variety of solvents and hence easier to separate than the parent proteins. The disadvantage is the tremendous increment in the number of species to be resolved. Since, after fractionation in coupled columns, the eluate is sent directly to MS instrumentation, the sample complexity might still be too high for proper analysis. While MS technology is rapidly improving, its dynamic range measurement capability is still two to three orders of magnitude less than the range of protein expression found within mammalian cells. In one of the earliest reports, Opiteck *et al.* (Opiteck, Lewis and Jorgenson, 1997; Opiteck *et al.*, 1998) described a 2D HPLC system that used size exclusion chromatography (SEC) followed by RP-HPLC for mapping of *E. coli* proteins. Perhaps one of the most successful approaches, though, is that of Yates and co-workers (1997; Link *et al.*, 1997; Washburn, Wolters and Yates, 2001), who developed an on-line 2D ion-exchange column coupled to RP-HPLC, termed MudPit, for separating tryptic digests of 80S ribosomes from yeast. The acidified peptide mixture was loaded onto a strong cation exchanger (SCX) column; discrete eluate fractions are then fed onto a RP column, whose effluent is fed directly into a mass spectrometer. This iterative process is repeated 12 times using increasing salt gradient elution from the SCX bed and an increasing organic solvent concentration from the RP beads (typically a C_{18} phase). In a total yeast lysate, the MudPit strategy allowed the identification of almost 1500 proteins (Washburn, Wolters and Yates, 2001). A similar procedure was used for separation of proteins and peptides in human plasma filtrates (Raida *et al.*, 1999), plasma (Richter *et al.*, 1999), blood ultrafiltrates (Schrader *et al.*, 1997) and human urines (Heine, Raida and Forssmann, 1997). The same set-up (SCX followed by C18-RP) was used by Davis *et al.* (2001) to resolve a protein digest derived from conditioned medium from human lung fibroblasts. Perhaps the most sophisticated instrumentation is that devised by Unger and co-workers (Wagner *et al.*, 2002) for processing proteins of $M_r < 20$ kDa. The set-up consists of two gradient HPLC instruments, two UV detectors, an isocratic pump, four RP columns, and ion-exchange column, four ten-port valves, an injection valve, two fraction collector stations and a work station to control such a fully automated system. This system was applied to mapping of human haemofiltrates as well as lysates from human foetal fibroblasts. There are also other, hybrid systems, consisting in coupling, e.g., an HPLC column to electrophoretic instrumentation, notably capillary electrophoresis or isoelectric focusing, or even isotachophoresis (Chen *et al.*, 2001; Wang *et al.*, 2002; Mohan and Lee, 2002). For these, we refer the reader to some recent reviews (Issaq *et al.*, 2002; Issaq, 2001; Shen and Smith, 2002). At times, bidimensionality can simply be achieved by coupling a single-dimension separation to powerful MS probes. A nice example of this comes from the work of Shen *et al.* (2001). The first-dimension separation is RPLC (reversed-phase liquid chromatography) (which discriminates on the basis of a hydrophobic parameter) coupled directly to FTICR-MS (Fourier transform ion cyclotron resonance mass spectrometry). With this coupled system, in a single run of about 3 h, more than 100 000 components could be resolved, a truly impressive achievement (see Figure 5.7). More than 1000 soluble yeast proteins could be identified from more than 9000 peptide database search 'hits' in a single,

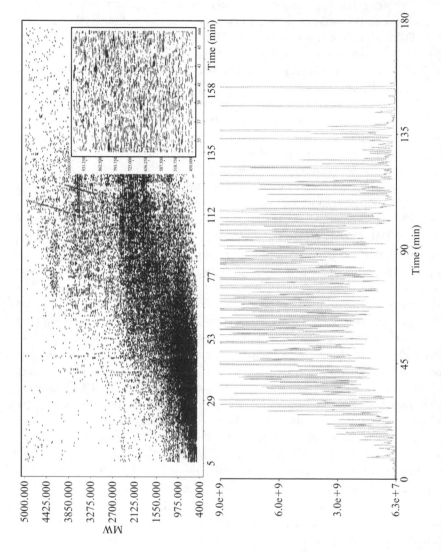

Figure 5.7 High-efficiency capillary RPLC-FTICR-MS of a cytosol tryptic digest and its super power resolution for extremely complex mixtures ($>10^5$ components detected) (reprinted, by permission, from Shen *et al.*, 2001)

high-efficiency capillary RPLC-FTICR-MS run of 3 h from a total of only 40 μg sample (Shen *et al.*, 2001).

5.7 Protein chip arrays

One of the latest technology platforms that have been developed for proteomics includes chip-based arrays. Such arrays have been elaborated for separating proteins based on surface chemistries or known protein ligands (e.g., antibodies), with subsequent identification by mass spectrometry (Borrebaeck, 2000; Lueking *et al.*, 1999; Borrebaeck *et al.*, 2001). Other promising applications of protein chip microarrays have been differential profiling (MacBeath and Schreiber, 2000) and high-throughput functional analysis (Fung *et al.*, 2001; Sawyer, 2001). An example of such protein chip arrays is shown in Figure 5.8. Such arrays can be divided into chemical and bioaffinity surfaces. In the first case (see the upper row in Figure 5.8) such surfaces function essentially like mini-chromatographic columns, in that they capture a given protein population by, e.g., hydrophobic interaction (reversed phase), metal chelation and different types of ion exchanger. Such chemical surfaces are not highly selective; nevertheless, they can be used in a cascade fashion, e.g. proteins captured on ion exchangers can be eluted and re-adsorbed onto a reversed phase, or on metal chelators so as to further sub-fractionate a very heterogeneous protein mixture as a total cell lysate. The bioaffinity surfaces (see the lower row in Figure 5.8) clearly work on a much higher-selectivity principle and allow capture of a narrow and well defined protein population, immediately ready for MS analysis. How the latter may be performed is shown schematically in Figure 5.9: once a selected protein population has been captured, the surface of the chip is bombarded with a laser beam, which will desorb and ionize the different proteins, ultimately identified via their precise molecular mass. The main difference between this kind of structural analysis and MALDI-TOF desorption ionization is that, in the latter case, the sample has to be manually transferred to a microwell plate, admixed with a special matrix (such as sinapinic acid) and then desorbed/ionized via pulse laser shots. Conversely, in the protein chips here illustrated (Figure 5.9) a stick containing eight (or multiples thereof) such affinity surfaces, with the captured protein, is directly inserted in the MS instrument and desorbed by surface-enhanced laser desorption/ionization (SELDI), in the absence of added matrix (the surface of such arrays being, in fact, already coated with energy absorbing polymers) (Voivodov, Ching and Hutchens, 1996). Such instrumentation is rapidly being adopted in many clinical chemistry laboratories around the world and might

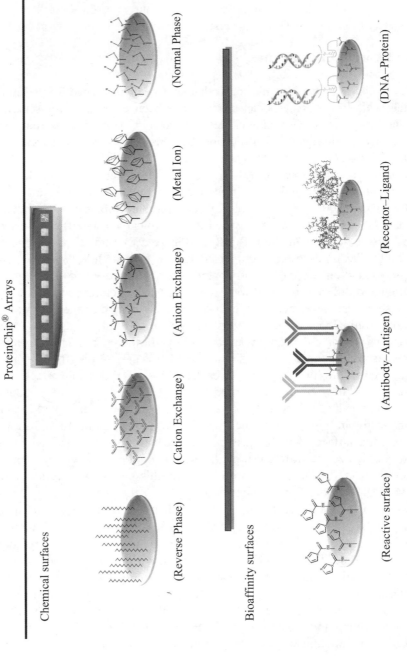

Figure 5.8 Scheme of different surfaces adopted in protein chip arrays (by courtesy of Ciphergen)

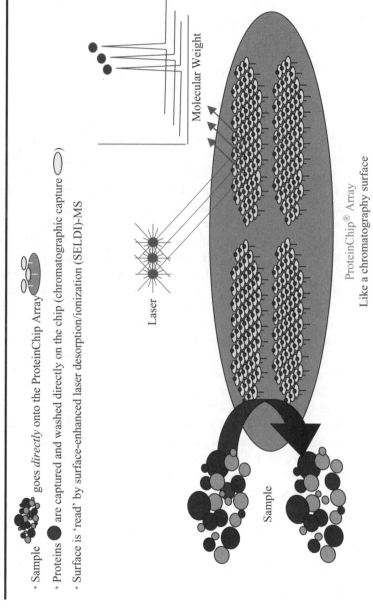

Figure 5.9 Scheme of processing/detection of protein species captured by different chemical surfaces (chromatographic adsorption) (by courtesy of Ciphergen)

soon become part of the standard instrumentation of such laboratories. The following is a (non-exhaustive) list of papers recently published on the field of arrays exploiting SELDI technology: Lin *et al.* (2001); Paweletz, Liotta and Petricoin (2001); Boyle *et al.* (2001); Wang *et al.* (2001); Thulasiraman *et al.* (2000); Austen, Frears and Davies (2000); Von Eggeling *et al.* (2000); Merchant and Weinberger (2000); Weinberger, Viner and Ho (2002).

5.8   Imaging mass spectrometry

It seems appropriate to end this chapter with an emerging, quite powerful MS technique: imaging mass spectrometry (IMS). In this newest development, MALDI-TOF-MS is used in profiling and imaging peptides and proteins directly from these tissue sections in order to obtain specific information on their local relative abundance and spatial distribution. Results from such imaging experiments yield a wealth of information, allowing investigators to measure and compare many of the major molecular components of the section in order to gain a deeper understanding of biomolecular processes involved. For instance, in tumour tissues, one could detect and map tumour-specific markers as well as determine their relative concentrations within a section (Stoeckli *et al.*, 2001). For clinicians, this would permit molecular assessment of tumour biopsies with the potential of identifying subpopulations that are not evident based on cellular phenotypes determined microscopically. In one application (Stoeckli *et al.*, 2001), key protein markers present in a human glioblastoma xenograph section (transplanted in a mouse hind limb) could be properly mapped, with a spatial resolution of about 30 μm for compounds in a mass range from 1000 to over 50 000 Da. Direct analysis of such tissue sections requires spotting or coating of the tissue with a matrix compound (typically sinapinic acid or other cinnamic acid analogues). A raster of this sample by the laser beam (usually obtained via 20–50 consecutive laser shots at each coordinate) and subsequent mass analysis of the desorbed ions can record molecular intensities throughout the section. A nice example of this technique is given in Figure 5.10, which shows the IMS of a cauda tissue section of mouse epididymis (Chaurand and Caprioli 2002). Figure 5.10(a) presents an optical image of the section prior to coating, outlining the general shapes of the cauda and of the tubules. Figures. 5.10(b)–(d) displays three images obtained after scanning the same section, each at specific m/z values. In this case, the images were acquired by using a 50 μm resolution step on the section with 20 laser shots per data point. From these data, it would appear that the peptides/proteins

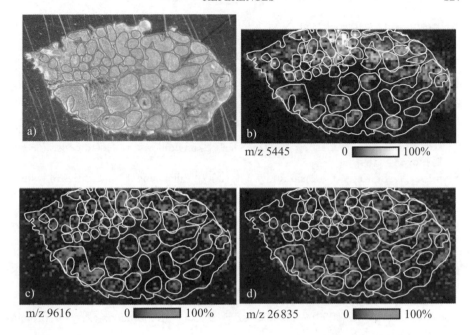

Figure 5.10 Imaging mass spectrometry. (a) Optical image of a 12 μm thin mouse cauda epididymis section: the outer perimeters of the section and tubules have been outlined. (b)–(d) Ion density maps obtained for three different *m/z* values from the section after MALDI-IMS (reprinted, by permission, from Chaurand and Caprioli, 2002)

at *m/z* 5445, 9616 and 26 835 are localized within the lumen of the epididymal tubules.

Acknowledgements

Supported in part by grants from ASI (Agenzia Spaziale Italiana), grant No. I/R/294/02, by the European Community, grant No. QLG2-CT-2001-01903, by GSK, grant No GEC 556-2000, and by a grant from AIRC (Associazione Italiana Ricerca sul Cancro).

References

Aebersold, R., and Leavitt, J. (1990) Sequence analysis of proteins separated by polyacrylamide gel electrophoresis: towards an integrated protein database. *Electrophoresis* **11**, 517–527.

Anderson, N. L., and Anderson, N. G. (1998) Proteome and proteomics: new technologies, new concepts, and new words. *Electrophoresis* **19**, 1853–1861.

Anderson, L., and Seilhamer, J. (1997) A comparison of selected mRNA and protein abundances in human liver. *Electrophoresis* **18**, 533–537.

Antonucci, F., Chilosi, M., Parolini C., Hamdan, M., Astner, H., and Righetti, P. G. (2003) Two-dimensional molecular profiling of mantle cell lymphoma. *Eletrophoresis* **24**, 2376–2358.

Austen, B. M., Frears, E. R., and Davies, H. (2000) The use of SELDI Protein Chip arrays to monitor production of Alzheimer's β-amyloid in transfected cells. *J. Pept. Sep.* **6**, 459–469.

Beardsley, R. L., Karty, J. A., and Reilly, J. P. (2000) Enhancing the intensities of lysine-terminated tryptic peptide ions in matrix assisted laser desorption/ionisation mass spectrometry. *Rapid Commun. Mass Spectrom.* **14**, 2147–2153.

Bjellqvist, B., Ek, K., Righetti, P. G., Gianazza, E., Görg, A., Postel, W., and Westermeier, R. (1982) Isoelectric focusing in immobilized pH gradients: principle, methodology and some applications. *J. Biochem. Biophys. Methods* **6**, 317–339.

Bordini, E., Hamdan, M., and Righetti, P. G. (1999) Probing the reactivity of S–S bridges to acrylamide in some proteins under high pH conditions by matrix-assisted laser-desorption/ionization. *Rapid Commun. Mass Spectrom.* **13**, 1818–1827.

Borrebaeck, C. A. K. (2000) Antibodies in diagnostics: from immunoassays to protein chips. *Immunol. Today* **21**, 379–382.

Borrebaeck, C. A. K., Ekstrom, S., Molmborg Hager, A. C., Nilson, J., Laurell, T., and Marko-Varga, G. (2001) Protein chips based on recombinant antibody fragments: a highly sensitive approach as detected by mass spectrometry. *BioTechniques* **30**, 1126–1132.

Boyle, M. D., Romer, T. G., Meeker, A. K., and Sledjeski, D. D. (2001) Use of surface-enhanced desorption ionization protein chip system to analyze streptococcal exotoxin B activity secreted by *Streptococcus pyogenes*. *J. Microbiol. Methods* **46**, 87–97.

Brancia, F., Oliver, S. G., and Gaskell, S. J. (2000) Improved matrix-assisted laser desorption/ionisation mass spectrometric analysis of tryptic hydrolysates of proteins following guanidination of lysine-containing peptides. *Rapid Commun. Mass Spectrom.* **14**, 2070–2073.

Cagney, G., and Emili, A. (2002) De novo peptide sequencing and quantitative profiling of complex protein mixtures using mass-coded abundance tagging. *Nat. Biotechnol.* **20**, 163–170.

Celis, J. E., Ostergaard, M., Jensen, N. A., Gromova, I., and Rasmussen, H. H. (1998) Human and mouse proteomic databases: novel resources in the protein universe. *FEBS Lett.* **430**, 64–72.

Chaurand, P., and Caprioli, R. M. (2002) Direct profiling and imaging of peptides and proteins from mammalian cells and tissue sections by mass spectrometry. *Electrophoresis* **23**, 3125–3135.

Chen, J., Lee, C. S., Shen, Y., Smith, R. D., and Baehrecke, E. H. (2002) Integration of capillary isoelectric focusing with capillary reversed-phase liquid chromatography for two-dimensional proteomics separations. *Electrophoresis* **23**, 3143–3148.

Chevallet, M., Santoni, V., Poinas, A., Rouquie, D., Fuchs, A., Kieffer, S., Rossignol, M., Lunardi, J., Garin, J., and Rabilloud, T. (1998) New zwitterionic detergents improve the analysis of membrane proteins by two-dimensional electrophoresis. *Electrophoresis* **19**, 1901–1909.

Davis, M. T., Beierle, J., Bures, E. T., McGinley, M. D., Mort, J., Robinson, J. H., Spahr, C. S., Yu, W., Luethy, R., and Pattersson, S. D. (2001) Automated LC–LC–MS–MS platform using binary ion-exchange and gradient reversed-phase chromatography for improved proteomic analyses. *J. Chromatogr. B* **752**, 281–291.

Figeys, D., and Ross, M. (Eds.) (2003) *Applied Proteomics*. New York, Wiley, in press.

Fountoulakis, M., Langen, H., Evers, S., Gray, C., and Takacs, B. (1997) Two-dimensional map of *Haemophilus influenzae* following protein enrichment by heparin chromatography. *Electrophoresis* **18**, 1193–1202.

Fountoulakis, M., Langen, H., Gray, C., and Takàcs, B. (1998) Enrichment and purification of proteins of *Haemophilus influenzae* by chromatofocusing. *J. Chromatogr. A* **806**, 279–291.

Fountoulakis, M., and Takàcs, B. (1998) Design of protein purification pathways: application to the proteome of *Haemophilus influenzae* using heparin chromatography. *Protein Expr. Purif.* **14**, 113–119.

Fountoulakis, M., Takàcs, M. F., Berndt, P., Langen, H., and Takacs, B. (1999) Enrichment of low abundance proteins of *Escherichia coli* by hydroxyapatite chromatography. *Electrophoresis* **20**, 2181–2195.

Fountoulakis, M., Takàcs, M. F., and Takàcs, B. (1999) Enrichment of low-copy-number gene products by hydrophobic interaction chromatography. *J. Chromatogr. A* **833**, 157–168.

Fung, E. T., Thulasiraman, V., Weibernger, S. R., and Dalmasso, E. A. (2001) Protein chips for differential profiling. *Curr. Opin. Biotechnol.* **12**, 65–69.

Galvani, M., Hamdan, M., Herbert, B., and Righetti, P. G. (2001a) Alkylation kinetics of proteins in preparation for two-dimensional maps: a MALDI-TOF mass spectrometry investigation. *Electrophoresis* **22**, 2058–2065.

Galvani, M., Rovatti, L., Hamdan, M., Herbert, B., and Righetti, P. G. (2001b) Proteins alkylation in presence/absence of thiourea in proteome analysis: a MALDI-TOF mass spectrometry investigation. *Electrophoresis* **22**, 2066–2074.

Gehanne, S., Cecconi, D., Carboni, L., Righetti, P. G., Domenici, E., and Hamdan, M. (2002) Quantitative analysis of two-dimensional gel-separated proteins using isotopically marked alkylating agents and matrix-assisted laser desorption/ionisation mass spectrometry. *Rapid Commun. Mass Spectrom.* **16**, 1692–1698.

Gharbi, S., Gaffney, P., Yang, A., Zvelebil, M. J., Cramer, R., Waterfield, M. D., and Timms, J. F. (2002) Evaluation of two-dimensional differential gel electrophoresis for proteomic expression analysis of a model breast cancer cell system. *Mol. Cell Proteomics* **2**, 91–98.

Goodlett, D. R., Keller, A., Watts, J. D., Newitt, R., Yi, E. C., Purvine, S., Eng, J. K., Haller, P., Aebersold, R., and Kolker, E. (2001) Differential stable isotope labelling of peptides for quantitation and de novo sequence derivation. *Rapid Commun. Mass Spectrom.* **15**, 1214–1221.

Gygi, S. P., Corthals, G. L., Zhang, Y., Rochon, Y., and Aebersold, R. (2000) Evaluation of two-dimensional gel electrophoresis-based proteome analysis technology. *Proc. Natl. Acad. Sci. USA* **97**, 9390–9395.

Gygi, S. P., Rist, B., Gerber, S. A., Turecek, F., Gelb, M. H., and Aebersold, R. (1999) Quantitative analysis of complex protein mixtures using isotope-coded affinity tags. *Nature Biotech* **17**, 994–999.

Hale, J. E., Butler, J. P., Knierman, M. D., and Becker, G. W. (2000) Increased sensitivity of tryptic peptides detection by MALDI-TOF mass spectrometry is achieved by conversion of lysine to homoarginine. *Anal. Biochem.* **287**, 110–117.

Hamdan, M., and Righetti, P. G. (2003) Assessment of protein expression by means of 2-D electrophoresis with and without mass spectrometry. *Mass Spectrom. Rev.* **22**, 272–284.

Han, D. K., Eng, J., Zhou, H., and Aebersold, R. (2001) Quantitative profiling of differentiation-induced microsomal proteins using isotope-coded affinity tags and mass spectrometry. *Nature Biotech.* **19**, 946–951.

Hanash, S. (1998) Two dimensional gel electrophoresis. In Hames, B. D. (Ed.). *Gel Electrophoresis of Proteins*. Oxford University Press, Oxford, 189–212.

Heine, G., Raida, M., and Forssmann, W. G. (1997) Mapping of peptides and protein fragments in human urine using liquid chromatography–mass spectrometry. *J. Chromatogr. A* **766**, 117–124.

Herbert, B., Galvani, M., Hamdan, M., Olivieri, E., McCarthy, J., Pedersen, S., and Righetti, P. G. (2001) Reduction and alkylation of proteins in preparation of two-dimensional map analysis: why, when and how? *Electrophoresis* **22**, 2046–2057.

Herbert, B., Hopwood, F., Oxley, D., McCarthy, J., Laver, M., Grinyer, J., Goodall, A., Williams, K., Castagna, A., and Righetti, P. G. (2003) *β*-elimination: an unexpected artefact in proteome analysis. *Proteomics*, **3**, 826–831.

Herbert, B. R., Molloy, M. P., Gooley, B. A. A., Walsh, J., Bryson, W. G., and Williams, K. L. (1998) Improved protein solubility in two-dimensional electrophoresis using tributyl phosphine as reducing agent. *Electrophoresis* **19**, 845–851.

Herbert, B., and Righetti, P. G. (2000) A turning point in proteome analysis: sample pre-fractionation via multicompartment electrolyzers with isoelectric membranes. *Electrophoresis* **21**, 3639–3648.

Hochstrasser, D. F., Harrington, M. G., Hochstrasser, A. C., Miller, M. J., and Merrill, C. R. (1988) Methods for increasing the resolution of two-dimensional protein electrophoresis. *Anal. Biochem.* **173**, 424–435.

Hooker, T. F., Jeffrey, D. J., and Jorgenson, J. W. (1998) Two dimensional separations in high-performance capillary electrophoresis. In: Khaledi MG, (Ed.). *High Performance Capillary Electrophoresis: Theory, Techniques and Applications*. Wiley, New York, 581–612.

Issaq, H. J. (2001) The role of separation science in proteomics research. *Electrophoresis* **22**, 3629–3638.

Issaq, H. J., Conrads, T. P., Janini, G. M., and Veenstra, T. D. (2002) Methods for fractionation, separation and profiling of proteins and peptides. *Electrophoresis* **23**, 3048–3061.

James, P. (Ed.) (2001) *Proteome Research: Mass Spectrometry.* Springer, Berlin.

Kellner, R., Lottspeich, F., and Meyer, H. E. (Eds.). (1999) *Microcharacterization of Proteins.* Wiley-VCH, Weinheim.

Keough, T., Lacey, M. P., and Youngquist, R. S. (2000) Derivatisation procedures to facilitate de novo sequencing of lysine-terminated tryptic peptides using post-source decay matrix assisted laser desorption/ionisation mass spectrometry. *Rapid Commun. Mass Spectrom.* **14**, 2348–2356.

Lahm, H. W., and Langen, H. (2000) Mass spectrometry: a tool for the identification of proteins separated by gels. *Electrophoresis* **21**, 2105–2114.

Lander, E. S., and Weinberg, R. A. (2000) Genomics: journey to the center of biology. *Science* **287**, 1777–1782.

Lin, S., Tornatore, P., King, D., Orlando, R., and Weinberger, S. R. (2001) Limited acid hydrolysis as a means for fragmenting proteins isolated upon ProteinChip array surfaces. *Proteomics* **1**, 1172–1184.

Link, A. J., Eng, J., Schieltz, D. M., Carmack, E., Mize, G. J., Morris, D. R., Garvik, B. M., and Yates, J. R. III (1997) Direct analysis of protein complexes using mass spectrometry. *Nature Biotech.* **17**, 676–682.

Liu, P., and Regnier, F. E. (2002) An isotope coding strategy for proteomics involving both amine and carboxyl group labelling. *J. Proteom. Res.* **1**, 443–450.

Lottspeich, F., Houthaeve, T., and Kellner, R. (1999) Chemical methods for protein synthesis analysis. In Kellner, R., Lottspeich, F., Meyer, H. E., (Eds.). *Microcharacterization of Proteins.* Wiley–VCH, Weinheim, 141–158.

Lueking, A. M., Horn, H., Eickhoff, H., Lehrach, H., and Walter, G. (1999) Protein microarrays for gene expression and antibody screening. *Anal. Biochem.* **270**, 103–111.

MacBeath, G., and Schreiber, S. L. (2000) Printing proteins as microarrays for high-throughput function determination. *Science* **289**, 1760–1763.

McCarthy, J., Hopwood, F., Oxley, D., Laver, M., Castagna, A., Righetti, P. G., Williams, K., and Herbert, B. (2003) Carbamylation of proteins in 2D electrophoresis – myth or reality? *J. Proteome Res.* **2**, 239–242.

Merchant, M., and Weinberger, S. R. (2000) Recent developments in surface-enhanced laser desportion/ionization (SELDI)-time of flight mass spectrometry. *Electrophoresis* **21**, 1164–1167.

Mohan, D., and Lee, C. S. (2002) On-line coupling of capillary isoelectric focusing with transient isotachophoresis-zone electrophoresis: a two dimensional separation system for proteomics. *Electrophoresis* **23**, 3160–3167.

Molloy, M. P. (2000) Two dimensional electrophoresis of membrane proteins using immobilized pH gradients. *Anal. Biochem.* **280**, 1–10.

Molloy, M. P., Herbert, B. R., Slade, M. B., Rabilloud, T., Nouwens, A. S., Williams, K. L., and Gooley, A. A. (2000) Proteomic analysis of the *Escherichia coli* outer membrane. *Eur. J. Biochem.* **267**, 1–12.

Musante, L., Candiano, G., and Ghiggeri, G. M. (1998) Resolution of fibronectin and other characterized proteins by 2D-PAGE with thiourea. *J. Chromatogr. B* **705**, 351–356.

O'Farrell, P. (1975) High resolution two dimensional electrophoresis of proteins. *J. Biol. Chem.* **250**, 4007–4021.

Opiteck, G. J., Lewis, K. C., and Jorgenson, J. W. (1997) Comprehensive LC/LC/MS of proteins. *Anal. Chem.* **69**, 1518–1524.

Opiteck, G. J., Ramirez, S. M., Jorgenson, J. W., and Moseley, M. A, III. (1998) Comprehensive two-dimensional high-performance liquid chromatography for the isolation of overexpressed proteins and proteome mapping. *Anal. Biochem.* **258**, 349–361.

Patterson, S. D., and Aebersold, R. (1995) Mass spectrometric approaches for the identification of gel-separated proteins. *Electrophoresis* **16**, 1791–1814.

Paweletz, C. P., Liotta, L. A., and Petricoin, E. F. III. (2001) New technologies for biomarker analysis of prostate cancer progression: laser capture microdissection and tissue proteomics. *Urology* **57**, 160–164.

Pedersen, S. K., Harry, J. L., Sebastian, L., Baker, J., Traini, M. D., McCarthy, J. T., Manoharan, A., Wilkins, M. R., Gooley, A. A., Righetti, P. G., Packer, N. H., Williams, K. L., and Herbert, B. R. (2003) Unseen proteome: mining below the tip of the iceberg to find low abundance and membrane proteins. *J. Proteome Res.* **2**, 303–311.

Peters, E. C., Horn, D. M., Tully, D. C., and Brock, A. (2001) A novel multifunctional labeling reagent for enhanced protein characterisation with mass spectrometry. *Rapid Commun. Mass Spectrom.* **15**, 2387–2392.

Rabilloud, T., Adessi, C., Giraudel, A., and Lunardi, J. (1997) Improvement of the solubilization of proteins in two-dimensional electrophoresis with immobilized pH gradients. *Electrophoresis* **18**, 307–316.

Rabilloud, T., and Chevallet, M. (2000) Solubilization of proteins in two dimensional electrophoresis. In Rabilloud, T. (Ed.). *Proteome Research: Two-Dimensional Gel Electrophoresis and Identification Methods*, Springer, Berlin, 9–29.

Rabilloud, T. (Ed.) (2000) Proteome Research: Two-Dimensional Gel Electrophoresis and Identification Methods. Springer, Berlin.

Raida, M., Schultz-Knape, P., Heine, G., and Forssmann, W. G. (1999) Liquid chromatography and electrospray mass spectrometric mapping of peptides from human plasma filtrate. *J. Am. Mass Spectrom.* **10**, 45–54.

Richter, R., Schultz-Knape, P., Schrader, M., Standker, L., Jurgens, M., Tammen, H., and Forssmann, W. G. (1999) Composition of the peptide fraction in human blood plasma: database of circulating human peptides. *J. Chromatogr. B* **726**, 25–35.

Righetti, P. G. (1990) *Immobilized pH Gradients: Theory and Methodology*. Elsevier, Amsterdam.

Righetti, P. G., Castagna, A., and Herbert, B. (2001) Prefractionation techniques in proteome analysis. *Anal. Chem.* **73**, 320A–326A.

Righetti, P. G., Stoyanov, A. V., and Zhukov, M. Y. (2001) The Proteome Revisited: Theory and Practice of All Relevant Electrophoretic Steps. Elsevier, Amsterdam.

Righetti, P. G., Tudor, G., and Gianazza, E. (1982) Effect of 2-mercaptoethanol on pH gradients in isoelectric focusing. *J. Biochem. Biophys. Methods* **6**, 219–227.

Righetti, P. G., Wenisch, E., Jungbauer, A., Katinger, H., and Faupel, M. (1990)

Preparative purification of human monoclonal antibody isoforms in a multicompartment electrolyzer with Immobiline membranes. *J. Chromatogr.* **500**, 681–696.

Rimpilainen, M., and Righetti, P. G. (1985) Membrane protein analysis by isoelectric focusing in immobilized pH gradients. *Electrophoresis* **6**, 419–422.

Santoni, V., Kieffer, S., Desclaux, D., Masson, F., and Rabilloud, T. (2000) Membrane proteomics: use of additive main effects with multiplicative interaction model to classify plasma membrane proteins according to their solubility and electrophoretic properties. *Electrophoresis* **21**, 3329–3344.

Sawyer, T. K. (2001) Proteomics – structure and function. *BioTechniques* **31**, 156–160.

Schrader, M., Jurgens, M., Hess, R., Schultz-Knape, P., Raida, M., and Forssmann, W. G. (1997) Matrix-assisted laser desorption/ionisation mass spectrometry guided purification of human guanylin from blood ultrafiltrate. *J. Chromatogr. A* **766**, 139–145.

Sechi, S. (2002) A method to identify and simultaneously determine the relative quantities of proteins isolated by gel electrophoresis. *Rapid Commun. Mass Spectrom.* **16**, 1416–1424.

Shen, Y., and Smith, R. D. (2002) Proteomics based on high-efficiency capillary separations. *Electrophoresis* **23**, 3106–3124.

Shen, Y., Tlic, N., Zhao, R., Pasa-Tolic, L., Li, L., Berger, S. J., Harkewicz, R., Anderson, G. A., Belov, M. E., and Smith, R. D. (2001) *Anal. Chem.* **73**, 3011–3021.

Stoeckli, M., Chaurand, P., Hallahan, D. E., and Caprioli, R. M. (2001) *Nat. Med.* **7**, 493–496.

Thulasiraman, V., McCutchens-Maloney, S. L., Motin, V. L., and Garcia, E. (2000) Detection and identification of virulence factors in *Yersinia pestis* using SELDI Protein Chip systems. *BioTechniques* **30**, 428–432.

Tonge, R., Shaw, J., Middleton, B., Rwlinson, R., Rayner, S., Young, J., Pognan, F., Hawkins, E., Currie, I., and Davison, M. (2001) Validation and development of fluorescence two-dimensional differential gel electrophoresis proteomics technology. *Proteomics* **1**, 377–396.

Unlu, M., Morgan, M. E., and Minden, J. S. (1997) Difference gel electrophoresis: a single gel method for detecting changes in protein extracts. *Electrophoresis* **18**, 2071–2077.

Voivodov, K. I., Ching, J., and Hutchens, T. W. (1996) Surface arrays of energy absorbing polymers enabling covalent attachment of biomolecules for subsequent laser-induced uncoupling/desorption. *Tetrahedron Lett.* **37**, 5669–5672.

Von Eggeling, F., Davies, H., Lomas, L., Fiedler, W., Junker, K., Claussen, U., and Ernst, G. (2000) Tissue-specific microdissection coupled with Protein Chip array technologies: application in cancer research. *BioTechiques* **29**, 1066–1070.

Wagner, K., Miliotis, T., Marko-Varga, G., Bischoff, R., and Unger, K. K. (2002) An automated on-line multidimensional HPLC system for protein and peptide mapping with integrated sample preparation. *Anal. Chem.* **74**, 809–820.

Wang, S., Diamond, D. L., Hass, G. M., Sokoloff, R., and Vessella, R. L. (2001) Identification of prostate-specific membrane antigen (PMSA) as the target of monoclonal antibody 107-1A4 by protein chip array surface-enhanced laser desorption/ionization (SELDI) technology. *Int. J. Cancer* **92**, 871–876.

Wang, H., Kachman, M. T., Schwartz, D. R., Cho, K. R., and Lubman, D. M. (2002) A protein molecular weight map of ES2 clear cell ovarian carcinoma cells using a two-dimensional liquid separation/mass mapping technique. *Electrophoresis* **23**, 3168–3181.

Washburn, M. P., Wolters, D., and Yates, J. R. (2001) Large-scale analysis of the yeast proteome by multidimensional protein identification technology. *Nature Biotech.* **19**, 242–247.

Weinberger, S., Viner, R. I., and Ho, P. (2002) Tagless extraction-retentate chromatography: a new global protein digestion strategy for monitoring differential protein expression. *Electrophoresis* **23**, 3182–3192.

Westermeier, R. (1997) High resolution 2D electrophoresis. In Westermeier, R. (Ed.). *Electrophoresis in Practice*. VCH, Weinheim, 213–228.

Wilkins, M. R., Gasteiger, E., Sanchez, J.C., Bairoch, A., and Hochstrasser, D. F. (1998) Two-dimensional gel electrophoresis for proteome projects: the effects of protein hydrophobicity and copy number. *Electrophoresis* **19**, 1501–1505.

Wilkins, M. R., Williams, K. L., Appel, R. D., and Hochstrasser, D. F. (Eds). (1997) *Proteome Research: New Frontiers in Functional Genomics*. Springer, Berlin.

Yates, J. R. III, McCormack, A. L., Schieltz, D., Carmack, E., and Link, A. (1997) Direct analysis of protein mixtures by tandem mass spectrometry. *J. Protein Chem.* **16**, 495–497.

Zhang, R., and Regnier, F. J. (2002) Minimizing resolution of isotopically coded peptides in comparative proteomics. *J. Proteom. Res.* **1**, 139–147.

Zhou, G., Li, H., DeCamps, D., Chen, S., Shu, H., Gong, Y., Flaig, M., Gillespie, J. W., Hu, N., Taylor, P. R., Emmert-Buck, M. R., Liotta, L.A., Petricon, E. F. III, and Zhao, Y. (2002) 2D differential in-gel electrophoresis for the identification of esophageal scans cell cancer-specific protein markers. *Proteomics* **2**, 117–124.

6

Mass Spectrometry in Proteomics

Pierre-Alain Binz

6.1   Introduction

Mass spectrometry has become a central analysis tool in the field of proteome analysis. Mass spectrometry has been used for many years to analyse proteins and peptides. However, since the origin of proteome analysis, which dates from about 9 years ago, the requirements have drastically evolved. To cope with the challenges of proteomics needs, MS instrument providers have been pushed to generate, and further optimize many different instruments. New requirements include higher sensitivity, higher precision, high-throughput capacities, unsupervised multi-mode selection etc. A summarized list updated at the end of 2002 has been published in the *Chemisch2Weekblad 98* (2002-18): 32–45. The list is also available online on http://www.hyphenms.nl. Within a couple of months of this publication, already a number of improvements had been made; new features had been added to existing instrumentation and even new instruments have been put on the market.

Here we will focus on the description of instruments used in proteomics. We will not describe the detailed architecture of all these instruments. We will describe the fundamental principles of the implemented technology. This means that we will describe the main ionization and analysis components of an MS instrument. Then we will summarize the set-up of the commonly used instruments with brief information on their functional properties.

The choice for the use of a particular MS instrument in a laboratory is closely related to the general experimental workflow designed for a particular study. Identifying all available proteins in a biological sample is different from

Genomics, Proteomics and Vaccines edited by Guido Grandi
© 2004 John Wiley & Sons, Ltd   ISBN 0 470 85616 5

specifically searching differentially expressed protein forms, or from looking for the presence of chosen post-translational modifications in a subset of proteins. Therefore we will also discuss the major proteomics workflows that include mass spectrometry as major analytical tool.

6.2 MS technology

A mass spectrometer is an instrument that aims to measure the mass of submitted analytes. In almost all set-ups, it produces charged particles (ions) from the molecules to be analysed. Various means are implemented to exert forces on the charged particles and to measure the molecular mass of the analytes.

According to http://masspec.scripps.edu/information/index.html J. J. Thomson built the first mass spectrometer in 1897. From then on techniques defining mass spectrometry have been improved over the years. J. H. Beynon first attempted to identify organic compounds with MS in 1956. Today, more than a dozen different instruments are commercialized by about half a dozen companies for proteomics usage. Even if these instruments might behave very differently in term of functionalities, one can distinguish a reduced number of common constituents to all of them. All mass spectrometers carry out four distinct functions:

- ionization

- ion analysis

- ion detection

- signal processing

6.2.1 Ionization

Various ionization methods include electron impact (EI), chemical ionization (CI), fast atom bombardment (FAB), field desorption (FD), electro-spray ionization (ESI) and derivatives and laser desorption (LD) and derivatives.

In proteomics, mainly ESI and matrix-assisted laser desorption ionization (MALDI) are implemented. We will therefore focus on these now.

Electro-spray ionization (ESI) was described by Dole *et al.* in 1968. It was revisited by Yamashita and Fenn in 1984. ESI produces molecular ions directly from samples in solution (Figure 6.1). The process transfers ions in solution

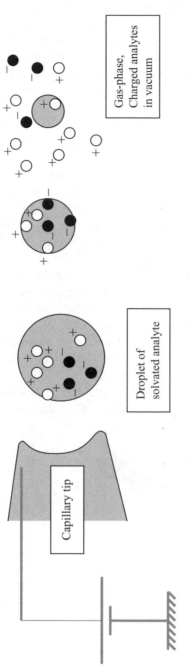

Capillary tip

Droplet of
solvated analyte

Gas-phase,
Charged analytes
in vacuum

Figure 6.1 Principle of the ESI process

into the gas phase. The process involves the application of an electric field to an inlet capillary tip. Liquid samples are delivered to the end of the high-potential capillary either under pressure from a pump or by electrostatic forces alone via a nano-electrospray source (Wilm and Mann, 1996). As the solution containing the analytes elutes from the tip to the high-voltage field fine droplets are formed. Hot nitrogen gas causes the solvent to evaporate and charged peptide ions are released into the gas phase. Ionization is generally obtained by protonation of the analytes. In this case, one or more protons can bind to the analyte. The analyte becomes singly or more often multiply charged. The flow rate ranges in the order of μl/min.

ESI is a continuous ionization method that is compatible with HPLC or capillary electrophoresis as interface or sample inlet. Multiply charged ions are usually produced. A nanospray is a special type of electro-spray, which uses a pulled and coated glass capillary to achieve low flows of the order 20–50 n/min. The use of a nanospray needle allows a microlitre sample to be available for measurement for several minutes.

The MALDI ionization was first described by Karas and Hillenkamp (1988) and by Tanaka *et al.* (1987, 1988). Microlitre quantities of dissolved samples are mixed with a so-called matrix solution and dried onto a stainless-steel-plated target. The mixture is co-crystallized. A pulsing laser (commonly N_2 emitting at 337 nm, selected to be absorbed by the matrix molecules) is used to irradiate the matrix-sample crystals and to generate molecular ions. Molecules are desorbed from the surface and ionized, usually by single protonation. An electric field between the sample target and a grid placed further down the ion path accelerates the ions in the direction of the ion analyser (Figure 6.2).

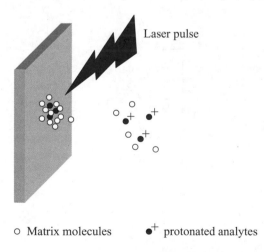

o Matrix molecules $\bullet^+$ protonated analytes

Figure 6.2   Principle of the MALDI process

Generally, the observed ions are singly charged (denoted $(M + H)^+$).

A limited number of matrix compounds are usually considered for the analysis of polypeptides. They are often aromatic organic acids. We can mention, among the most commonly used, alpha-cyano-4-hydroxy-cinnamic acid, 2,5-dihydroxybenzoic acid, sinapinic acid and 2-(4-hydroxyphenylazo)-benzoic acid.

The ionization mechanism is not completely understood and is still subject to debate. However, it is accepted that the matrix acts in absorbing radiatively the energy of the laser. Then the available energy is used to sublimate and ionize the molecules present, including the analytes.

ESI and MALDI can be considered as complementary ionization methods. In ESI, the sample must be soluble, stable in solution, polar and relatively clean (free of non-volatile buffers, detergents, salts etc.). Conversely, MALDI is less sensitive to salt concentration (Qin *et al.*, 1997). It is to be noted that ESI and MALDI do not highlight identical ionization efficiencies. In fact, some peptides are 'easier to see' with MALDI and some others with ESI.

In standards workflows aiming to identify protein entries in a database, the choice of a MALDI or ESI ionization source is not the most crucial part of the process. The workflow and its instrumental implementation define whether the sample will be more convenient to analyse with one or the other ionization. As the MS instruments are still pretty expensive and constitutively cannot usually be installed with two types of ionization source, the choice might be a decision by default.

6.2.2 Ion analysis

Once the ions are formed, they are separated and treated in the analyser part of the mass spectrometer. Different types of analyser have been developed. In proteomics, the preferences go to the time of flight (TOF), the quadrupole, the ion-trap, the FT-ICR (Fourier transform ion cyclotron resonance) and combinations of these. The main quantity to be measured is the mass-to-charge ratio (m/z) rather than the mass alone, where m is the mass in Daltons and z is the exact integer multiple of elementary charges on the ion (unit values 1, 2, 3 etc.). In proteomics instrumentation, protonation is the usual ionization process applied in ESI and MALDI. Protonated ions are denoted $(M + H)^+$, $(M + 2H)^{2+}$, etc.

We will now briefly describe the principle of the main mass analysers imbedded in MS instrumentation used in proteomics. Good reviews and tutorials can be found at http://www.ionsource.com or at http://www.spectroscopynow.com.

Time of Flight

The principle description of a TOF analyser (Figure 6.3) can almost be reduced to a stop watch and linear tube in vacuum, ended by an ion source on one side and a detector on the other. All ions accelerated in the ion source have all the same kinetic energy but different masses; they will therefore travel with different velocities in a tube in high vacuum. One of the simplest forms of mass analysis takes advantage of these differing velocities. After a pulsed acceleration, the time of flight of the ions can be measured. A direct relation between time of flight and m/z can be calculated.

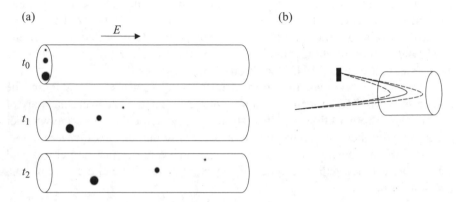

Figure 6.3 Principle of the TOF analyser. (a) Linear TOF analyser. At t_0, all ions are accelerated from the source with the same energy. At t_1 and t_2, the ions with different m/z will fly with different velocities and reach the detector at different times. (b) The reflectron applies an electromagnetic field to the entering ions and deviates them on a reflectron detector. Identical ions entering with slightly different velocities will be refocused on the detector

The TOF flight tube of currently used high-resolution instruments usually includes a so-called reflectron (Mamyrin *et al.*, 1973). The reflectron is a device designed to reverse the ions' travel path as they near the end of the flight tube with an opposing electric field gradient. At some point in the reflectron, the ions are deviated and then accelerated back out. The ions may return through the flight tube or along a slightly different trajectory to their initial one and reach a second detector. Highest resolution can be obtained by combining the use of a reflectron with a delayed extraction device (Takach *et al.*, 1997). The delayed extraction allows us to homogenize the initial velocity of the ions at the entrance of the TOF analyser.

In the linear mode (high mass, low resolution) m/z can be measured to >300 000 (in rare cases) with a resolution of about 100–1000 whereas in the reflectron mode (lower mass, higher resolution) the upper bound for m/z is of 6000–10 000 with a resolution reaching 20 000. The resolution is described here as the ratio between the width of a signal at half of its height, divided by the m/z value of its maximum.

Quadrupole

The quadrupole (Q) mass analyser is a so-called mass filter. A quadrupole mass filter consists of four parallel metal rods (Figure 6.4). Two opposite rods have an applied potential of $(U + V\cos(wt))$ and the other two rods have a potential of $-(U + V\cos(wt))$, where U is a direct current voltage and $V\cos(wt)$ is an alternative current voltage (known as *RF potential*). The applied voltages influence the trajectory of ions travelling down the flight path centred between the four rods. For given DC and AC voltages, only ions of a certain m/z ratio fly through the quadrupole filter and reach the detector. All other ions are ejected from their original path. A mass spectrum is obtained by monitoring the ions that pass through the quadrupole filter, as the voltages on the rods are varied. In fact, voltages are scanned so that ever-increasing m/z can find a successful path through the rods to the detector. As a large portion of the ions are ejected out of the filter during an m/z scan, the quadrupole displays a lower sensitivity than the TOF, which provides a signal for all flying ions. Standard m/z ranges can reach about 3000. Masses that are higher than 3 kDa can be measured with a quadrupole if the number of charges is high enough. As ESI may generate multiply charged ions and therefore low mass/charge ratios, proteins up to few

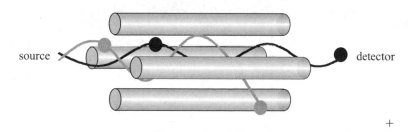

Figure 6.4 Principle of the quadrupole mass filter. Black ions: ions that are in resonance with the field applied in the quadrupole. They fly from the source to the detector. The m/z of these ions can be detected. White ions: ions that are filtered out of the quadrupole

dozen KDa can be analysed with ESI-Q MS. The resolution typically reaches 5000–10 000.

Ion trap

The ion trap mass analyser consists of three hyperbolic electrodes (Figure 6.5): the ring electrode, the entrance endcap electrode and the exit endcap electrode. These electrodes limit a small volume cavity in which ions can be trapped and selectively ejected to a detector. Both endcap electrodes have a small hole in their centres through which the ions can travel. The ring electrode is located halfway between the two endcap electrodes. A mass spectrum is obtained by changing the electrode voltages to eject the ions from the trap. Electric and magnetic fields can be configured within the trap volume so that ions can be held in stable orbits for a period of time long enough to perform useful measurements on them.

Ions are focused in the ion trap using an electrostatic lens system. An electrostatic ion gate pulses to inject ions into the trap. The pulsing of the ion gate differentiates ion traps from 'beam' instruments such as quadrupoles, where ions continuously enter the mass analyser. The time during which ions are allowed into the trap is a function of the total ion capacity of the trap and is

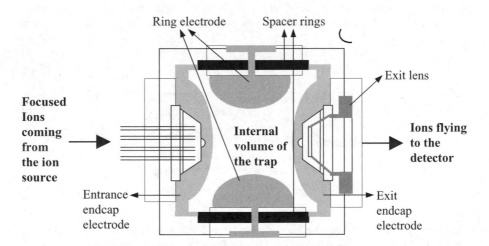

Figure 6.5   Architecture of the ion trap analyser

set to maximize the signal while minimizing space-charge effects (ion–ion repulsion). These effects reduce both resolution and sensitivity.

Unlike the quadrupole, scanning the voltage range will selectively eject the ions to be analysed while keeping all the others available in the trap. Furthermore, helium gas within the trapping volume is used to improve the mass resolution of the instrument. The ion trajectories are concentrated towards the centre of the trap and the kinetic energy of the ions is reduced. The formed ion packet is ejected faster and more efficiently than a diffuse cloud of ions, thus improving resolution.

The specific features of the ion-trap mass spectrometer include compact size and the ability to trap and accumulate ions to increase the signal-to-noise ratio of a measurement. Furthermore, the ion-trap technique is able to perform multiple stages of mass spectrometry. Up to three or four stages of tandem mass spectrometry can be performed by using an ion trap, thus greatly improving the amount of structural information available for a given analyte.

Fourier transform ion cyclotron resonance (FT-ICR)

The basis of an FT-ICR analyser is an ion trap that allows ions to be accumulated and stored for periods as long as minutes (Figure 6.6). During this time, reactions of the ions with neutral molecules can be followed. FT-ICR takes advantage of ion-cyclotron resonance to optimize ion selection and detection.

Ion cyclotron resonance MS (ICR MS) is a technique in which ions are subjected to a simultaneous radio-frequency electric field and a uniform magnetic field, causing them to follow spiral paths in an analyser chamber. In a static magnetic field, the ions move in cyclotron orbits within the cell and do not generate any signal on the detectors. However, when an oscillating electrical field is applied, a selected packet of ions of a given m/z is excited. If the frequency of the applied field is the same as the cyclotron frequency of the ions, the ions absorb energy and increase their velocity and orbital radius. They can then be detected. By scanning the radio-frequency or magnetic field, the ions can be detected sequentially. In order to generate small enough orbital radii, very high magnetic fields are required.

The major advantage of FT-ICR MS is that it allows many different ions to be determined at once, instead of one at a time. The method has the highest resolving power in mass spectrometry, a high upper mass limit, high sensitivity, non-destructive detection and high accuracy for mass measurement.

Looking at all these advantages, one might ask why this technology is not the mainly used method. The main reason is still price, as those instruments

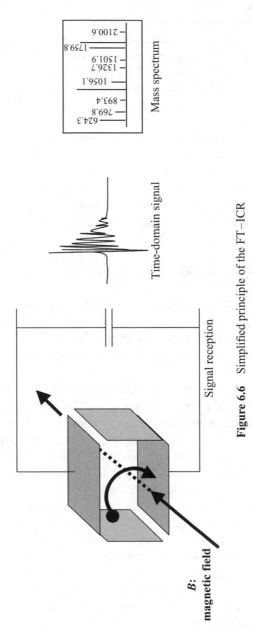

Figure 6.6 Simplified principle of the FT–ICR

available on the market are out of the financial range of many laboratories. In addition, they are more difficult to operate than others.

6.2.3 Tandem MS, MS/MS

Analysers can obviously measure m/z of the submitted analytes. In addition to this 'straightforward' and 'basic' functionality, a particular mode of operation of some of them, or combinations of analysers, allows for the generation of more structurally related information on the considered analytes. This feature implies the selective fragmentation of an analysed compound in the mass spectrometer. This measurement mode is called tandem MS or MS/MS.

The high-level description of the principle of MS/MS is always the same, independent from the architectural set-up of the instrument. The process can be divided into three steps.

(1) *Parent ion selection.* From a full MS spectrum, the analyser is able to select only a small m/z range to be further processed. The ions that are selected are named parent ions or precursor ions.

(2) *Fragmentation.* The considered parent ions are fragmented into smaller pieces within the analyser(s).

(3) *Measurement of the fragments.* The instrument will then measure a spectrum of the fragments only (which may be named daughter ions or product ions).

The technique requires either two or more analysers in series or relies on the intrinsic characteristics of some spectrometers such as the ion trap (Hoffmann, 1996; Busch, Glish and McLuckey, 1988). When several analysers are involved, precursor ions are separated according to their m/z values in the first analyser and selected for fragmentation. Fragments are generated in a collision cell and analysed in another analyser. As we will see later, MS/MS has become very popular in proteomics. A number of applications and instrumentations have been specifically developed that combine MS and MS/MS modes to improve the identification of proteins by adding structural information to the parent ions.

6.2.4   Instrumentation

From here on, we can briefly describe the instrumental set-ups currently available on the proteomics market. They reflect the possibilities to virtually combine, a little bit like a Lego toy, various ionization sources, analysers and fragmentation devices (James, 1997; Yates, 1998).

- MALDI–TOF (with delayed extraction and reflectron). This instrument is mainly dedicated to rapid peptide mass fingerprinting measurements (see below for the description of the principle). The mass range is well suited for tryptic peptides in reflectron mode. The instrument can be used in linear mode to measure very high molecular weights (up to a few hundred kiloDaltons) with a reduced resolution. MS/MS can be performed in post-source-decay mode. However, this mode of operation is difficult to handle and generally does not compete with other more standard types of fragmentation. The SELDI MS is a quasi-MALDI–linear TOF instrument equipped with a short flight tube. It is particularly suited for profiling approaches, and is linked with a specific sample preparation procedure (see below).

- MALDI–TOF–TOF (a MALDI source coupled with a short linear TOF, followed by a CID fragmentation cell and then a TOF equipped with a reflectron). Some operation modes on commercial instruments also allow in-source decay or post-source decay as an MS/MS feature. The MALDI–TOF–TOF is a true MS/MS-capable MALDI-source-based high-through-put instrument. In MS mode, only the TOF equipped with reflectron is activated.

- ESI triple quadrupole. This is today a pretty old set-up, still interesting for specific measurement modes, such as parent scanning or neutral loss. It is routinely used for screening of fragmentation products, a little less for protein identification and systematic characterization purposes.

- ESI–Q–TOF (the ESI source is coupled with a quadrupole mass filter followed by a CID fragmentation cell sending the fragment ions to a reflectron–TOF analyser). This high-resolution instrument is routinely used, coupled upstream with an LC system for LC–MS/MS runs. It is possible to replace the ESI source by a MALDI source on some commercial instruments.

- ESI–ion trap (the ESI source is coupled with an ion trap). The ion trap acts as mass analyser and as fragmentation chamber. It allows fragmentation in the same space as the mass analysis. Therefore it is capable of MS^n measurements. Coupled with an LC system, it is capable of high acquisition rate. It is highly sensitive, but generally displays lower resolution compared to a Q-TOF analyser. There have also been a number of attempts to couple a MALDI source to this analyser.

- ESI Q-trap (the analyser is similar to a triple quadrupole, with the third quadrupole capable of working as a linear trap). This recent set-up displays interesting features that will further evolve in the future.

- FT-ICR (the source can be MALDI, ESI or others, the main analyser is a kind of trap working with cyclotron properties). This is a very high-resolution, high-accuracy, high-sensitivity, wide-mass-range instrument. It is still not very much used mainly due to its high price.

For more information on the description and availability of the instrumentation, see www.hyphenms.nl, www.ionsource.com and the instrument providers' websites.

6.3   Principle of protein identification based on MS data

Mass spectrometers are generically able to measure masses, or more precisely mass-to-charge ratios (m/z). This means that the signals obtained from mass spectra are to be interpreted. Each raw data is processed and the instrument software provide a list of signals in the form of peak lists. Each peak is represented as at least an m/z value, often with a measure of signal intensity. Less frequently, one can also obtain information on the signal shape (resolution, width, area, measure of fit with a signal theoretical model, etc.).

Lists of m/z and related intensity values have then to be matched with peptide sequences or with protein database entries. How does this work?

6.3.1   PMF and MS/MS

In order to identify a protein entry, or a list of protein entries, one needs to generate and measure a set of experimental protein attributes that is specific for this/these proteins. As mass spectrometry measures mass-related values, what are the possible attributes that represent the analysed protein with high specificity and sensitivity? The exact mass of the protein? Other masses? The amino acid composition? The amino acid sequence? The isoelectric point? The hydrophobicity index? The 3D structure? Other properties?

Certainly, mass spectrometry is not able to measure all of these attributes. Proteomics today focuses mainly on two methods: peptide mass fingerprint (PMF) and MS/MS.

PMF

In most of the proteomics workflows, mass spectrometry is involved in the analysis of relatively short peptides and is mainly not used to measure the molecular mass of entire proteins. Generally, protein samples are treated with cleaving agents, such as trypsin, and less often chymotrypsin, Lys-C, endo-protease V8 or CNBr. Therefore specific peptides are formed and subjected to MS measurement.

The product of the MS measurement of the generated peptides is a spectrum called the peptide mass fingerprint (Henzel *et al.*, 1993; James *et al.*, 1993; Mann, Hojrup and Roepstorff, 1993; Pappin, Jojrup and Bleasby, 1993; Yates *et al.*, 1993) and represents the masses and intensities of the generated peptides (Figure 6.7). The experimentally measured mass list is then sent to a dedicated PMF identification tool. This tool generates theoretical fingerprints from all the proteins included in a protein sequence database and compares them with the experimental data. The result of such a search is a list of database protein entries that match the experimental data with decreasing scores. The best score(s) therefore correspond(s) to the potentially identified protein entry(ies).

PMF performed using currently available rapid and high-resolution MALDI-TOF instruments provide high-throughput capabilities. As a spectrum can be acquired within a few seconds, the instrument is in principle able to process a few hundred if not a few thousand samples a day. This capacity is very rarely implemented, as the processing and validation steps usually do not reach the instrument throughput. PMF approaches however reveal some efficiency limitations and sometimes give ambiguous results:

- If the PMF results have to be searched against large sequence databases; as the specificity of the method is based on statistics, the larger the database, the higher the chance of randomly matching unwanted protein entries.

- If the proteins are post-translationally or post-transcriptionally modified; *m/z* values are mainly used to match unmodified peptides, i.e. made up of the 20 natural amino acids only. Post-translational modifications modify

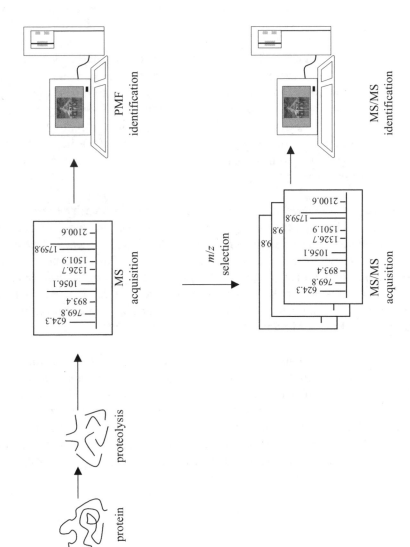

Figure 6.7   Principle of PMF and MS/MS-based protein identification process

the mass of the peptides, therefore reducing the number of possible matched mass values for a protein. Similarly, non-annotated alternative splicing products, unconsidered processing events and mutations also reduce the number of matches, therefore also reducing the score of the protein.

- In the case of extreme protein molecular masses (very small or very large) very small proteins also produce a very small number of peptides to be analysed. It is possible that these few peptides might not all be present in a spectrum. In this case, the required minimum number of matching peptides might not be reached. Similarly, a fragment of a protein might be difficult to identify. In the case of large proteins (above ~150 kDa) the number of theoretical peptides is so huge that a portion of them can randomly match nearly every spectrum.

- If the protein sequence under investigation is not known by the protein sequence databases; here either (1) random matches are not correlating with a high score and no significant hit is obtained, which means identification is not successful, (2) there is a protein match that corresponds to a homologous and similar protein; here additional knowledge such as sample species or tissue will help to sort; or (3) there is a random match that displays a significant score. Here the user should appreciate the quality of the hit, according to additional biological knowledge of the sample.

- If an excised spot contains several proteins; in this case, the spectrum can be highly complex and might contain a high number of signals. The danger exists that significant scores might be obtained for proteins that randomly match m/z values from other proteins present (for instance, if two proteins A and B are present and match for n and m m/z values, respectively, a protein C might by chance match some of the n values from A and some of the m values from B and obtain a significant score).

In these cases, obviously, more detailed and accurate information is needed.

MS/MS

Thanks to the possibility of current MS instrumentation to specifically select and fragment peptides, spectra acquired either by post-source decay (PSD) or more often by collision-induced dissociation (CID) can be used to determine sequence tags or the complete sequence of a peptide, with the help of computer algorithms. This additional peptide sequence information makes protein identification less ambiguous and can be used to search expressed sequence tag (EST) databases or nucleotide sequence databases if the protein is not yet listed in a protein database, or to identify with higher confidence peptides from various proteins in a mixture. A series of protein identification tools is able to handle these so-called MS/MS data (or tandem-MS data). These include SEQUEST (Eng, McCormack and Yates, 1994), Mascot (Perkins *et al.*, 1999) and SONAR (Fenyo, 2000), among others.

In the so-called MS/MS or tandem-MS approach (McCormack *et al.*, 1997; Ducret *et al.*, 1998; Perkins *et al.*, 1999) (Figure 6.7) proteins are usually processed using a specific proteolytic step, as in the PMF approach. The peptides are then analysed separately by MS instrumentation that can work in a more elaborate mode than for PMF. The peptides usually enter the MS process either as a mixture or after an online RP-HPLC separation. The instrument will then individually isolate the entering peptides and physically fragment them into smaller pieces. The obtained spectra are specific for each individual peptide. These data can be interpreted by dedicated MS/MS identification tools in order to identify peptides and therefore protein entries in sequence databases.

Tandem MS provides structural information by establishing relationships between precursor ions and their fragmentation products. Different types of fragment ions observed in an MS/MS spectrum depend on the peptide primary sequence, the amount of internal energy, how the energy was introduced, the charge state etc. Fragment ions are named according to an accepted nomenclature (Figure 6.8).

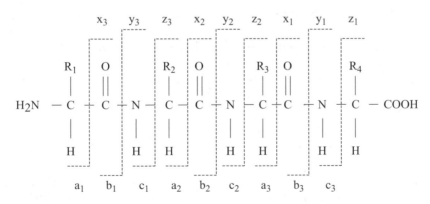

Figure 6.8 Peptide fragmentation in MS/MS mode and fragment nomenclature

In a mass spectrometer, peptide ions do not fragment at random positions. The peptide backbone preferentially carries the bond breaking locations under the condition generated in MS instruments used in proteomics (Hunt *et al.*, 1986, Kinter and Sherman, 2000). As for the parent ion signals, only fragments carrying at least one charge will be detected. In a generally accepted nomenclature, ions are labelled as type a, b or c if the charge is retained on the N-terminal fragment, and as x, y or z if the charge is retained on the C terminal (Roepstorff and Fohlman, 1984; Biemann, 1988). A subscript indicates the number of residues in the fragment. The most common position of fragmentation is the C(O)–N amide bond that links two adjacent residues. This fragmentation type generates primarily b and y ions. A fragmentation spectrum might ideally contain series of b and/or y ions. Is such a spectrum, the mass difference between two adjacent b or y ions corresponds to that of an amino acid residue. The peptide sequence can ideally be extracted and 'read' from series of mass differences (Figure 6.9).

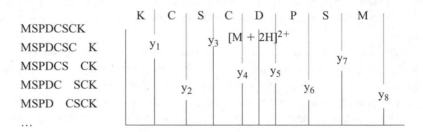

Figure 6.9   Principle of MS/MS spectrum interpretation

6.3.2   Protein identification tools

Protein identification implies the comparison of experimentally measured protein attributes with those that could be generated *in silico* from a database. A number of tools have been developed that specifically consider these different attributes, such as amino acid composition, sequence tags or mass-spectrometry derived data. All tools aim to identify a protein by providing the best match to a protein entry in a protein database.

Table 6.1 describes and gives references for a list of protein identification tools available on the web.

PMF and MS/MS-based protein identification tools are able to annotate/ interpret a portion of the experimental signals only. Currently available tools are able to look for the presence of peptides as amino acid sequences. Some

Table 6.1 List of some protein identification tools available on the web

Protein attribute (PMF, MS/MS, seq. Info.)	Program name	Search database	Internet URL address	References
PMF	ALDentE	SWISS-PROT, TrEMBL	http://www.expasy.org/tools/aldente/	Tuloup et al., 2002
PMF, MS/MS, seq.	Mascot	OWL*, NCBInr, SWISS-PROT, MSDB, EST	http://www.matrixscience.com	Perkins et al., 1999
PMF seq., MS/MS	Protein Prospector suite: MS-Fit, MS-Tag, MS-Seq	SWISS-PROT, Genepept, dbEST, OWL*, NCBInr	http://prospector.ucsf.edu/	Clauser et al., 1995
PMF	ProFound (Part of Prowl)	NCBInr, SWISS-PROT	http://129.85.19.192/profound_bin/WebProFound.exe/	Zhang and Chait, 2000
MS/MS	PepFrag (Part of Prowl)	SWISS-PROT, nr, dbEST and others	http://129.85.19.192/prowl/pepfragch.html	Fenyo et al., 1998
PMF, seq	PepSea	nrdb	http://195.41.108.38/PepSeaIntro.html	Mann, Hojrip and Roepstorff, 1993
PMF	PeptideSearch	NCBInr	http://www.mann.embl-heidelberg.de/GroupPages/PageLink/peptidesearchpage.html	
PMF	PeptIdent	SWISS-PROT, TrEMBL	http://www.expasy.org/tools/	Binz et al., 1999b
PMF MS/MS	PepMAPPER Sonar MS/MS	SWISS-PROT NCBInr, SWISS-PROT, dbEST	http://wolf.bms.umist.ac.uk/mapper/ http://knxs.bms.umist.ac.uk/prowl/knexus.html	

* OWL was last updated in May 1999 (release 31.4) and is no longer maintained.

modifications can be considered. One can include those that are known to be experimentally induced, such as alkylation of cysteine residues, modification with ICAT reagent; those that are supposed to happen artefactually, such as oxidation of methionine or tryptophan residues; those that are intrinsically potentially present in the native peptides, such as protein N-terminal formylation N-term acetylation or serine/threonine phosphorylation. Some tools do also search unspecifically cleaved peptides or amino acid mutated peptides. Users have to handle such parameters with care, as the inherent combination does increase the number of peptide matches for a number of protein entries; however, it does not guarantee high confidence, due to an increased rate of random (and therefore also incorrect) matches.

It is often preferred to perform identification with a restricted number of such 'loose' parameters. The yield of identification might be reduced, the number of matched peptides is not as high, the sequence coverage is also reduced, but the overall result is of higher quality. Then, as a second step, it might be recommended to look for the presence of modified peptides that are not annotated as in databases. To do this, users can perform queries to the same tools with looser parameters, or decide to use tools that are specifically designed for such tasks. This is where the characterization tools enter the game.

Protein characterization tools aim to interpret experimental information that cannot be matched in a database using the protein identification tools only, such as non-annotated post-translational modifications or mutations. Differing from the identification tools, they try to match experimental data against one database entry. The characterization step often takes place after a protein entry has been associated to experimental data: the protein has been identified and needs to be further characterized. These tools are closely associated with the identification software and are part of a 'package'. As an example, the ExPASy server provides PeptIdent and Aldente as PMF identification tools and FindMod (http://www.expasy.org/tools/findmod/) (Wilkins et al., 1999), FindPept (http://www.expasy.org/tools/findpept.html) (Gattiker et al., 2002) and GlycoMod (http://www.expasy.org/tools/glycomod/) (Cooper et al., 2001) as characterization tools. In brief, FindMod uses a protein entry and a PMF spectrum and searches MS signals matching post-translationally modified peptides or amino acid mutated peptides. FindPept looks similarly for the presence of signals corresponding to unspecifically cleaved peptides, signals matching autolysis products of the cleaving agent and signals matching contamination products such as keratins. GlycoMod predicts the possible oligosaccharide structures that occur on proteins from PMF data.

Analysis tools are being developed to help in the interpretation of identification and/or characterization tools and simulate experimental data. They do not necessarily require experimental data to be computed. Some examples available on the ExPASy server include PeptideMass (http://www.expasy.org/ tools/peptide-mass-html) (Wilkins *et al.*, 1997), which calculates masses of peptides and their post-translational modifications for a Swiss-Prot or TrEMBL entry or for a user sequence, PeptideCutter (http://www.expasy.org/tools/ peptidecutter/), which predicts potential protease and cleavage sites and sites cleaved by chemicals in a given protein sequence, or BioGraph (http://www.ex-pasy.org/tools/BioGraph/), which graphically represents the annotation of a PMF spectrum after a process launched with PeptIdent, FindMod, FindPept or Aldente.

6.4   Proteomics workflows

An increasing number of workflows and procedures that use mass spectrometry as central tool have been developed and published in the last decade. These workflows intend to describe various aspects of proteomics information. The type of biological question varies from one to another. One can discriminate two major tasks:

- to identify the list of the proteins present in a given biological sample

- to quantify different expression levels of proteins or peptides.

The major families of proteomics workflows can be described as follows:

- the classical gel-based approach

- the molecular scanner

- the MuDPIT approach

- the ICAT approach

- the SELDI profiling approach

- other technologies.

Before we begin the description of these workflows, it is to be noticed that they should be considered as complementary, partially overlapping, but not as one definitely better than another. Even if the original author intends to exalt the arguments that make his/her approach more efficient than another's, it is important to understand what are the possibilities and the limits of the technologies. These could be described as a range of molecular weights, a range of physico-chemical properties, a range of concentration, or other means to help the end user in choosing the most appropriate methodology for his/her needs.

There have been many discussions between the 'European' gel-based approach and the 'American' non-gel-based approach to separate and identify the highest number of proteins (Corthals *et al.*, 2000). As we will see, the type of information one can extract from one approach or another is partially identical, but also partially different. This will make these approaches definitely complementary.

6.4.1 The classical gel-based approach

Historically, proteomics has been closely associated with large-scale protein identification using two-dimensional gel electrophoresis (2D PAGE) as separation method and mass spectrometry as subsequent analytical tool. This method is still used in quite a large number of laboratories, although it cannot cover the entire range of proteins found in complex biological samples. In this approach (Figure 6.10), protein samples are subjected to the high-resolution separation capacity of 2D PAGE, independently developed by O'Farrel (1975), Anderson and Anderson (1982), Scheele (1975) and Klose (1975). This technique has been shown to be able to separate up to 10 000 polypeptides in a single analysis, using a microgram-to-milligram amount of sample. The proteins are denatured and then subjected to a first separation step named isoelectrofocusing (IEF). They are separated according to their isoelectric point (pI). Simplistically described, the proteins are focused, on a pH gradient gel, at the pH value corresponding to a net charge of zero. The IEF gel is then placed on top of a second dimension sodium dodecyl sulphate polyacrylamide gel electrophoresis (SDS-PAGE), on which the proteins are electrophoretically separated according to relative molecular mass (M_r) (Lämmli, 1970). The 2D-PAGE patterns are classically visualized using ultra-sensitive staining methods, such as Coomassie brilliant blue (Neuhoff *et al.*, 1988), silver (Rabilloud, 1990) or negative zinc staining (Fernandez-Patron, Castellanos-Serra and Rodriguez, 1992). The obtained two-dimensional polyacrylamide coloured gel can be digitized and analysed computationally. Several image analysis programs have been developed to perform tasks such as 2D-PAGE image visualization, spot detection, spot quantitation, image and spot matching, image comparison, clustering analysis and data management. Many software packages have been developed over the years and cover some or all of the above-mentioned functionalities. We can mention TYCHO (Anderson *et al.*, 1981), ELSIE (Olson and Miller, 1988), GELLAB (Lemkin, Wu and Upton, 1993) (Wu, Lemkin and Upton, 1993),

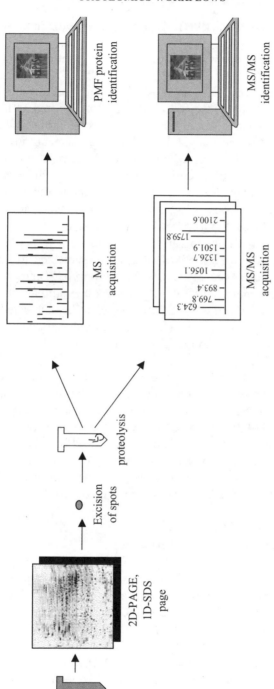

Figure 6.10 Classical gel-based approach

PDQUEST (Monardo *et al.*, 1994), Z3 (Smilansky, 2001), and Melanie (Appel *et al.*, 1997a,b).

The next step is the protein identification process that employs mass spectrometry. Protein spots of interest are excised from 2D-PAGE gels and submitted to endoproteolytic digestion. The most commonly used enzyme is trypsin. Typically, proteins are identified from PMF spectra acquired on high-resolution MALDI-TOF instruments. MALDI-TOF instruments are still widely used as they can measure PMF spectra with high-throughput capabilities. The extracted PMF peaklists are then submitted to dedicated identification tools, such as Aldente (Tuloup *et al.*, 2002), PeptIdent (Binz *et al.*, 1999b), MS-FIT (Clauser *et al.*, 1995), and Mascot (Perkins *et al.*, 1999), among others. If PMF results do not provide unambiguous answers, or if no proteins are identified, peptide mixtures can be subjected to an MS/MS measurement, with or without the use of prefractionation or chromatographic separation of the peptides before the MS measurement. Then the MS/MS spectra are subjected to dedicated tools able to work with this type of data, such as Mascot (Perkins *et al.*, 1999), Sequest (Eng, McCormack and Yates, 1994) or other similar tools. In the case of unsuccessful identification, MS/MS data can be subjected to *de novo* sequencing tools, that aim to extract sequence information directly from the MS/MS spectra, without comparing the experimental data with a sequence database content.

Due to some limitations inherent to the technologies, to the complexity of the sample, to the properties of the proteins or to the instrumentation available in different laboratories, a number of modifications can be described.

Instead of 2D PAGE, 1D-SDS gel electrophoresis can be used for less complex samples (up to about 15 proteins per band) (Scherl *et al.*, 2002). Some extreme proteins, such as very acidic or very basic or membrane proteins, can be handled with 1D-SDS electrophoresis as they are usually outside the standard possibilities of the 2D-PAGE technology (Corthals *et al.*, 2000).

6.4.2   The molecular scanner approach

The classical approach is characterized by a very highly resolutive method to separate proteins and to keep them available for analyses for months. However, one bottleneck of the classical approach is clearly linked to the heavy sample handling to perform in a systematic identification of all components in a 2D PAGE. In the molecular scanner approach (Binz *et al.*, 1999a; Bienvenut *et al.*, 1999) (Figure 6.11), proteins separated by 1D SDS or 2D PAGE are simultaneously digested and electrotransferred onto a collecting membrane. On the membrane, the digested peptides are concentrated into spots whose positions

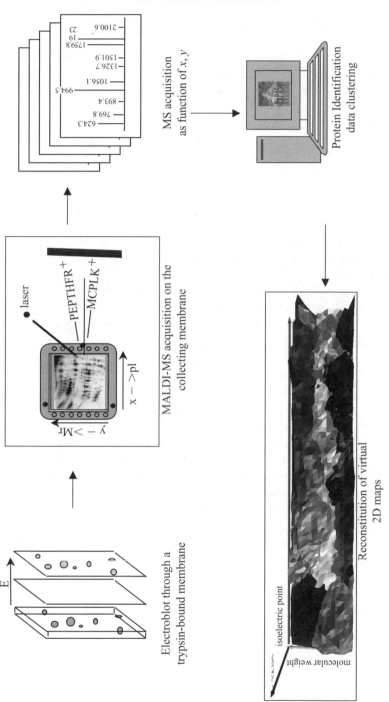

Figure 6.11 Molecular scanner

correspond to their 'parent' protein spots on the original gel. After deposition of a MALDI matrix solution on its surface, the collecting membrane is inserted into a MALDI–TOF MS instrument and scanned on a grid much finer than the average spot size, producing a peptide mass fingerprint for each grid point. All data points can be used for individual protein identification. Thanks the redundancy in the data due to the high resolution of the grid scan (more than one MS spectrum is measured per spot), synthetic spectra can be extracted that are purged of noise and peptide masses from overlapping proteins (Muller et al., 2002a,b). The method requires no chemical staining procedure. It is fully parallel (the digestion of all proteins is performed in one single step). The sample handling is reduced to the minimum (one single digestion operation and one single sample transfer from the gel to the MS target). Potentially, as the target surface is stable over a long period of time, result-dependent acquisition can be designed; for instance running MS/MS acquisition on specific position of the membrane, to validate identification or to perform a de novo sequencing experiment.

The technology can be appended with various modifications. Instead of 2D PAGE as the main separation system, a combination of LC and 1D-SDS gel electrophoresis can be appended. As the sample remains stable on the collecting membrane, spots/or bands of particular interest can be subjected to MS/MS acquisition in order to either confirm identifications or to characterize further the original protein content of the analysed sample.

6.4.3   The MuDPIT approach

In the face of some of the difficulties encountered by the 2D-PAGE and 1D-SDS technologies to recover very small, very large, very hydrophobic or very basic proteins, Yates and co-workers (Wolters, Washburn and Yates, 2001) have developed a non-gel-based approach to identify as many proteins as possible using a so-called shotgun approach. The multi-dimensional protein identification technology (MuDPIT) principle can be described as follows (Figure 6.12). A whole complex protein mixture is submitted to trypsin digestion. The obtained peptides are stepwise separated, in a first dimension on a cation exchange column and then, in a second dimension on a reverse phase column. The series of two columns is set up in line with an ESI–MS instrument capable of MS/MS acquisition. In its first description of the analysis of a yeast proteome, Yates and co-workers evaluated that, starting from an estimate of 200 000 different tryptic peptides subjected to the system, the peptides are entering the mass spectrometer at an average rate of over 500 per minute (Washburn, Wolters and Yates, 2001). Only a portion of them can be analysed

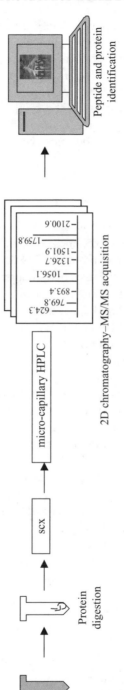

Figure 6.12 MuDPIT approach

in MS/MS mode. This technique requires less time and work to identify many components of a full proteome in a first approach. It does not, however, provide any information on the relative abundance of proteins that differ by post-translational modifications. It also does not inform on other processing events such as protein fragmentation, and hardly on splicing variants and in general on the entire proteins. The sensitivity is limited by the capacity of the columns. It is also not appropriate for comparative studies due to the current low reproducibility of picking the same peptides between two runs.

6.4.4 The ICAT approach

Aebersold and coworkers have approached the simultaneous identification and relative quantitation aspects of proteomics by developing an isotope labelling technique included in a gel-free proteomics workflow (Gygi, *et al.*, 1999) (Figure 6.13). They proposed the use of a reagent that is available in two isotopically discriminated forms, the isotopically coded affinity tag (ICAT) reagent. The structure of the reagent contains a biotin head, a linker portion and an iodo-activated carbonyl group that reacts specifically with cysteine side chains. The heavy form differs from the light form by the replacement of eight hydrogen atoms by eight deuterium atoms in the linker portion.

The approach is designed to compare a pair of biological samples. Two protein samples (cell state 1 and cell state 2) are individually labelled with d0-ICAT and d8-ICAT reagents, respectively, then mixed and digested with trypsin. The peptide mixture is subjected to biotin–streptavidin affinity chromatography. This step simplifies the peptide mixture by isolating and then eluting the cysteine-containing peptides. The elution step is coupled with a MS and MS/MS capable instrument. Pairs of differentially labelled peptides elute at approximately the same time and can be recognized as two MS signals separated by 8 Da. As the ionization properties of the d0- and of the d8-modified peptides are supposed to be identical, relative quantitation can be performed on each pair. Once a pair is found as displaying 'differentially expressed' peptides, MS/MS can be specifically performed to identify the considered peptides. This technique generates fewer data than the MuDPIT approach, and allows direct relative quantitation of peptides for binary comparison. However, 14 per cent of eukaryotic proteins lack cysteine residues, making this technology not compatible for these proteins. It has similar advantages to MuDPIT over 2D-PAGE-based approaches for the difficult small, large, acidic, basic and hydrophobic proteins. It also shares some disadvantages, such as the lack of information on entire proteins, on processing events, on PTMs and on relative abundance of PTM modified proteins in a mixture.

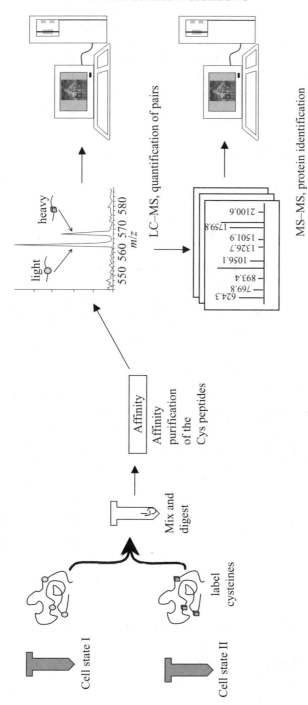

Figure 6.13 ICAT approach

In order to cope with the advantages of the technique and of those of 2D PAGE, a new reagent has been developed, which is called cleavable ICAT and is compatible with use as a binary mixture in 2D PAGE. The reagent is commercialized by Applied Biosystems.

6.4.5   The SELDI profiling approach

The surface-enhanced laser desorption/ionization (SELDI) technology is based on the MALDI technique for its measurement part. SELDI refers to the process of affinity capture on special chemical surfaces, followed by mass analysis using laser desorption/ionization-based detection (Merchant and Weinberger, 2000).

The SELDI Process may be described in four parts.

1.   Affinity capture of one or more proteins on a protein chip array, by spotting a complex mixture on a chip.

2.   Wash away the material that does not bind to the chip surface.

3.   Read one or more of the target protein(s) retained by the SELDI–TOF MS instrument that provides directly the molecular weight of the targets.

4.   Process the spectra by considering the spectra as profiles. This analysis resembles the comparison of 2D-PAGE gel images as for the discovery of up–down-regulated signals.

Biomolecules bind to surfaces through hydrophobic, electrostatic or coordinate covalent bonds or Lewis acid–base interaction. Arrays with biochemical surfaces may be created using any molecule of interest covalently linked to the surface (antibodies, receptors, enzymes, DNA, ligands, lectins etc.).

The technology is able to measure entire protein molecular masses up to a few tens of kiloDaltons. It is particularly well suited for protein profiling approaches and the associated software performs clustering analyses. In principle, identification of the retained proteins could be processed by trypsin treatment and PMF and/or MS/MS analysis of the peptides on the chip. There are some sensitivity and complexity issues to take advantage of this approach still.

6.4.6 Other technologies

Other workflows and technologies are developed for proteomics purposes that are not yet using mass spectrometry as the main analytical tool. Microarray and microchip technologies (Cahill, 2001; Jenkins and Pennington, 2001) will doubtless play an important role in proteomics research in the future.

References

Anderson, N. L., Taylor, J., Scandora, A. E., Coulter, B. P., and Anderson, N. G. (1981) The TYCHO system for computer analysis of two-dimensional gel electrophoresis patterns. *Clin. Chem.* **27**, 1807–1820.

Anderson, N. G., and Anderson, L. (1982) The Human Protein Index. *Clin. Chem.* **28**, 739.

Appel, R. D., Palagi, P. M., Walther, D., Vargas, J. R., Sanchez, J.-C., Ravier, F., Pasquali, C., and Hochstrasser, D. F., (1997a) Melanie II – a third-generation software package for analysis of two-dimensional electrophoresis images: I. Features and user interface. *Electrophoresis* **18**(15), 2724–2734.

Appel, R. D., Vargas, J. R., Palagi, P. M., Walther, D., and Hochstrasser, D. F. (1997b) Melanie II – a third-generation software package for analysis of two-dimensional electrophoresis images: II. Algorithms. *Electrophoresis* **18**(15), 2735–2748.

Beynon, J. H. (1956) The use of the mass spectrometer for the identification of organic compounds. *Mikrochim. Acta*, 437–453.

Biemann, K. (1988) Contribution of mass spectrometry to peptide and protein structure. *Biomed. Environ. Mass Spectrom.* **16**, 99–100.

Bienvenut, W. V., Sanchez, J.-C., Karmime, A., Rouge, V., Rose, K., Binz, P.-A., and Hochstrasser, D. F. (1999) Toward a clinical molecular scanner for proteome research: parallel protein chemical processing before and during western blot. *Anal. Chem.* **71**, 4800–4807.

Binz, P.-A., Müller, M., Walther, D., Bienvenut, W. V., Gras, R., Hoogland, C. *et al.* (1999a) A molecular scanner to automate protcomic research and to display proteome images. *Anal. Chem.* **71**, 4981–4988.

Binz, P. A., Wilkins, M., Gasteiger, E., Bairoch, A., Appel, R., and Hochstrasser, D. F. (1999b) Internet resources for protein identification and characterization. In: Kellner, R., Lottspeich, F., and Meyer, H. (Eds.). *Microcharacterization of Proteins*. Wiley–VHC, 277–299.

Busch, K. L., Glish, G. L., and McLuckey, S. A. (1988) *Mass Spectrometry/Mass Spectrometry*. VCH, New York.

Cahill, D. J. (2001) Protein and antibody arrays and their medical applications. *J. Immunol. Meth.* **250**, 81–91.

Clauser, K. R., Hall, S. C., Smith, D. M., Webb, J. W., Andrews, L. E., Tran, H. M., Epstein, L. B., and Burlingame, A. L. (1995) Rapid mass spectrometric peptide sequencing and mass matching for characterization of human melanoma proteins isolated by two-dimensional PAGE. *Proc. Natl. Acad. Sci. USA* **92**, 5072.

Cooper, C. A., Gasteiger, E., and Packer, N. (2001) GlycoMod – a software tool for determining glycosylation compositions from mass spectrometric data. *Proteomics.* **1**, 340–349.

Corthals, G. L., Wasinger, V. C., Hochstrasser, D. F., and Sanchez, J.-C. (2000) The dynamic range of protein expression: a challenge for proteomic research. *Electrophoresis* **21**, 1104–1115.

Dole, M., Mack, L. L., Hines, R. L., Mobley, R. C., Ferguson, L. D., and Alice, M. B. (1968) Molecular beams of macroions. *J. Chem. Phys.* **49**(5), 2240–2249.

Ducret, A., Van Oostveen, I., Eng, J. K., Yates, J. R. III, and Aebersold, R. (1998) High throughput protein characterization by automated reverse-phase chromatography/electrospray tandem mass spectrometry. *Protein Sci.* **7**(3), 706–719.

Eng, J. K., McCormack, A. L., and Yates, J. R. (1994) *J. Am. Soc. Mass. Spectrom.* **5**, 976.

Fenyo, D. (2000) Identifying the proteome: software tools. *Curr. Opin. Biotechnol.* **11**, 391–395.

Fenyo, D., Qin, J., and Chait, B. T. (1998) Protein identification using mass spectrometric information. *Electrophoresis* **19**, 998–1005.

Fernandez-Patron, C., Castellanos-Serra, L., and Rodriguez, P. (1992) Reverse staining of sodium dodecyl sulfate polyacrylamide gels by imidazole-zinc salts: sensitive detection of unmodified proteins. *Biotechniques* **12**, 564.

Gattiker, A., Bienvenut, W. V., Bairoch, A., and Gasteiger, E. (2002) FindPept, a tool to identify unmatched masses in peptide mass fingerprinting protein identification. *Proteomics* **2**, 1435–1444.

Gygi, S. P., Rist, B., Gerber, S. A., Turecek, F., Gelb, M. H., and Aebersold, R. (1999) Quantitative analysis of complex protein mixtures using isotop-coded affinity tags. *Nat. Biotechnol.* **17**, 994.

Henzel, W. J., Billeci, T. N., Stults, J. T., Wong, S. C., Grilmey, C., and Watanabe, C. (1993) Identifying proteins from two-dimensional gels by molecular mass searching of peptide fragments in protein sequence databases. *Proc. Natl. Acad. Sci. USA* **90**, 5011.

Hoffmann, E. (1996) Tandem mass spectrometry: a primer. *J. Mass Spectrometry* **31**, 129–137.

Hunt, D. F., Yates, J. R. III, Shabanowitz, J., Winston, S., and Hauer, C. R. (1986) Protein sequencing by tandem mass spectrometry. *Proc. Natl. Acad. Sci. USA* **83**, 6233–6237.

James, P. (1997) Protein identification in the post-genome era: the rapid rise of proteomics. *Q. Rev. Biophys.* **30**, 279–331.

James, P., Quadroni, M., Carafoli, E., and Gonnet, G. (1993) Protein identification by mass profile fingerprinting. *Biochem. Biophys. Res. Commun.* **195**, 58.

Jenkins, R. E., and Pennington, S. E. (2001) Arrays for protein expression profiling:

towards a viable alternative to two-dimensional gel electrophoresis? *Proteomics* **1**, 13–29.

Karas, M., and Hillenkamp, F. (1988) Laser desorption ionization of proteins with molecular masses exceeding 10,000 daltons. *Anal. Chem.* **60**(20), 2299–2301.

Kinter, M., and Sherman, N. E. (2000) *Protein Sequencing and Identification Using Tandem Mass Spectrometry*. Wiley–Interscience, New York.

Klose, J. (1975) Protein mapping by combined isoelectric focusing and electrophoresis of mouse tissues. A novel approach to testing for induced point mutations in mammals. *Humangenetik* **26**, 231–243.

Lämmli, U. K. (1970) Cleavage of structural proteins during the assembly of the head of bacteriophage T4. *Nature* **277**, 680.

Lemkin, P. F., Wu, Y., and Upton, K. (1993) An efficient disk based data structure for rapid searching of quantitative two-dimensional gel databases. *Electrophoresis* **14**, 1341–1350.

Mamyrin, B. A., Karatajev, V. J., Shmikk, D. V., and Jagulin, V. (1973) *Sou. Phys. – JETP* **37**, 45.

Mann, M., Hojrup, P., and Roepstorff, P. (1993) Use of mass spectrometric molecular weight information to identify proteins in sequence databases. *Biol. Mass Spectrum* **22**, 338.

McCormack, A. L., Schieltz, D. M., Goode, B., Yang, S., Barnes, G., Drubin, D., and Yates, J. R. III. (1997) Direct analysis and identification of proteins in mixtures by LC/MS/MS and database searching at the low-femtomole level. *Anal. Chem.* **69**(4), 767–776.

Merchant, M., and Weinberger, S. R. (2000) Recent advancements in surface-enhanced laser desorption/ionization-time of flight-mass spectrometry. *Electrophoresis* **21**, 1164–1177.

Monardo, P. J., Boutell, T., Garrels, J. I., and Latter, G. I. (1994) A distributed system for two-dimensional gel analysis. *Comput. Appl. Biosci.* **10**, 137–143.

Muller, M., Gras, R., Appel, R. D., Bienvenut, W. V., and Hochstrasser, D. F. (2002a) Visualization and analysis of molecular scanner peptide mass spectra. *J. Am. Soc. Mass. Spectrom.* **13**(3), 221–231.

Muller, M., Gras, R., Binz, P. A., Hochstrasser, D. F., and Appel, R. D. (2002b) Molecular scanner experiment with human plasma : improving protein identification by using intensity distributions of matching peptide masses. *Proteomics* **2**(10), 1413–1425.

Neuhoff, V., Arold, N., Taube, D., and Ehrhardt, W. (1988) Improved staining of proteins in polyacrylamide gels including isoelectric focusing gels with clear background at nanogram sensitivity using Coomassie Brilliant Blue G-250 and R-250. *Electrophoresis* **9**, 255.

O'Farrell, P. H. (1975) High resolution two-dimensional electrophoresis of proteins. *J. Biol. Chem.* **250**, 4007.

Olson, A. D., and Miller, M. J. (1988) Elsie 4: quantitative computer analysis of sets of two-dimensional gel electrophoretograms. *Anal. Biochem.* **169**, 49–70.

Pappin, D. J. C., Hojrup, P., and Bleasby, A. J. (1993) *Curr. Biol.* **3**, 327.

Perkins, D. N., Pappin, D. J., Creasy, D. M., and Cottrell, J. S. (1999) Probability-based protein identification by searching sequence databases using mass spectrometry data. *Electrophoresis* **20**(18), 3551–3567.

Qin, J., Fenyo, D., Zhao, Y., Hall, W. W., Chao, D. M., Wilson, C. J., Young, R. A., and Chait, B. T. (1997) A strategy for rapid, high-confidence protein identification. *Anal. Chem.* **69**, 3995–4001.

Rabilloud, T. (1990) *Electrophoresis* **11**, 785.

Roepstorff, P., and Fohlman, J. (1984) Proposal for a common nomenclature for sequence ions in mass spectra of peptides. *Biomed. Mass Spectrom.* **11**, 601.

Scheele, G. A. (1975) Two-dimensional gel analysis of soluble proteins. Characterization of guinea pig exocrine pancreatic proteins. *J. Biol. Chem.* **250**, 4007–4021.

Scherl, A., Coute, Y., Deon, C., Calle, A., Kindbeiter, K., Sanchez, J. C., Greco, A., Hochstrasser, D., and Diaz, J. J. (2002) Functional proteomic analysis of human nucleolus, *Mol. Biol. Cell.* **13**(11), 4100–4109.

Smilansky, Z. (2001) Automatic registration for images of two-dimensional protein gels. *Electrophoresis* **22**(9),1616–1626.

Takach, E. J., Hines, W. M., Patterson, D. H., Juhasz, P., Falick, A. M., Vestal, M. L., and Martin, S. A. (1997) Accurate mass measurements using MALDI–TOF with delayed extraction. *J. Protein Chem.* **16**, 363–369.

Tanaka, K., Ido, Y., Akita, S., and Yoshida, T. (1987) *Second Japan–China Symposium on Mass Spectrometry*, Osaka, Japan.

Tanaka, K., Waki, H., Ido, Y., Akita, S., Yoshida, Y., and Yoshida, T. (1988) Protein and polymer analysis up to *m/z* 100,000 by laser ionisation time-of-flight mass spectrometry. *Rapid Commun. Mass Spectrom.* **2**, 151.

Tuloup, M., Hoogland, C., Binz, P.-A., and Appel, R. D. (2002) A new peptide mass fingerprinting tool on ExPASy: ALDentE (abstract). *Swiss Proteomics Society 2002 Congress: Applied Proteomics*, Lausanne.

Washburn, M. P., Wolters, D., and Yates, J. R. III. (2001) Large-scale analysis of the yeast proteome by multidimensional protein identification technology. *Nat. Biotechnol.* **19**(3), 242–247.

Wilkins, M. R., Gasteiger, E., Gooley, A., Herbert, B., Molloy, M. P., Binz, P. A., Ou, K., Sanchez, J.-C., Bairoch, A.,Williams, K. L., and Hochstrasser, D. F. (1999) High-throughput mass spectrometric discovery of protein post-translational modifications. *J. Mol. Biol.* **289**, 645–657.

Wilkins, M. R., Lindskog, I., Gasteiger, E., Bairoch, A., Sanchez, J.-C., Hochstrasser, D. F., and Appel, R. D. (1997) Detailed peptide characterisation using PEPTIDE-MASS – a World-Wide Web accessible tool. *Electrophoresis* **18**, 403–408.

Wilm, M., and Mann, M. (1996) Analytical properties of the nanoelectrospray ion source. *Anal. Chem.* **6**, 1–8.

Wolters, D. A., Washburn, M. P., and Yates, J. R. III. (2001) An automated multidimensional protein identification technology for shotgun proteomics. *Anal. Chem.* **73**(23), 5683–5690.

Wu, Y., Lemkin, P. F., Upton, K., (1993) A fast spot segmentation algorithm for two-dimensional gel electrophoresis analysis. *Electrophoresis* **14**, 1351–1356.

Yamashita, M., and Fenn, J. B. (1984) Another variation on the free-jet theme. *J. Phys. Chem.* **88**(20), 4451–4459.

Yates J. R. III. (1998) Mass spectrometry and the age of the proteome. *J. Mass Spectrom.* **33**, 1–19.

Yates, J. R. III, Speicher, S., Griffin, P. R., and Hunkapiller, T. (1993) *Anal. Biochem.* **214**, 397.

Zhang, W., and Chait, B. T. (2000) ProFound: an expert system for protein identification using mass spectrometric peptide mapping information. *Anal. Chem.* **72**, 2482–2489.

7

High-Throughput Cloning, Expression and Purification Technologies

Andreas Kreusch and **Scott A. Lesley**

7.1 Introduction

The ongoing transition from genomics to a more proteomics-oriented approach in the biological sciences has made it necessary to develop and implement new and improved methods for the efficient and successful production of proteins. The complexity of any proteomic project demands, from the beginning, a clear understanding of the required protein quantity and quality. This essentially divides a proteomic approach into application-driven strategies. For instance, only microgram amounts of soluble and properly folded proteins are required for most analytical and functional purposes, whereas milligram quantities will often be required for immunization or structural studies using X-ray and NMR techniques.

This 'application-driven' approach to the field of proteomics results in one very important consequence largely absent from genomics research: not one single method or system for the expression and purification of proteins can be successfully employed, but many different technologies need to be developed and integrated synergistically (Kenyon *et al.*, 2002). One obvious example for this is the expression of higher-eukaryotic proteins, which often require post-translational modifications, necessary for their proper function. Since bacterial expression systems are not able to provide these post-translational modifications, expression of mammalian proteins is likely to be more amenable to baculovirus, yeast and mammalian systems such as Chinese hamster ovary and

Genomics, Proteomics and Vaccines edited by Guido Grandi
© 2004 John Wiley & Sons, Ltd   ISBN 0 470 85616 5

NS0 murine myeloma cell lines. Another important, but until now largely neglected, field in the proteomic space is the development of adequate expression systems for integral membrane proteins and membrane-associated proteins. Considering that these proteins not only constitute roughly 30–40 per cent of all human proteins but also represent a major target class for pharmaceutical drug development, it will be vital to a truly successful proteomic effort to develop and validate methods for their expression and purification. Obviously, such proteins also offer tremendous utility for vaccine development (Jeannin *et al.*, 2002).

Considerable progress in the field of molecular biology has contributed to an ever growing arsenal of options for protein expression and purification. It is beyond the scope of this review to give a detailed overview of the available expression and purification systems, but a number of excellent papers regarding protein expression using yeast (Lin Cereghino and Cregg, 2001; Lin Cereghino *et al.*, 2001), mammalian expression systems (Werner *et al.*, 1998; Schlesinger and Dubensky, 1999; Koller *et al.*, 2001; Andersen and Krummen, 2002), cell-free expression systems (Kigawa, Yabuki and Yokoyama, 1999a; Kigawa *et al.*, 1999b; Shimizu *et al.*, 2001) and microbial expression systems (Baneyx, 1999; Swartz, 2001) have been published.

In this review, we will describe one route that our laboratory has taken to implement a flexible, high-throughput protein expression and purification pipeline, designed for large-scale expression and purification of 96 proteins in parallel using a bacterial expression system. The design of any high-throughput system needs to conform to the intended target application in terms of desired protein quantity and quality. Our goal was to produce milligram quantities of pure and homogeneous protein in a highly automated way, amenable for structural studies using X-ray and NMR methods. Such high-quality protein is in general also sufficient for most other proteomics applications.

7.2   Gene cloning

The first interface in the translation of genomic information into something of proteomic utility comes at the stage of cloning for expression. A vast amount of genomic information is generally available through common databases. This information can be mined for predicted open reading frames (ORFs), which can then be classified based on sequence similarity to predict function. However, in order to test any of these hypotheses, the difficult task of obtaining a physical clone of a predicted ORF and converting it to a form that is readily expressed must be undertaken. For many prokaryotes, this can be as simple as designing amplification primers from predicted ORFs and amplifying directly from genomic DNA. Most eukaryotic proteins, however, first require the production of an appropriate cDNA library, which can often be difficult. Attention to the predicted biology is important for the success of this first step, as cDNAs are dependent on the active expression of a gene. For example, genes involved in the cell cycle, long coding regions or those expressed at very low levels can be very difficult to clone without custom libraries and substantial luck. Beyond issues of abundance, integrity of the predicted gene is critical. Rarely is a predicted gene product recreated in exact integrity to its database entry. Splice variations, clonal diversity and other mutations often result in multiple amplifers and variants that force a choice as to what is the true relevant target for expression.

The desired final application can also play a major role in decision making at this point. Functional activities, desired biophysical attributes and immunogen cross-reactivity lead to one of the first critical decisions in the expression pathway. The choice of expressing full length versus a domain can have a huge impact on vector and expression system choices, as well as the expected success rate. One of the best applications of bioinformatics to proteomics is in the classification of proteins and prediction of domains. For a comprehensive list of the growing numbers of databanks that classify sequences, the reader is referred to an article by Baxevanis (2003). Most proteins fall into functional categories that can be predicted directly from sequence similarity. Due to the large and ever-increasing amount of structural information available, structural domains can usually be predicted within a protein target. Many tools such as DALI (Dietmann *et al.*, 2001), SCOP (Lo Conte *et al.*, 2000) and CATH (Pearl *et al.*, 2003) are publicly available for this purpose. For practical purposes, it is important at this stage to distinguish between functional domain prediction and structural domain prediction. While they often overlap, it is structural domain prediction that is key to the successful design of expression constructs. These tools are not exact, and even variants of a single amino acid can make the difference between success and failure when it comes to soluble expression of a target. We take a combined empirical and predictive approach to construct design whereby various permutations around a domain prediction are created in parallel and then tested directly for appropriate expression properties.

Once the desired ORF has been identified and cloned, the next step is to convert this clone into something that can easily be expressed. At this point there are a bewildering number of options, all with proposed benefits and utilities. Unfortunately, there is no panacea for generic expression of proteins. Unlike DNA technologies, which often take advantage of the narrow range of DNA biophysical properties, proteomics technologies must take into account a huge range of shape, stability, charge and hydrophobicity. To provide some measure of consistency, purification tags are normally included when purifying recombinant proteins. A number of tags are routinely utilized (Baneyx, 1999; Kapust and Waugh, 1999; Lesley, 2001). Since each has its own advantages, cloning methods based on homologous recombination have been promoted to rapidly evaluate the various expression options (Landy, 1989; Liu et al., 1998; Ghosh and Van Duyne, 2002). While these systems offer substantial convenience when shuttling between vectors, they also have limitations in vector design and often result in the fusion of irrelevant sequence onto the ORF of interest. We routinely use a more traditional cloning approach utilizing a blunt vector with a downstream eight-base restriction site to provide directionality. This conventional approach to cloning is implemented in a high-throughput way by parallel processing and simple PCR screens to provide a robust and generic cloning approach and has been enhanced further with the incorporation of commercial robotics systems.

Regardless of the cloning approach, the resulting clones need verification and reorganization for protein expression. While such tasks as sequencing and reracking of clones are trivial for individual clones, they become complex and expensive when applied in a proteomic fashion. Most sequence errors are introduced by the oligonucleotides used for PCR rather than the amplification reaction itself, which can be done with proofreading enzymes. It is essential that the cloning junctions be sequence verified for each clone.

An example of how such an approach has been utilized for a proteomics effort is demonstrated by the successful processing of a bacterial proteome as part of a structural genomics initiative (Lesley et al., 2002). Using relatively traditional cloning methods applied in a parallel and high-throughput fashion, 73 per cent of the predicted Thermotoga maritima proteome was cloned into a single expression vector and sequence verified in a matter of months using minimal human resources.

7.3   Protein expression

After expression-ready clones become available, they can be tested on a small scale using a 96-well format for expression of target protein in order to select the clones that produce soluble protein. While high-throughput expression and purification of proteins in the microgram range can be achieved by either traditional methods (Doyle et al., 2002) or commercially available robotics (Lesley, 2001), our intended large-scale approach required the development of custom-designed automation. Expression of milligram quantities of proteins is traditionally achieved by fermentation of at least 500–1000 ml cultures, which is not only labour intensive and time consuming but also requires sufficient incubator space and large quantities of culture media if applied to the parallel expression of 96 proteins. We therefore designed a compact, high-performance 96-tube fermenter that uses culture volumes of 65 ml per tube (Figure 7.1). Optimized aeration and media conditions result in a high-density cell growth that can reach 40 OD (600 nm), a sufficient cell mass to produce the milligram quantities of proteins required for most applications. The relatively low cell culture volume is also considerably more cost effective for special applications such as the use of ^{15}N-enriched media (for NMR studies) or selenomethionine enriched media (for X-ray crystallography purposes) than conventional fermentation procedures. A complete fermenter run from the time the precultures are added to the media until the cell cultures are harvestable usually takes 6 h.

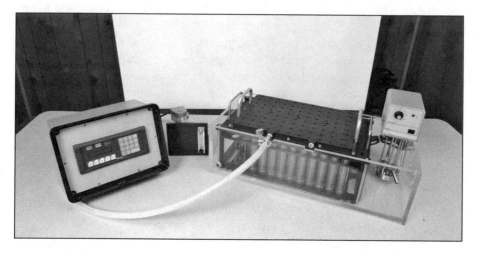

Figure 7.1   Fermenter for parallel expression of 96 proteins

7.4   High throughput protein purification

In order to achieve a maximum of protein purity while at the same time avoiding too many labour intensive purification steps, it has become customary to attach purification tags to the protein of interest. A comprehensive list of available fusion tags can be found in an article by Stevens (2000). In general, larger tags such as thioredoxin (La Vallie *et al.*, 1993), maltose-binding protein (di Guan, Riggs and Inouye, 1988) and glutathione-S-transferase (Smith and Johnson, 1988) have the distinct advantage that they can often increase the solubility and expression level of otherwise difficult to express proteins (Kapust and Waugh, 1999). They are also preferable for use with small peptides, for instance the expression of haptens, which can then be used to raise antibodies. One disadvantage of these protein tags is the inability to use denaturing purification conditions, i.e. for the purification of proteins expressed as inclusion bodies. A second disadvantage of large tags is that they can interfere with the subsequent application explicitly in structural analysis of the target protein. To circumvent the latter problem, most large tags can be removed by proteolysis. However, this requires at least one additional purification step and is more difficult to implement in a high-throughput pipeline since the success rate of proteolytic digests can vary widely, often resulting in a significant reduction of yield. Therefore, small tags such as polyarginine (Smith *et al.*, 1984), biotin acceptor peptide (Morag, Bayen and Wilchek, 1996) and polyhistidine tags (Hochuli *et al.*, 1988), which introduce only 6–13 additional residues at the N- or C-terminus of the target protein, represent a more convenient choice for standardized and automated parallel purification. In addition, these small tags do not usually perturb the function and structure of the target protein. For our expression and purification pipeline, we implemented a hexa-His tag because of its proven track record for successful use in structural biology (Bucher, Evdokimov and Waugh, 2002). However, preliminary tests in our laboratory indicated that in a significant number of cases the overall expression levels of His-tag fusion proteins were lower than for proteins fused to larger tags. We therefore explored the possibility of improving target protein expression by introducing a short translational leader sequence N-terminal to our hexa-His tag. The addition of six residues, representing the N-terminal sequence of thioredoxin (MGSDKI), considerably improved protein expression and has been implemented in all of our bacterial expression vectors.

Parallel protein purification on a microgram scale has become feasible simply by adopting instrumentation and modification of methods that were originally designed for nucleic acid separation and purification. However, commercially accessible robotics for handling larger cell culture volumes are not available and required us to develop custom designed instrumentation for our purification pipeline. Based on our 96-tube fermenter design we developed a purification robot (Figure 7.2), which is able to perform the basic steps in protein extraction and purification, including centrifugation, aspiration and dispensing of buffers, and cell lysis via sonication for 96 cell cultures at a time. The resulting supernatants, which contain the expressed and soluble proteins of interest, are automatically loaded onto chromatography columns containing pre-equilibrated Ni-NTA resins. The purification robot then runs through a standard wash and elution protocol. All steps are temperature controlled to minimize protein aggregation and degradation. Obviously, proteins that are

Figure 7.2 Robot for parallel purification of 96 proteins

expressed as inclusion bodies will not be captured by this particular purification scheme. In order to be able to collect the insoluble proteins for refolding studies, the protocol can be modified to dissolve inclusion bodies and purify those proteins using denaturing conditions. Such inclusion bodies have been shown to be effective immunogens when an appropriate fusion tag is used (Knuth *et al.*, 2000). A set of 96 individual target proteins can be purified within 6 hours using this automation. Daily fermentation and protein purification of 96 individual proteins including media preparation, sterilization of equipment and analytical SDS gels to assess the protein purity and quantity can be accomplished with a minimum of dedicated personnel. The amount of purified protein usually exceeds 6–7 mg and can reach 40 mg per sample. The protein purity after the one-step affinity chromatography is above 80 per cent, which is sufficient for most applications such as antibody production or target-based high-throughput screens using small molecule libraries. More demanding applications such as structural studies require additional purification steps. To meet higher quality standards, we have implemented a standardized secondary purification protocol using ion-exchange chromatography on multiple HPLC systems, equipped with autosampler. A final, optional purification step involves gel-filtration chromatography, which is also convenient for buffer exchanges.

The parallel expression and purification of 96 proteins per day requires several quality checkpoints in order to minimize and identify errors during fermentation and purification. To monitor the success of a fermenter run in terms of growth characteristics we use as a control a cell culture that overexpresses GFP protein. If a fermenter run fails due to problems caused by improper media, aeration, induction or temperature, GFP will not be expressed.

A second checkpoint is the apparent molecular weights of the purified proteins according to SDS gel electrophoresis. This can identify errors associated with wrongly labelled or mishandled tubes and errors in indexing by the purification robot. Finally, the most important check is being done at the end of the protein purification: analytical mass spectrometry of a trypsinized sample of each purified protein is performed for the unambiguous identification of the purified protein.

7.5   Validation of the pipeline and outlook

In order to test the robustness of our expression and purification pipeline we have chosen to process the complete proteome of the thermophilic bacteria *Thermotoga maritima* with its 1877 open reading frames, 46 per cent of which have no assigned function (Nelson *et al.*, 1999). From the 1877 targets, 1376 expression clones (73 per cent) could be derived and channelled through the fermentation and purification process. 542 of those clones (542/1376 = 40 per cent) resulted in the expression of milligram amounts of soluble protein and are currently in structure determination trials (Lesley *et al.*, 2002). The distribution of soluble expressed proteins among the different functional classes in the *Thermotoga* genome shows that our generic approach to protein expression captures at best 50 per cent of all proteins within one protein class (Table 7.1). This number drops significantly for cell envelope proteins and transport and binding proteins, categories that include a large number of membrane and membrane associated proteins. In general, for protein expression on a proteomic scale, considerable attrition has to be expected at each step of the gene-to-protein pipeline. The most severe problem is the lack of solubility of the expressed proteins. This is simply due to our currently very limited knowledge in terms of which proteins require specific binding partners or cofactors for proper folding. That it is possible to systematically identify multiprotein complexes on a genome-wide scale has recently been demonstrated for yeast (Gavin *et al.*, 2002). In addition, membrane-associated proteins and integral membrane proteins are not likely to be suitable for the expression and purification strategy we have chosen.

However, the possibility to efficiently express and purify nearly a hundred proteins a day in a highly automated manner allows us to approach the attrition problem by generating and testing an unprecedented number of different gene constructs for expression positive clones. This includes the use of different

Table 7.1 Functional assignment of soluble expressed proteins

Protein classification[1]	T. maritima proteins	soluble expression	(%)
Amino acid biosynthesis	72	34	47
Biosynthesis of cofactors and prosthetic groups	31	14	45
Cell envelope	73	15	21
Cellular processes	49	14	28
Central intermediary metabolism	44	23	52
DNA metabolism	54	15	28
Energy metabolism	195	88	45
Fatty acid and phospholipid metabolism	15	6	40
Hypothetical proteins	774	188	24
Other categories	13	4	31
Protein fate	48	6	13
Protein synthesis	106	31	29
Purines, pyrimidines, nucleosides and nucleotides	45	20	44
Regulatory functions	70	18	26
Transcription	16	8	50
Transport and binding proteins	188	30	16
Unknown function	83	28	34
Total	1877	542	29

[1] According to Nelsen *et al.* (1999).

affinity tags, varying the size and borders of gene constructs and, in the case of multidomain-containing proteins, the expression of single, stable domains. Once protein partners have been identified, these proteins can be coexpressed and copurified with their interacting partner. Orthologous proteins from many different species can be tested simultaneously and the best behaving proteins will then be selected for further examination (Savchenko *et al.*, 2003). Similarly, protein engineering by introducing random or directed single point mutations can be used to improve protein properties like stability and solubility.

7.6   Conclusion

If we want to take advantage of the vast number of data that have been accumulated through genomics, the need to develop methods and strategies for the efficient expression and purification of proteins will represent one of the crucial steps in the proteomic era. In this review we have presented one of the possible routes to efficiently express and purify proteins in milligram quantities by using a highly automated gene-to-protein pipeline. This approach makes it possible to process complete proteomes of organisms in an unpreceded short time as compared to traditional methods. For now, our protein expression strategy is centred around a bacterial expression system that is adequate at this early stage of the proteomic timescale. In the near future, however, improved methods for the parallel expression of eukaryotic proteins, as well as membrane proteins, will need to be developed in order to truly meet the challenges of proteomics research.

References

Andersen, D. C., and Krummen, L. (2002) Recombinant protein expression for therapeutic applications. *Curr. Opin. Biotechnol.* **13**, 117–123.

Baneyx, F. (1999) Recombinant protein expression in *Escherichia coli*. *Curr. Opin. Biotechnol.* **10**, 411–421.

Baxevanis, A. D. (2003) The molecular biology database collection: 2003 update. *Nucleic Acids Res.* **31**, 1–12.

Bucher, M. H., Evdokimov, A. G., and Waugh, D. S. (2002) Differential effects of short affinity tags on the crystallization of *Pyrococcus furiosus* maltodextrin-binding protein. *Acta Crystallogr.* **D 58**, 392–397.

Dietmann, S., Park, J., Notredame, C., Heger, A., Lappe, M., and Holm, L. (2001) A fully automatic evolutionary classification of protein folds: DALI domain dictionary version 3. *Nucleic Acids Res.* **29**, 55–57.

di Guan, C., Li, P., Riggs, P. D., and Inouye, H. (1988) Vectors that facilitate the expression and purification of foreign peptides in *Escherichia coli* by fusion to maltose-binding protein. *Gene* **67**, 21–30.

Doyle, S. A., Murphy, M. B., Massi, J. M., and Richardson, P. M. (2002) High-throughput Proteomics: A flexible and efficient pipeline for protein production. *J. Proteome Res.* **1**, 531–536.

Gavin, A. C., Bosche, M., Krause, R., Grandi, P., Marzioch, M., Bauer, A., Schultz. J., Rick, J. M., Michon, A. M., Cruciat, C. M., Remor, M., Hofert, C., Schelder, M., Brajenovic, M., Ruffner, H., Merino, A., Klein, K., Hudak, M., Dickson, D., Rudi, T., Gnau, V., Bauch, A., Bastuck, S., Huhse, B., Leutwein, C., Heurtier, M. A., Copley, R. R., Edelmann, A., Querfurth, E., Rybin, V., Drewes, G., Raida, M., Bouwmeester, T., Bork, P., Seraphin, B., Kuster, B., Neubauer, G., and Superti-Furga, G. (2002) Functional organization of the yeast proteome by systematic analysis of protein complexes. *Nature* **415**, 141–147.

Ghosh, K., and Van Duyne, G. D. (2002) Cre-loxP biochemistry. *Methods* **28**, 374–383.

Hochuli, E., Bannwarth, W., Dobeli, H., Gentz, R., and Stuber, D. (1988) Genetic approach to facilitate purification of recombinant proteins with a novel metal chelate absorbent. *BioTechnology* **6**, 1321–1325.

Jeannin, P., Magistrelli, G., Goetsch, L., Haeuw, J. F., Thieblemont, N., Bonnefoy, J. Y., and Delnest, Y. (2002) Outer membrane protein A (OmpA): a new pathogen-associated molecular patten that interacts with antigen presenting cells – impact on vaccine strategies. *Vaccine* **20** (Suppl. 4), A23–A27.

Kapust, R. B., and Waugh, D. S. (1999) *Escherichia coli* maltose-binding protein is uncommonly effective at promoting solubility of polypeptides to which it is fused. *Prot. Sci.* **8**, 1668–1674.

Kenyon, G. L., DeMarini, D. M., Fuchs, E., Galas, D. J., Kirsch, J. F., Leyh, T. S., Moos, W. H., Petsko, G. A., Ringe, D., Rubin, G. M., and Sheahan, L. C. (2002) Defining the mandate of proteomics in the post-genomics era: workshop report. *Mol. Cell Proteomics* **1**, 763–780.

Kigawa, T., Yabuki, T., and Yokoyama, S. (1999a) Large-scale protein preparation using cell-free synthesis. *Tanpakushitsu Kakusan Koso* **44**, 598–605.

Kigawa, T., Yabuki, T., Yoshida, Y., Tsutsui, M., Ito, Y., Shibata, T., and Yokoyama, S. (1999b) Cell-free production and stable isotope labeling of milligram quantities of proteins. *FEBS Lett.* **442**, 15–19.

Knuth, M. W., Okragly, A. J., Lesley, S. A., and Haak-Frendscho, M. (2000) Facile generation and use of immunogenic polypeptide fusions to a sparingly soluble non-antigenic protein carrier. *J. Immunol. Methods* **236**, 53–69.

Koller, D., Ruedl, C., Loetscher, M., Vlach, J., Oehen, S., Oertle, K., Schirinzi, M., Deneuve, E., Moser, R., Kopf, M., Bailey, J. E., Renner, W., and Bachmann, M. F. (2001) A high-throughput alphavirus-based expression cloning system for mammalian cells. *Nat. Biotechnol.* **19**, 851–855.

La Vallie, E. R., DiBlasio, E. A., Kovacic, S., Grant, K. L., Schendel, P. F., and McCoy, J. M. (1993) A thioredoxin gene fusion expression system that circumvents inclusion body formation in the *E. coli* cytoplasm. *BioTechnology* **11**, 187–193.

Landy, A. (1989) Dynamic, structural, and regulatory aspects of lambda site-specific recombination. *Annu. Rev. Biochem.* **58**, 913–949.

Lesley, S. A. (2001) High-throughput proteomics: protein expression and purification in the postgenomic world. *Prot. Expression Purification* **22**, 159–164.

Lesley, S. A., Kuhn, P., Godzik, A., Deacon, A. M., Mathews, I., Kreusch, A., Spraggon, G., Klock, H. E., McMullan, D., Shin, T., Vincent, J., Robb, A., Brinen, L. S., Miller, M. D., McPhillips, T. M., Miller, M. A., Scheibe, D., Canaves, J. M., Chittibabu, G., Jaroszewski, L., Selby, T. L., Elsliger, M. A., Wooley, J., Taylor, S. S., Hodgson, K. O., Wilson, I. A., Schultz, P. G., and Stevens, R. C. (2002) Structural genomics of the *Thermotoga maritima* proteome implemented in a high-throughput structure determination pipeline. *Proc. Natl. Acad. Sci.* **99**, 11 664–11 669.

Lin Cereghino, G. P., Sunga, A. J., Lin Cereghino, J., and Cregg, J. M. (2001) Expression of foreign genes in the yeast *Pichia pastoris*. In Setlow, J. K. (Ed.). *Genetic Engineering: Principles and Methods*. Kluwer–Plenum, London, Vol. 23, 157–169.

Lin Cereghino, J., and Cregg, J. M. (2001) Heterologous protein expression in the methylotrophic yeast *Pichia pastoris*. *FEMS Microbiol. Rev.* **24**, 45–66.

Liu, Q., Li, M. Z., Leibham, D., Cortez, D., and Elledge, S. J. (1998) The univector plasmid-fusion system, a method for a rapid construction of recombinant DNA without restriction enzymes. *Curr. Biol.* **8**, 1300–1309.

Lo Conte, L., Ailey, B., Hubbard, T. J., Brenner, S. E., Murzin, A. G., and Chothia, C. (2000) SCOP: a structural classification of proteins database. *Nucleic Acids Res.* **28**, 257–259.

Morag, E., Bayer, E. A., and Wilchek, M. (1996) Immobilized nitro-avidin and nitro-streptavidin as reusable affinity matrices for application in avidin–biotin technology. *Anal. Biochem.* **243**, 257–263.

Nelson, K. E., Clayton, R. A., Gill, S. R., Gwinn, M. L., Dodson, R. J., Haft, D. H., Hickey, E. K., Peterson, J. D., Nelson, W. C., Ketchum, K. A., McDonald, L., Utterback, T. R., Malek, J. A., Linher, K. D., Garrett, M. M., Stewart, A. M., Cotton, M. D., Pratt, M. S., Phillips, C. A., Richardson, D., Heidelberg, J., Sutton, G. G., Fleischmann, R. D., Eisen, J. A., White, O., Salzberg, S. L., Smith, H. O., Venter, J. C., and Fraser, C. M. (1999) Evidence for lateral gene transfer between Archaea and bacteria from genome sequence of *Thermotoga maritima. Nature* **399**, 323–329.

Pearl, F. M., Bennett, C. F., Bray, J. E., Harrison, A. P., Martin, N., Shepherd, A. J., Sillitoe, I., Thornton, J., and Orengo, C. A. (2003) The CATH database: an extended protein family resource for structural and functional genomics. *Nucleic Acids Res.* **31**, 452–455.

Savchenko, A., Yee, A., Khachatryan, A., Skarina, T., Evdokimova, E., Pavlova, M., Semesi, A., Northey, J., Beasley, S., Lan, N., Das, R., Gerstein, M., Arrowsmith, C. H., and Edwards, A. M. (2003) Strategies for structural proteomics of prokaryotes: quantifying the advantages of studying orthologous proteins and of using both NMR and x-ray crystallography approaches. *Proteins* **50**, 392–399.

Schlesinger, S., and Dubensky, T. Jr. (1999) Alphavirus vectors for gene expression and vaccines. *Curr. Opin. Biotechnol.* **10**, 434–439.

Shimizu, Y., Inoue, A., Tomare, Y., Suzuki, T., Yokogawa, T., Nishikawa, K., and Ueda, T. (2001) Cell-free translation reconstituted with purified components. *Nat. Biotechnol.* **19**, 751–755.

Smith, J. C., Derbyshire, R. B., Cook, E., Dunthorne, L., Viney, J., Brewer, S. J., Sassenfeld, H. M., and Bell, L. D. (1984) Chemical synthesis and cloning of a poly(arginine)-coding gene fragment designed to aid polypeptide purification. *Gene* **32**, 321–327.

Smith, D. B., and Johnson, K. S. (1988) Single-step purification of polypeptides expressed in *Escherichia coli* as fusions with glutathione-S-transferase. *Gene* **67**, 31–40.

Stevens, R. C. (2000) Design of high-throughput methods of protein production for structural biology. *Structure* **8**, R177–R185.

Swartz, J. R. (2001) Advances in *Escherichia coli* production of therapeutic proteins. *Curr. Opin. Biotechnol.* **12**, 195–201.

Werner, R. G., Noe, W., Kopp, K., and Schulter, M. (1998) Appropriate mammalian expression systems for biopharmaceuticals. *Drug Res.* **48**, 870–880.

Part 3
Applications

8

Meningococcus B: from Genome to Vaccine

Davide Serruto, Rino Rappuoli and Mariagrazia Pizza

8.1 Meningococcus, a major cause of bacterial meningitis

8.1.1 Microbiological features and pathogenesis

Meningococcal meningitis and septicaemia are devastating diseases caused by the infection of *Neisseria meningitidis* (meningococcus), a Gram-negative, capsulated bacterium, which belongs to the phylogenetic order of the B subgroup of Proteobacteria. Humans are the only natural reservoir of *N. meningitidis*, and the mucosal epithelium of the nasopharynx is the site from which meningococci can be transmitted. Upon colonization, the organism often assumes a carrier status, with 5–15 per cent of healthy individuals carrying meningococci in their nasopharynx. Under particular conditions and in susceptible individuals meningococcus is able to actively invade the respiratory tract epithelia, reach and multiply within the bloodstream, causing sepsis. From the bloodstream, the bacterium then crosses the blood–brain barrier and spreads to the cerebrospinal fluid, causing meningitis (Rosenstein *et al.*, 2001) (Figure 8.1).

Under endemic situations, the disease can reach an incidence of one to five cases per 100 000 population, while during epidemics the cases reported are 15–500 per 100 000 (Schwartz, Moore and Broome, 1989; Achtman, 1995; Schuchat *et al.*, 1997). Currently, antibiotics represent the elected treatment for meningitis, with penicillin generally administered to infected patients, while rifampicin is mainly used prophylactically to prevent disease in carriers (Broome, 1986). Although antimicrobial therapy has contributed to reduce the

Genomics, Proteomics and Vaccines edited by Guido Grandi
© 2004 John Wiley & Sons, Ltd   ISBN 0 470 85616 5

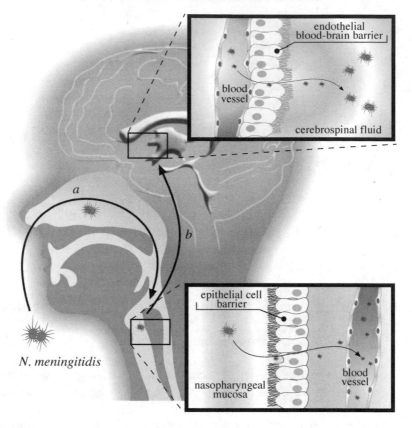

Figure 8.1 Colonization steps of *N. meningitidis*: from the nasopharynx to the bloodstream (a) and from the bloodstream to the cerebrospinal fluid (b)

mortality rate of 65–80 per cent of the preantibiotic era to the current figure of 4–15 per cent, cases of fulminant meningococcaemia are unfortunately generally hopeless. Fatality rates reported for such cases are of 15–71 per cent (Peltola, 1983). For all these reasons, prevention by vaccination is being regarded and actively pursued as the gold standard method to control meningitis. Unfortunately, despite numerous efforts, the problem of producing a general vaccine against this alarming pathogen has not yet been solved (Morley and Pollard, 2001).

A schematic representation of bacterial compartments and of the main membrane components is reported in Figure 8.2. The bacterium is surrounded by a polysaccharide capsule that is essential for pathogenicity as it confers resistance to phagocytosis and complement-mediated lysis and offers protection

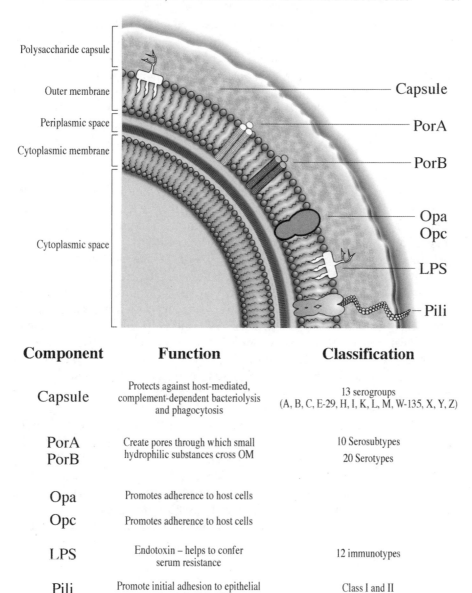

Component	Function	Classification
Capsule	Protects against host-mediated, complement-dependent bacteriolysis and phagocytosis	13 serogroups (A, B, C, E-29, H, I, K, L, M, W-135, X, Y, Z)
PorA PorB	Create pores through which small hydrophilic substances cross OM	10 Serosubtypes 20 Serotypes
Opa	Promotes adherence to host cells	
Opc	Promotes adherence to host cells	
LPS	Endotoxin – helps to confer serum resistance	12 immunotypes
Pili	Promote initial adhesion to epithelial and endothelial cells	Class I and II

Figure 8.2   Schematic representation of the different bacterial compartments. The main components of the outer membrane, their function and their contribution to meningococcus classification are reported

against environmental insults. As a consequence, acapsulated bacteria are inactive *in vivo* since they cannot persist in the blood. The outer membrane is composed by many proteins, some of which enable the organism to adhere to, and to invade, host cells. Pili, composed of pilin subunits, are filamentous structures essential in mediating the interaction between meningococci and both epithelial and endothelial cells, and are also associated with other phenotypes such as a high level of competence, bacterial autoagglutination and twitching motility. Pilins are classified into class I and class II based on their differences in antigenicity (Nassif, 1999). Opa and Opc proteins are capable of mediating adhesion to and invasion into eukaryotic cells, but this effect is visible only in the absence of the capsule (Virji *et al.*, 1993, 1992). Outer membrane also encloses lipopolysaccharide (LPS), which helps in serum resistance and plays a key role in the pathogenesis of meningococcal disease (Vogel and Frosch, 1999; McNeil, Virji and Moxon, 1994).

Meningococci can be classified on the basis of the chemical composition of distinctive polysaccharides into 13 serogroups: A, B, C, H, I, K, L, M, W135, X, Y, Z and 29E. Among them, only five, A, B, C, Y and W135, have been associated with the disease and are thus considered as pathogenic serogroups. A further classification uses serological methods and is based on the antigenicity of the most abundant membrane proteins PorA and PorB, whose sequence defines the serotype and serosubtype, respectively. Moreover, on the basis of the lipopolysaccharide structure is possible to define 12 immunotypes (Morley and Pollard, 2001; Frasch, Zollinger and Poolman, 1985; Scholten *et al.*, 1994; Kogan *et al.*, 1997).

New methods of classification have been developed more recently. Multi-locus enzyme electrophoresis (MLEE) uses the electrophoretic mobility of various cytoplasmic enzymes to classify meningococci into clonal families with similar characteristics (Caugant *et al.*, 1986a,b). Multi-locus sequence typing (MLST) is based on a similar principle but makes use of molecular sequence data generated from a number of different housekeeping genes (Maiden *et al.*, 1998). Both classification methods have allowed to classify the strains into a small number of hypervirulent lineages: electrophoretic types ET-5, ET-37, cluster A4, lineage 3, subgroup I, III, IV-1 and sequence types ST11, ST32, ST8 ST41, ST1, ST5 and ST4, respectively. A better understanding of clonal changes within meningococci is essential in vaccine design.

8.1.2   Prevention: state of the art on Neisseria vaccines

The meningococcal vaccine currently in use is a tetravalent vaccine composed of purified capsular polysaccharides of serogroups A, C, Y and W135. Although efficacious in adolescent and adults, this vaccine is poorly immunogenic in infants and fails to induce immunological memory. Experience with *Haemophilus influenzae* conjugate vaccine has shown that the immunogenicity of polysaccharides can be improved by chemical conjugation to a protein carrier, which is able to elicit a T-cell-dependent antibody response. After the introduction of this vaccine into clinical practice in 1995, the incidence of *H. influenzae*-associated meningitis decreased to five per cent of total cases (Lindberg, 1999). In 1995 meningococcus became the major cause of meningitis for all ages in the United States and the second leading cause worldwide (Schuchat *et al.*, 1997).

Conjugate vaccines against serogroups A and C have been developed and tested in clinical trials (Costantino *et al.*, 1992; Anderson *et al.*, 1994; Fairley *et al.*, 1996; Lieberman *et al.*, 1996). The first trials conducted in the United Kingdom with the meningococcus C conjugate showed a dramatic decline in the incidence of serogroup C disease in all age groups that received the vaccine (Borrow *et al.*, 2000; Miller, Salisbury and Ramsay, 2001) with an efficacy of 97 and 92 per cent for teenagers and toddlers, respectively (Ramsay *et al.*, 2001). Similar conjugate vaccines against Y and W135 serogroups are being developed.

Today, the most critical target for vaccination remains meningococcus B, which is responsible for 32 per cent of all cases of meningococcal disease in the United States and for 45–80 per cent or more of the cases in Europe (Scholten *et al.*, 1993), and for which conventional biochemical and microbiological approaches have so far failed to produce a vaccine able to induce broad protection. A polysaccharide-based vaccine approach cannot be used for group B meningococcus because the MenB capsular polysaccharide is a polymer of $\alpha(2\text{-}8)$-linked N-acetylneuranimic acid that is also present in mammalian tissues. This means that it is almost completely non-immunogenic as it is recognized as a self-antigen. Attempts to break tolerance to induce immunity to this polysaccharide are likely to lead to autoimmunity (Hayrinen *et al.*, 1995; Finne *et al.*, 1987).

An alternative approach to MenB vaccine development is based on the use of surface-exposed proteins contained in outer membrane preparations (outer membrane vesicles, OMVs). The first OMV vaccines were developed in Norway and Cuba and showed efficacy in humans ranging from 50 to 80 per cent (Tappero *et al.*, 1999). However, while each vaccine was shown to induce good protection against the homologous strain, it failed to induce protection against heterologous strains (Rosenstein, Fischer and Tappero, 2001). The major protective antigen in both these vaccines is PorA, the most abundant outer membrane protein, which is known to be highly variable across different isolates of serogroup B *N. meningitidis*.

To develop an OMV vaccine able to induce a broader protection, one approach has been the use of six different PorA. This hexavalent PorA vaccine is based on the OMVs of two isogenic meningococcal strains each expressing three different PorA. This vaccine has been tested in clinical trials and shown to induce antibodies with bactericidal activity higher in toddlers than in children (de Kleijn *et al.*, 2001). However, the main issue of this kind of vaccine is the number of different PorA to include in order to obtain protection against all MenB strains. In the last few years, new antigens have been identified as possible vaccine candidates, and most of them are currently in different phases of development or under clinical evaluation.

Transferring binding proteins A and B (TbpA and TbpB, respectively) are involved in iron acquisition from the host during infection (Schryvers and Morris, 1988; Schryvers and Stojiljkovic, 1999). Functional antibodies against these antigens could inhibit the uptake of iron and, consequently, inhibit bacterial growth. TbpB is a surface-exposed lipoprotein, is recognized by antibodies in human convalescent sera, indicating that it is immunogenic *in vivo*, during infection, and is protective in the mouse infection model (Lissolo *et al.*, 1995). In a recent phase I clinical trial, adults who were immunized with a recombinant TbpB vaccine developed high titres of antibody but with low bactericidal activity (Danve *et al.*, 1998). This vaccine has been improved by the addition of more different variants of TbpB or by the inclusion of TbpA. Phase I clinical trials are ongoing.

Neisseria surface protein NspA has been described as the first antigen with high levels of sequence conservation among different meningococcus strains. NspA is able to elicit serum bactericidal antibodies that confer passive protection in the infant rat model (Martin *et al.*, 1997; Cadieux *et al.*, 1999). However, there are large differences in antibody accessibility of NspA on the surface of different encapsulated meningococcal strains (Moe, Tan and Granoff, 1999; Moe *et al.*, 2001). Therefore, a vaccine based only on NspA might not be able to confer protection against multiple strains.

8.2   Group B meningococcus as an example of reverse vaccinology

As discussed above, conventional biochemical and microbiological approaches used for more than four decades have been unsuccessful in providing an universal vaccine against meningococcus B. As described in Chapter 2, the availability of whole genome sequences has entirely changed the approach to vaccine development. The genome represents a list of virtually all the protein antigens that the pathogen can express at any time. It becomes possible to choose potentially surface-exposed proteins in a reverse manner, starting from the genome rather than from the microorganism. This novel approach has been named 'reverse vaccinology' (Rappuoli, 2000, 2001). Here we will describe the application of this genome-based approach to the discovery of novel vaccine candidates of *N. meningitidis* serogroup B.

8.2.1 The complete genome sequence of *N. meningitidis* serogroup B

In collaboration with TIGR, we sequenced the genome of a virulent MenB strain (MC58) using the random shotgun strategy (Tettelin *et al.*, 2000). The genome contains 2 272 351 base pairs with an average content of G + C of 53 ˙ per cent. Using open reading frame (ORF) prediction algorithms and whole genome homology searches, 2158 putative ORFs have been identified. Biological roles have been assigned to 1158 ORFs (53.7 per cent) on the basis of homologies with proteins of known function. 345 (16 per cent) predicted coding sequences matched gene products of unknown function from other species and 532 (24.7 per cent) had no database match.

8.2.2 Experimental strategy

To identify novel vaccine antigens we designed a strategy aimed to select, among the more than 2000 predicted proteins, those that were predicted to be surface exposed or secreted and tested them for their potential to induce protection against disease. A schematic representation of the 'reverse vaccinology' approach applied to *N. meningitidis* serogroup B is reported in Figure 8.3.

Candidate antigen prediction by bioinformatic analysis

N. meningitidis is essentially an extracellular pathogen and the major protective response relies on circulating antibody. Complement-mediated bactericidal activity is the accepted correlate for the *in vivo* protection and as such is the surrogate endpoint in clinical trials of potential meningococcal vaccines. On the basis of this evidence, we worked on the assumption that protective antigens are more likely to be found among surface-exposed or secreted proteins. Hence the initial selection of candidate analysis is based on computer predictions of surface location or secretion. In the preliminary selection, all *N. meningitidis* predicted ORFs were searched using computer programs such as PSORT and

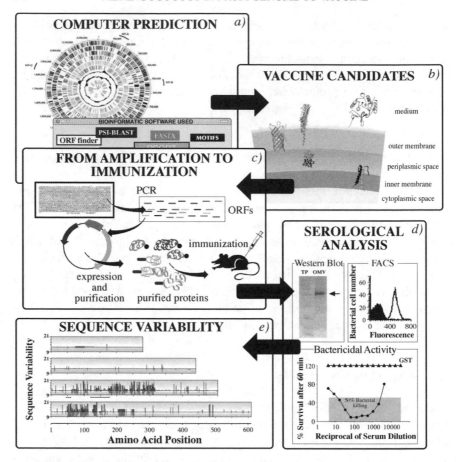

Figure 8.3 *N. meningitidis* serogroup B as an example of reverse vaccinology. Use of different bioinformatic software to analyse the genome (a) to identify potential vaccine candidates (b). Selected ORFs are amplified, cloned in expression vectors, purified and used to immunize mice (c). Mice immune sera are analysed in western blot and FACS to verify whether the antigens are expressed and surface exposed; the bactericidal assay is used to evaluate the complement-mediated bacterial killing activity of antibodies (d). Antigen conservation is evaluated by sequencing the genes in a panel of *Neisseria* strains (e) (Pizza *et al.*, 2000)

SignalP, which predict signal peptide sequences. Moreover, proteins containing predicted membrane spanning regions (using TMPRED) or lipoprotein signature and proteins homologous to surface-exposed proteins in other microorganisms were also selected. Finally, proteins with homology to known virulence factors or protective antigens from other pathogens were added to the list (Figure 8.3(a)).

Of the 2158 predicted ORFs in the *N. meningitidis* genome, 600 were selected by these criteria and could therefore represent new potential vaccine candidates.

These ORFs have been classified according to their topological features. The higher proportion is represented by inner membrane proteins and transporters (34 per cent), followed by periplasmic proteins (27 per cent) and lipoproteins (20 per cent). The smallest groups include predicted outer membrane and secreted proteins (13 per cent) and proteins that have been selected exclusively on the basis of sequence homology criteria (6 per cent) (Figures 8.3(b) and 8.4).

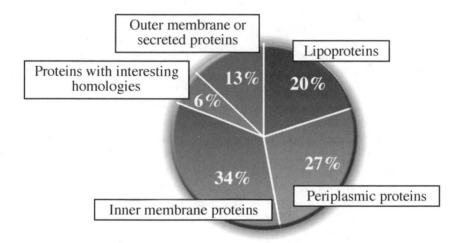

Figure 8.4 Percentage distribution of the novel MenB vaccine candidates, identified by genome mining, according to their topological features

The 600 ORFs identified were also analysed for presence and conservation in *N. meningitidis* serogroup A and *N. gonorrhoeae*, whose partial genome sequences were available on the databases: The Sanger Centre (http://www.sanger.ac.uk/Projects/N_meningitidis/) and the Advanced Center for Genome Technology at The University of Oklahoma (http://www.genome.ou.edu/gono.html), respectively. This analysis indicated that 98.8 per cent and 90 per cent of the selected *N. meningitidis* ORFs were present and conserved in serogroup A and in *N. gonorrhoeae*, respectively.

Candidate screening

The 600 ORFs selected were amplified, cloned and analysed for expression in a heterologous system as either C-terminal His-tag or N-terminal glutathione S-transferase (GST) fusion proteins. These two expression systems were chosen to achieve the highest level of expression and the easiest purification procedure by a single chromatography step. Of the 600 ORFs, 350 were successfully cloned in *Escherichia coli* and purified in a sufficient amount for mice immunizations (Figure 8.3(c)). Most of the failures, both in cloning and in expression, were related to proteins with more than one transmembrane spanning region. This is likely to be due to toxicity for *E. coli* or to their intrinsic insolubility.

Serological analysis

Each purified recombinant protein was used to immunize CD1 mice in the presence of Freund's adjuvant. Immune response was analysed by western blot analysis on both total-cell extracts and on purified OMV to verify whether the protein was expressed in meningococcus and localized on the membrane, and by enzyme-linked immunosorbent assay (ELISA) and flow cytometry on whole cells to verify whether the antigen was surface exposed. Finally, the bactericidal assay was used to evaluate the complement-mediated killing activity of the antibodies, since this property correlates with vaccine efficacy in humans (Figure 8.3(d)) (Goldschneider, Gotschlich and Artenstein, 1969).

Of the 91 proteins found to be positive in at least one of these assays, 28 were able to induce antibodies with bactericidal activity (Pizza *et al.*, 2000).

Gene variability analysis on vaccine candidates

To evaluate whether the 28 antigens able to induce a bactericidal response were also conserved in sequence among different strains, we analysed nine of them in a panel of different strains. As already mentioned, the analysis of a sequence conservation of a given antigens is a fundamental aspect in its evaluation as vaccine candidate. Meningococcus uses a broad range of microevolutionary tools (like phase variation and antigenic variability) to escape the human immune response, and the best example of this mechanism is represented by PorA, the most abundant and variable antigen of meningococcus. To approach this issue, a panel of serogroup B *N. meningitidis* strains was selected, as representative of the meningococcus diversity (Maiden *et al.*, 1998). In addition, a few strains of *N. meningitidis* serogroups A, C, Y and W135 and the species *N. cinerea*, *N. lactamica* and *N. gonorrhoeae* were also included in the analysis. Each of these strains were analysed by PCR and Southern blot to evaluate the presence of the genes coding for each of the nine antigens. Remarkably, eight out of the nine genes were found to be present in all strains of serogroup B *N. meningitidis* and most of them were also present in the other serogroups, in *N. cinerea*, *N. lactamica* and *N. gonorrhoeae*.

To evaluate the sequence variability, the genes were sequenced and compared. Surprisingly, although surface-exposed proteins are generally variable because they undergo selective pressure by the human immune system, most of the antigens analysed were found to be conserved in sequence and not mosaic in structure (Figure 8.3(e)) (Pizza *et al.*, 2000). This result is particularly important since the meningococcus antigens identified in the last four decades, with only the exception of NspA (Martin *et al.*, 1997), are antigenically variable.

Overall, these data demonstrate that the antigens identified by the reverse vaccinology approach are good candidates for the clinical development of a vaccine against MenB (Pizza *et al.*, 2000; Jodar *et al.*, 2002).

8.2.3   Functional characterization of novel candidates

The analysis of a genome can allow the discovery of a high number of novel proteins that could be not only promising vaccine candidates but also important virulence factors. It is therefore crucial to further characterize the novel molecules, evaluating their potential as vaccine candidates and understanding their function and their possible role in virulence and pathogenesis. Many of the novel outer membrane or surface-exposed proteins identified in MenB share interesting homology to known virulence factors (Tettelin *et al.*, 2000; Pizza *et al.*, 2000). Among these newly identified antigens GNA33, GNA992, NadA, App and GNA1870 have been further characterized from the biochemical, immunological and functional points of view (Figure 8.5). They deserve mention here.

GNA33, a novel mimetic antigen

GNA33 (Genome-derived *Neisseria* Antigen) is a lipoprotein highly conserved among meningococcus serogroup B strains, other meningococcal serogroups and gonococcus. GNA33 shows 33 per cent identity to a membrane bound lytic transglycolase (MltA) from *E. coli*. Biochemical analysis confirmed that the

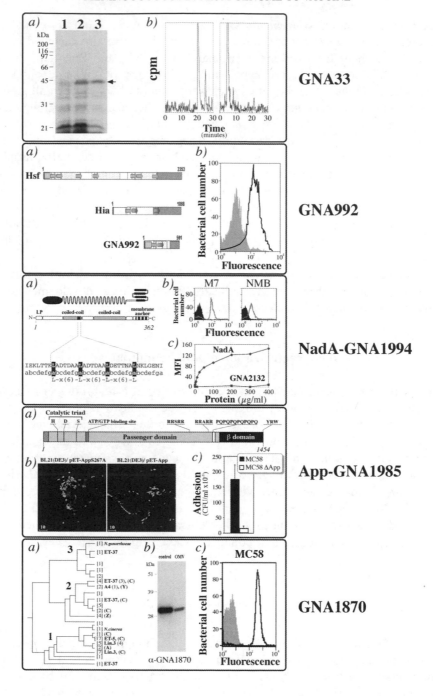

molecule is a murein hydrolase of the lytic transglycosylase class, as it is capable of degrading both insoluble murein sacculi and unsubstituted glycan strand (Jennings *et al.*, 2002; Figure 8.5). It has been shown that the recombinant GNA33 elicits antibodies that are bactericidal and able to confer passive protection in the infant rat model against strains of serosubtype P1.2, by mimicking a surface-exposed epitope on loop 4 of Por A. Epitope mapping of a bactericidal anti-GNA33 monoclonal antibody identified a short motif (QTP) present in GNA33 that is essential for antibody recognition. The QTP motif is also present in loop 4 of PorA and was also found to be essential but not sufficient for the binding of the Mab to PorA (Granoff *et al.*, 2001).

Figure 8.5   Main properties of some of the novel antigens identified. **GNA33**. (a) Autoradiography showing GNA33 lipidation when expressed as a full-length gene in *E. coli*; (b) HPLC analysis demonstrating hydrolysis of glycan strands in the presence or absence of GNA33 (left- and right-hand panels, respectively) (Jennings *et al.*, 2002). **GNA992**. (a) Schematic representation of the sequence relationship between Hsf, Hia and GNA992 proteins. The C-terminal homology is indicated in dark grey; arrows indicate the number of repeats. (b) Surface localization of GNA992 protein assayed by FACS analysis using a mouse polyclonal anti-GNA92 antiserum. Shaded and non-shaded profiles indicate binding of preimmune and immune sera, respectively (Scarselli, Rappuoli and Scarlato, 2001). **NadA-GNA1994**. (a) Schematic representation of NadA topology and the corresponding domain structure: the globular amino terminus, the amphipatic COOH terminus, the coiled-coil structure and the membrane anchor domain. (b) FACS analysis showing binding of mouse polyclonal anti-NadA antiserum to noncapsulated (M7) or capsulated (NMB) *N. meningitidis* strains. Grey profiles show binding of preimmune serum; non-shaded profiles show binding of immune serum. (c) Concentration-dependent binding of NadA to human Chang cells expressed as net mean fluorescence intensity (Comanducci *et al.*, 2002). **App-GNA1985**. (a) Schematic representation of App features. The N-terminal leader peptide (hatched), the putative passenger domain (grey) and the predicted C-terminal β-domain (black) are indicated. The predicted serine protease catalytic triad, a putative ATP/GTP binding site, two arginine-rich regions and a proline-rich region are also reported. (b) Immunofluorescence microscopy showing bacterial adherence and aggregation of recombinant *E. coli* strains expressing the wild-type App and the mutant form AppS267A (right- and left-hand panels respectively). (c) Adhesion of the wild-type *N. meningitidis* MC58 strain and the isogenic mutant strain MC58ΔApp to human Chang cells (Serruto *et al.*, 2003). **GNA1870**. (a) Dendrogram showing the strain clustering according to GNA1870 protein distances. 1, 2 and 3 indicate the three variants. Numbers in square brackets indicate the strains with identical sequence present in each branch of the dendrogram. (b) Western blot of the OMV vaccine from the Norwegian Institute of Public Health and of the recombinant GNA1870 (used as positive control) probed with anti-GNA1870 serum. (c) FACS analysis of capsulated MC58 using the serum against GNA1870. The shaded profile is the negative control obtained by reaction with the preimmune serum. The non-shaded profile shows the reaction with immune sera (Masignani *et al.*, 2003).

This case of antigen mimicry is not only interesting from the biological point of view, but could be also important for vaccine development. Although anti-GNA33 antibodies only recognize P1.2 PorA, this serosubgroup accounts for 8 per cent of disease-causing strains of *N. meningitidis* (Tondella *et al.*, 2000) and GNA33 could hence be considered as an antigen in a combination vaccine. The enormous advantage would be that, unlike PorA, GNA33 can be produced in *E. coli* and easily purified in a soluble form.

GNA992, a possible adhesin

GNA992 is surface exposed in meningococcus as shown by FACS analysis and western blot on outer membrane vesicles. Antibodies elicited by GNA992 are bactericidal against a subgroup of MenB strains and therefore this antigen is being regarded as a possible constituent of a multi-component protein-based vaccine. It appears to be strictly related to two adhesins encoded by *H. influenzae*, Hsf and its allelic variant Hia, both involved in the formation of type b fibrils (St Geme, Cutter and Barenkamp, 1996). The elevated amino acid sequence similarity of GNA992 with Hsf and Hia (57 and 51 per cent identity, respectively) and the similar topology of the three proteins suggest that they could share a common role in the mechanism of adherence (Figure 8.5).

The three adhesins are in fact composed of a different number of repeats. The core of these repetitive units has been identified and the conserved motif is present also in other adhesive molecules of *H. influenzae* (HMW1) as well as in human proteins belonging to the family of cell adhesion molecules (CAMs), such as NB-2 (Ogawa *et al.*, 1996). On the basis of these common features, GNA992 has been postulated to promote adherence of meningococcus to host cells by mimicking the cell–cell recognition phenomena that occur at the neural level (Scarselli, Rappuoli and Scarlato, 2001).

NadA: a novel Neisseria strain-specific antigen

GNA1994 was identified during the MenB genome screening as one of the proteins able to elicit the highest bactericidal response. Homology searches revealed a structural and topological similarity to virulence-associated proteins UspA2 of *Moraxella catarrhalis* (Helminen *et al.*, 1994) and *Yersinia* adhesin A (YadA) (El Tahir and Skurnik, 2001; Hoiczyk *et al.*, 2000). In spite of the low level of amino acid sequence homology, the three proteins show a well conserved secondary structure. This protein has been named NadA (Neisseria adhesin A) (Comanducci *et al.*, 2002). The protein has a C-terminal beta barrel domain necessary for anchoring to the outer membrane, and a central portion mostly characterized by a coiled-coil structure, which is possibly involved in oligomerization (Figure 8.5).

Interestingly, the NadA gene is not ubiquitous in *Neisseria* strains, being present only in 50 per cent of the strains analysed, but it is always present in strains of hyper-virulent clonal complexes ET5, ET37 and A4, whereas it is absent from Lineage III, from gonococcus and from *N. cinerea* and *N. lactamica*. The gene clusters in three well conserved alleles, whose overall identity ranges from 96 to 99 per cent. NadA is exposed on the surface of the meningococcus, where, in accordance with the described topological features, it is also able to form high-molecular-weight oligomers. Furthermore, NadA is able to bind to human cells *in vitro*, suggesting that the structural homology with YadA is also conserved at the functional level. The results suggest a possible role of this novel MenB antigen in the pathogenesis of *N. meningitidis*, particularly associated with the hypervirulent strains (Comanducci *et al.*, 2002).

App: a new adhesin of Neisseria meningitidis

GNA1985 was also identified by H. Abdel Hadi *et al.* (2001) and named adhesion and penetration protein (App). App is highly homologous to the *Haemophilus* adhesion and penetration protein (Hap) from *H. influenzae*. Both Hap and App belong to the autotransporter family that comprises proteins from Gram-negative bacteria characterized by a distinct mechanism of secretion. App protein is exported to the outer membrane, cleaved and released in the culture supernatant. App protein was functionally characterized using *E. coli* as an expression system. Mutation of serine S267 to alanine abolishes processing and secretion of App, which remains cell associated, and suggests that App has a serine protease activity. Moreover, App is an adhesin able to bind to human epithelial but not to endothelial cells. Deletion of the *app* gene in a virulent serogroup B capsulated strain (MC58) significantly reduces its adherence capability compared to the wild-type strain (Figure 8.5). The results provide evidence that App is a new adhesin, which may play a role in *N. meningitidis* colonization of the nasopharyngeal mucosa (Serruto *et al.*, 2003).

GNA1870: a novel lipoprotein as vaccine candidate

GNA 1870 is a lipoprotein of 26.9 kDa. The protein has been detected on the surface of meningococci by FACS analysis on whole-cell bacteria and western blot on OMV. Analysis of the expression on different strains has shown that protein is expressed at different levels in the different strains, which have been therefore classified as high, intermediate and low expressors. The gene is present in all strains analysed and the sequence analysis have shown that it is present as three variants, namely variants 1, 2 and 3 (Figure 8.5). Amino acid identity between variant 1 and variant 2 is 74.1 per cent, that between variant 1 and variant 3 is 62.8 per cent, and 84.7 per cent between variant 2 and variant 3. Sequences within each variant are well conserved, the most distant showing 91.6 per cent, 93.4 and 93.2 per cent identity to their type strains, respectively. The protein is able to induce antibodies with high bactericidal activity against each strain carrying the same variant even when the amount of protein expressed is low. The activity is low or absent against strains of the other variants. Furthermore, antibodies are able to induce protection in the infant rat model (Masignani *et al.*, 2003).

8.3   Conclusions

All the antigens identified as potential vaccine candidates using conventional approaches are highly variable and capable of inducing protection only against the homologous strain. By reverse vaccinology, based on the expression and evaluation as vaccine candidates of all potentially surface-exposed or secreted antigens, it has been possible to identify, in only two years, several conserved proteins capable of inducing bactericidal antibodies against different *N. meningitidis* strains. It is most likely that a combination of some of these antigens will provide an efficacious vaccine able to protect the population from this devastating disease.

Furthermore, the genome screening has also allowed the identification of novel virulence factors that are able to mediate adhesion of meningococci to epithelial cells. So far, the complete genome sequence of over 80 bacteria is available on the database, indicating that the reverse vaccinology approach can be used on a wide scale. The success of reverse vaccinology in the development of a novel vaccine against *N. meningitidis* indicates that this approach is likely to be successful for the development of vaccines against other human pathogens. Following the MenB paradigm, other groups have used this approach for the identification of vaccine antigens against major human pathogens. These antigens are currently under investigation as components of vaccines able to prevent each related disease.

In conclusion the analysis of the genomic repertoire of different pathogens can allow the identification of novel vaccine antigens and/or antigens that play key roles in virulence and pathogenesis.

References

Achtman, M., Epidemic spread and antigenic variability of *Neisseria meningitidis*. *Trends Microbiol.* **3**(5), 186–192.

Anderson, E. L. *et al.* (1994) Safety and immunogenicity of meningococcal A and C polysaccharide conjugate vaccine in adults. *Infect. Immun.* **62**(8), 3391–3395.

Borrow, R. *et al.* (2000) Reduced antibody response to revaccination with meningococcal serogroup A polysaccharide vaccine in adults. *Vaccine* **19**(9–10), 1129–1132.

Broome, C. V. (1986) The carrier state, *Neisseria meningitidis. J. Antimicrob. Chemother.* **18**(Suppl. A), 25–34.

Cadieux, N. *et al.* (1999) Bactericidal and cross-protective activities of a monoclonal antibody directed against *Neisseria meningitidis* NspA outer membrane protein. *Infect. Immun.* **67**(9), 4955–4959.

Caugant, D. A. *et al.* (1986a) Intercontinental spread of a genetically distinctive complex of clones of *Neisseria meningitidis* causing epidemic disease. *Proc. Natl. Acad. Sci. USA* **83**(13), 4927–4931.

Caugant, D. A. *et al.* (1986b) Multilocus genotypes determined by enzyme electrophoresis of *Neisseria meningitidis* isolated from patients with systemic disease and from healthy carriers. *J. Gen. Microbiol.* **132**(Pt 3), 641–652.

Comanducci, M. *et al.* (2002) NadA, a novel vaccine candidate of *Neisseria meningitidis. J. Exp. Med.* **195**(11), 1445–1454.

Costantino, P. *et al.* (1992) Development and phase 1 clinical testing of a conjugate vaccine against meningococcus A and C. *Vaccine* **10**(10), 691–698.

Danve, B. *et al.* (1998) Safety and immunogenicity of a *Neisseria meningitidis* group B transferrin binding protein vaccine in adults. In 11th International Pathogenic Neisseria Conference. Nice.

de Kleijn, E. *et al.* (2001) Serum bactericidal activity and isotype distribution of antibodies in toddlers and schoolchildren after vaccination with RIVM hexavalent PorA vesicle vaccine. *Vaccine* **20**(3/4), 352–358.

El Tahir, Y., and Skurnik, M. (2001) YadA, the multifaceted *Yersinia* adhesin. *Int. J. Med. Microbiol.* **291**(3), 209–218.

Fairley, C. K. *et al.* (1996) Conjugate meningococcal serogroup A and C vaccine, reactogenicity and immunogenicity in United Kingdom infants. *J. Infect. Dis.* **174**(6), 1360–1363.

Finne, J. *et al.* (1987) An IgG monoclonal antibody to group B meningococci cross-reacts with developmentally regulated polysialic acid units of glycoproteins in neural and extraneural tissues. *J. Immunol.* **138**(12), 4402–4407.

Frasch, C. E., Zollinger, W. D., and Poolman, J. T. (1985) Serotype antigens of *Neisseria meningitidis* and a proposed scheme for designation of serotypes. *Rev. Infect. Dis.* **7**(4), 504–510.

Goldschneider, I., Gotschlich, E. C., and Artenstein, M. S. (1969) Human immunity to the meningococcus. I. The role of humoral antibodies. *J. Exp. Med.* **129**(6), 1307–1326.

Granoff, D. M. *et al.* (2001) A novel mimetic antigen eliciting protective antibody to *Neisseria meningitidis. J. Immunol.* **167**(11), 6487–6496.

Hadi, H. A. *et al*. (2001) Identification and characterization of App, an immunogenic autotransporter protein of *Neisseria meningitidis*. *Mol. Microbiol.* **41**(3), 611–623.

Hayrinen, J. *et al*. (1995) Antibodies to polysialic acid and its N-propyl derivative, binding properties and interaction with human embryonal brain glycopeptides. *J. Infect. Dis.* **171**(6), 1481–1490.

Helminen, M. E. *et al*. (1994) A large, antigenically conserved protein on the surface of *Moraxella catarrhalis* is a target for protective antibodies. *J. Infect. Dis.* **170**(4), 867–872.

Hoiczyk, E. *et al*. (2000) Structure and sequence analysis of *Yersinia* YadA and *Moraxella* UspAs reveal a novel class of adhesins. *EMBO J.* **19**(22), 5989–5999.

Jennings, G. T. *et al*. (2002) GNA33 from *Neisseria meningitidis* serogroup B encodes a membrane-bound lytic transglycosylase (MltA). *Eur. J. Biochem.* **269**(15), 3722–3731.

Jodar, L. *et al*. (2002) Development of vaccines against meningococcal disease. *Lancet* **359**(9316), 1499–1508.

Kogan, G. *et al*. (1997) Structural basis of the *Neisseria meningitidis* immunotypes including the L4 and L7 immunotypes. *Carbohydr. Res.* **298**(3), 191–199.

Lieberman, J. M. *et al*. (1996) Safety and immunogenicity of a serogroups A/C *Neisseria meningitidis* oligosaccharide–protein conjugate vaccine in young children. A randomized controlled trial. *J. Am. Med. Assoc.* **275**(19), 1499–1503.

Lindberg, A. A. (1999) Glycoprotein conjugate vaccines. *Vaccine* **17**(Suppl 2), S28–36.

Lissolo, L. *et al*. (1995) Evaluation of transferrin-binding protein 2 within the transferrin-binding protein complex as a potential antigen for future meningococcal vaccines. *Infect. Immun.* **63**(3), 884–890.

Maiden, M. C. *et al*. (1998) Multilocus sequence typing, a portable approach to the identification of clones within populations of pathogenic microorganisms. *Proc. Natl. Acad. Sci. USA* **95**(6), 3140–3145.

Martin, D. *et al*. (1997) Highly conserved *Neisseria meningitidis* surface protein confers protection against experimental infection. *J. Exp. Med.* **185**(7), 1173–1183.

Masignani, V. *et al*. (2003) Vaccination against *Neisseria meningitidis* using three variants of the lipoprotein GNA1870. *J. Exp. Med.* **197**(6), 789–799.

McNeil, G., Virji, M., and Moxon, E. R. (1994) Interactions of *Neisseria meningitidis* with human monocytes. *Microb. Pathog.* **16**(2), 153–163.

Miller, E., Salisbury, D., and Ramsay, M. (2001) Planning, registration, and implementation of an immunisation campaign against meningococcal serogroup C disease in the UK, a success story. *Vaccine* **20**(Suppl 1), S58–67.

Moe, G. R., Tan, S., and Granoff, D. M. (1999) Differences in surface expression of NspA among *Neisseria meningitidis* group B strains. *Infect. Immun.* **67**(11), 5664–5675.

Moe, G. R. *et al*. (2001) Functional activity of anti-Neisserial surface protein A monoclonal antibodies against strains of *Neisseria meningitidis* serogroup B. *Infect. Immun.* **69**(6), 3762–3771.

Morley, S. L., and Pollard, A. J. (2001) Vaccine prevention of meningococcal disease, coming soon? *Vaccine* **20**(5–6), 666–687.

Nassif, X. (1999) Interaction mechanisms of encapsulated meningococci with eucaryotic cells, what does this tell us about the crossing of the blood–brain barrier by *Neisseria meningitidis? Curr. Opin. Microbiol.* **2**(1), 71–77.

Ogawa, J. *et al.* (1996) Novel neural adhesion molecules in the Contactin/F3 subgroup of the immunoglobulin superfamily, isolation and characterization of cDNAs from rat brain. *Neurosci. Lett.* **218**(3), 173–176.

Peltola, H. (1983) Meningococcal disease, still with us. *Rev. Infect. Dis.* **5**(1), 71–91.

Pizza, M. *et al.* (2000) Identification of vaccine candidates against serogroup B meningococcus by whole-genome sequencing. *Science* **287**(5459), 1816–1820.

Ramsay, M. E. *et al.* (2001) Efficacy of meningococcal serogroup C conjugate vaccine in teenagers and toddlers in England. *Lancet* **357**(9251), 195–196.

Rappuoli, R., (2000) Reverse vaccinology. *Curr. Opin. Microbiol.* **3**(5), 445–450.

Rappuoli, R., (2001) Reverse vaccinology, a genome-based approach to vaccine development. *Vaccine* **19**(17–19), 2688–2691.

Rosenstein, N. E. *et al.* (2001) Meningococcal disease. *N. Engl. J. Med.* **344**(18), 1378–1388.

Rosenstein, N. E., Fischer, M., and Tappero, J. W. (2001) Meningococcal vaccines. *Infect. Dis. Clin. North Am.* **15**(1), 155–169.

Scarselli, M., Rappuoli, R., and Scarlato, V. (2001) A common conserved amino acid motif module shared by bacterial and intercellular adhesins, bacterial adherence mimicking cell cell recognition? *Microbiology* **147**(Pt 2), 250–252.

Scholten, R. J. *et al.* (1993) Meningococcal disease in The Netherlands, 1958–1990, a steady increase in the incidence since 1982 partially caused by new serotypes and subtypes of *Neisseria meningitidis*. *Clin. Infect. Dis.* **16**(2), 237–246.

Scholten, R. J. *et al.* (1994) Lipo-oligosaccharide immunotyping of *Neisseria meningitidis* by a whole-cell ELISA with monoclonal antibodies. *J. Med. Microbiol.* **41**(4), 236–243.

Schryvers, A. B., and Morris, L. J. (1988) Identification and characterization of the transferrin receptor from *Neisseria meningitidis*. *Mol. Microbiol.* **2**(2), 281–288.

Schryvers, A. B., and Stojiljkovic, I. (1999) Iron acquisition systems in the pathogenic *Neisseria*. *Mol. Microbiol.* **32**(6), 1117–1123.

Schuchat, A. *et al.* (1997) Bacterial meningitis in the United States in 1995. Active Surveillance Team. *N. Engl. J. Med.* **337**(14), 970–976.

Schwartz, B., Moore, P. S., and Broome, C. V. (1989) Global epidemiology of meningococcal disease. *Clin. Microbiol. Rev.* **2**(Suppl.), S118–124.

Serruto, D. *et al.* (2003) *Neisseria meningitidis* App, a new adhesin with autocatalytic serine protease activity. *Mol. Microbiol.* **48**(2), 323–334.

St Geme, J. W. III, Cutter, D., and Barenkamp, S. J. (1996) Characterization of the genetic locus encoding *Haemophilus influenzae* type b surface fibrils. *J. Bacteriol.* **178**(21), 6281–6287.

Tappero, J. W. *et al.* (1999) Immunogenicity of 2 serogroup B outer-membrane protein meningococcal vaccines, a randomized controlled trial in Chile. *J. Am. Med. Assoc.* **281**(16), 1520–1527.

Tettelin, H. *et al.* (2000) Complete genome sequence of *Neisseria meningitidis* serogroup B strain MC58. *Science* **287**(5459), 1809–1815.

Tondella, M. L. *et al.* (2000) Distribution of *Neisseria meningitidis* serogroup B serosubtypes and serotypes circulating in the United States. The Active Bacterial Core Surveillance Team. *J. Clin. Microbiol.* **38**(9), 3323–3328.

Virji, M. *et al.* (1992) Expression of the Opc protein correlates with invasion of epithelial and endothelial cells by *Neisseria meningitidis*. *Mol. Microbiol.* **6**(19), 2785–2795.

Virji, M. *et al.* (1993) Meningococcal Opa and Opc proteins, their role in colonization and invasion of human epithelial and endothelial cells. *Mol. Microbiol.* **10**(3), 499–510.

Vogel, U., and Frosch, M. (1999) Mechanisms of neisserial serum resistance. *Mol. Microbiol.* **32**(6), 1133–1139.

9

Vaccines Against Pathogenic Streptococci

John L. Telford[1], Immaculada Margarit y Ros,
Domenico Maione, Vega Masignani, Hervé Tettelin,
Giuliano Bensi and Guido Grandi

9.1 Introduction

Streptococci are responsible for a substantial amount of morbidity and mortality, particularly in children and in elderly adults. There are three major human pathogens among the streptococci, *S. pneumoniae*, *S. agalactiae* and *S. pyogenes*, which, although related, colonize different mucosal niches and cause quite different pathologies. *S. pneumoniae* colonizes the upper respiratory tract and sinuses and is the most frequent cause of inner ear infections in children but can also descend to the lungs causing bacterial pneumonia, particularly in the elderly (Cartwright, 2002). On the other hand, *S. agalactiae*, also known as group B streptococcus (GBS), colonizes the digestive tract and vaginal passage and is today the major cause of neonatal sepsis (Schuchat, 1999). GBS is also responsible for substantial morbidity in the elderly, particularly among people with underlying diseases such as diabetes and cancer. *S. pyogenes* (group A streptococcus) is however probably the most feared of the pathogenic streptococci, causing a range of pathologies from simple inflammation of the throat and of the tonsils (strep throat) to life threatening infections such as streptococcal toxic shock syndrome (STSS) and necrotizing fasciitis (Cunningham,

[1] Correspondence to: John L. Telford. Email address: john_telford@chiron.it

Genomics, Proteomics and Vaccines edited by Guido Grandi
© 2004 John Wiley & Sons, Ltd   ISBN 0 470 85616 5

2000). GAS can also result in the autoimmune sequelae involved in rheumatic fever and glomerulonephritis. Complete genome sequences for each of these pathogens have been determined (Baltz *et al.*, 1998; Beres *et al.*, 2002; Dopazo *et al.*, 2001; Ferretti *et al.*, 2001; Glaser *et al.*, 2002; Smoot *et al.*, 2002; Tettelin *et al.*, 2002, 2001), leading to novel strategies to prevent these diseases. Recently a novel vaccine against *S. pneumoniae*, based on glycoconjugates of the major capsular polysaccharide serotypes, was introduced, resulting in a 90 per cent reduction in invasive pneumococcal disease caused by the serotypes in the vaccine (Shinefield and Black, 2000). Vaccines against GBS and GAS are however still not available. This chapter will focus on the use of genomic sequence information to identify novel vaccine candidates for these two remaining pathogens.

9.2   Comparative genomics of streptococci

As mentioned above, complete genome sequences are available for each of the three species of pathogenic streptococci. In fact, the genome sequence has been determined for three strains of *S. pneumoniae* (Baltz *et al.*, 1998; Dopazo *et al.*, 2001; Tettelin *et al.*, 2001), two strains of *S. agalactiae* (Glaser *et al.*, 2002; Tettelin *et al.*, 2002) and three strains of *S. pyogenes* (Beres *et al.*, 2002; Ferretti *et al.*, 2001; Smoot *et al.*, 2002). Perhaps the most remarkable observation is that the gene complement of these three species is widely different. While each genome contains around 2000 open reading frames (ORFs), the three species share only 1060 genes (Figure 9.1). One may consider these 1060 genes to be the 'core' genome and together to define the biological nature of the streptococci. However, it cannot be concluded that these genes would be sufficient to support a viable bacterium. It is likely that some species-specific genes are essential for viability in that species whose function is fulfilled or substituted by other genes specific to the other species. Nevertheless, genes coding for functions involved in the specific colonization, invasion and disease characteristics of each species are more likely to be found among the species-specific genes. In addition, quite a number of genes have been duplicated in each species after the separation from the others (130 in GBS) (Tettelin *et al.*, 2002). Among these may be genes which have evolved a species-specific function after duplication.

GBS shares 225 genes with GAS and 174 genes with *S. pneumoniae*, whereas the latter two species share only 74 genes, suggesting that GBS is more closely related to GAS. This is supported by the level of gene synteny between GBS and GAS. In fact, 828 genes in 35 regions are arranged in the

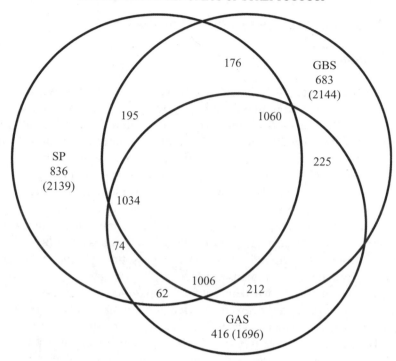

Figure 9.1 *In silico* comparisons between streptococci (modified from Tettelin *et al.*, 2002). The protein sets of *S. agalactiae* (group B *Streptococcus*, GBS), *S. pneumoniae* (SP) and *S. pyogenes* (group A *Streptococcus*, GAS) were compared by using FASTA3. Numbers under the species name indicate genes that are not shared with the other species; value in parentheses are the number of proteins in each species (excluding frameshifted and degenerated genes). Numbers in the intersections indicate genes shared by two or three species. Numbers in any given intersection are slightly different depending on the species used as the query due to gene duplications in some species

same order in GBS and GAS, whereas only 128 genes in nine regions are syntenic between GBS and *S. pneumoniae*. It is noteworthy that many of the species-specific genes are in regions of the genome with characteristics of mobile genetic elements, indicating that they may have been acquired by horizontal transfer of DNA from other bacteria, including other genera (Tettelin *et al.*, 2002).

9.3   A vaccine against group B streptococcus

9.3.1   Rationale

As mentioned above, GBS is the major cause of neonatal sepsis in the industrialized world, accounting for 0.5–3.0 deaths/1000 live births. Eighty per cent of the GBS infections in newborns occur within the first 24–48 h after delivery (Schuchat, 1998). This group is known as early onset disease and is generally caused by direct transmission of the bacteria from the mother to the baby during passage through the vagina. A second peak of infections, which begins a week after birth and continues through the first month of life, is known as late onset disease and is not associated with colonization of the mother. This latter group is often caused by infection acquired from other sources in the clinic. Protection against invasive GBS disease correlates with high titres of maternal anti-capsule antibodies, which can pass through the placenta and protect the child in the first months of life.

Experiments in mice have demonstrated that glycoconjugates of capsular polysaccharide with tetanus toxoid carrier protein can induce an immune response in pregnant females which can confer protection in the pups against lethal GBS challenge (Paoletti *et al.*, 1994). These data suggest that a strategy of immunizing women before they become pregnant could effectively prevent the majority of invasive GBS disease in newborns. Unfortunately, there are at least nine capsular serotypes and antibodies against any one of these fail to confer protection against the other serotypes (Berg *et al.*, 2000; Davies *et al.*, 2001; Hickman *et al.*, 1999; Lin *et al.*, 1998; Suara *et al.*, 1998). Hence a vaccine based on capsular polysaccharide would need to contain several serotype-specific antigens. Furthermore, although conjugation of the polysaccharide with a protein carrier helps to induce immunologic memory (Paoletti *et al.*, 1996), the antibody titres obtained are usually not maintained for periods of time as long as those against classical protein antigens. This would likely mean that women immunized at adolescence would require a booster immunization shortly before or during pregnancy. For these reasons we decided to try to identify protective antigens among the proteins expressed on the surface of the GBS bacteria.

Our strategy to identify vaccine candidates was to start from the available genome sequence information and to use the approach termed 'reverse vaccinology', which is described in detail elsewhere in this book. In collaboration with TIGR, we determined the complete genome sequence of serotype V strain 2603v/r of GBS (Tettelin *et al.*, 2002). A circular representation of the genome is shown in Figure 9.2. Modern computer algorithms, based on current knowledge of microbial biology, were used to identify in the genome genes coding

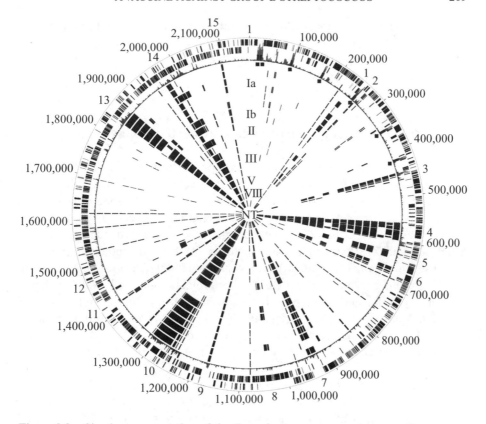

Figure 9.2 Circular representation of the *S. agalactiae* genome and comparative genome hybridizations using microarrays (modified from Tettelin *et al.*, 2002). Outer circle: predicted coding regions on the plus strand. Second circle: predicted coding regions on the minus strand. Third circle: black, atypical nucleotide composition curve; squares indicate rRNAs. Circles 4–22: comparative genome hybridizations of strain 2603V/R with 19 *S. agalactiae* strains. Cy3/Cy5 (2603 signal/test strain) ratio cut-offs were defined arbitrarily as Cy3/Cy5 = 1.0–3.0, gene present in test strain (not displayed); >3.0, gene absent or highly diverged in test strain (black line). Circles 4–9: type Ia strains 090, 515, A909, Davis, DK1, DK8; 10, 11: type Ib (S7) 7357b and H36B; 12, 13: type II 18RS21 and DK21; 14–18: type III COH1, COH31, D136C, M732 and M781; 19: type V strain CJB111; 20, 21: type VIII strains SMU014 and JM9130013; 22: non-typable (NT) strain CJB110. Varying regions of five or more consecutive genes are indicated by numbers

for proteins that are likely to be secreted from the bacteria or associated with the bacterial surface. These genes were then cloned in *E. coli* in order to obtain recombinant proteins. The expression vectors were chosen to produce fusion proteins with either an N-terminal tag of six histidine residues or a C-terminal domain corresponding to glutathione synthase transferase (GST) to facilitate rapid purification on IMAC or glutathione affinity columns, respectively. Each of the expressed proteins was then used to immunize mice to assay their capacity to induce protective immunity.

9.3.2   *In silico* identification of surface-exposed proteins

The complete genome sequence of a serotype V strain of GBS (2603v/r) was determined (Tettelin *et al.*, 2002). The genome consists of 1 160 267 base pairs and contains 2175 genes. Biological roles were assigned to 1333 (61 per cent) of the predicted proteins according to the classification scheme adapted from Riley (1993). A further 623 matched genes of unknown function in the databases and 219 had no database matches. A series of computer programs was used to identify potential signal peptides (PSORT, SignalP), transmembrane spanning regions (TMPRED), lipoproteins and cell wall anchored proteins (Motifs) and homology to known surface proteins in other bacteria (FastA). Approximately 650 surface-predicted proteins were identified by these means. In addition, genes coding for proteins homologous to known virulence factors in other bacterial species were selected. A total of 681 genes were thus selected.

From our previous experience in similar projects involving the expression in *E. coli* of large numbers of recombinant proteins, we knew that proteins that contain several transmembrane spanning regions are extremely difficult if not impossible to express. For this reason, we excluded from our list genes that were predicted to code for five or more transmembrane regions. The final list thus contained 473 GBS genes.

9.3.3   High-throughput cloning and expression

Each of the 473 genes was amplified by PCR and cloned in parallel into *E. coli* expression vectors containing sequences coding either for an N-terminal 6xHis tag or for a C-terminal GST domain. The reason for the parallel cloning in the two distinct vectors was to maximize the possibility of obtaining soluble recombinant proteins. Our strategy for selecting which clones to use was based on production of soluble antigen. If the 6xHis tagged protein was soluble, it was used directly. If not, we tested the GST fusion version of the protein, which, if soluble, was used. If neither of the fusion proteins were soluble, we attempted to obtain soluble material by resuspending the 6xHis tagged recombinant with guanidinium hydrochloride then stepwise dialysis to remove the denaturant. The rate of success of each of the steps in this strategy is shown in Figure 9.3.

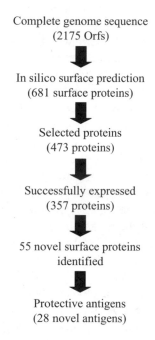

Complete genome sequence
(2175 Orfs)

In silico surface prediction
(681 surface proteins)

Selected proteins
(473 proteins)

Successfully expressed
(357 proteins)

55 novel surface proteins
identified

Protective antigens
(28 novel antigens)

Figure 9.3 Schematic representation of the stepwise approach to identification of GBS protective antigens

9.3.4   Primary screening of antigens

In order to obtain specific immune sera, each of the 357 successfully expressed recombinant proteins was used to immunize groups of four CD1 mice. This strain of mice was chosen because it is outbred and thus there was less risk of lack of immune response due to the MHC restriction of inbred strains. The sera obtained were then assayed in immunoblot of GBS protein extracts to identify the natural protein, by flow cytometry to assess surface exposure and in a mouse protection model.

Of the 357 sera produced, 139 recognized a predominant band corresponding to the expected molecular weight in immunoblots of protein extracts of GBS. Many others recognized multiple bands on the gels or bands with electrophoretic mobility incompatible with the predicted protein product of the genes expressed. This discrepancy in results from those expected may be due to postranslational modification of the proteins in their natural host or the recognition of multiple members of related proteins. Of course, artifactual cross-reaction cannot be ruled out. The lack of recognition of any band in some cases may be due to the limits of detection of the assay or may indicate that the corresponding genes are not expressed in the laboratory culture conditions used to grow the bacteria. Of the 139 sera which recognized a specific band in immunoblot, 55 also gave a significant signal in indirect fluorescent flow cytometry. Here the discrepancy may likewise be due to the limits of detection but may also indicate masking of the surface-associated proteins by cell wall and capsule. Nevertheless, we have identified 55 proteins in GBS that are clearly associated with the surface and exposed to the interaction with specific antibodies (Tettelin *et al.*, 2002).

The primary model used to screen the antigens for their capacity to induce immunological protection was passive serological protection of newborn mice against a lethal challenge of virulent GBS of serotype III (COH1). This strain was selected because it is a well characterized strain and is very virulent in mice. It should be noted that the recombinant antigens used were derived from PCR reactions using the sequenced strain, which is a serotype V strain. Thus, already in the primary screen, we were testing the antigens for their capacity to elicit cross-protection against challenge with a different serotype, a prerequisite for a broadly protective vaccine. Briefly, newborn mice within the first 24 h of life were injected subcutaneously with a small quantity of antisera raised against the recombinant proteins. Four hours later, the mice were challenged, again subcutaneously but at a different site, with a dose of live GBS calculated to kill around 75 per cent of the mice. As controls, preimmune sera from each group were used. Of the 357 immune sera tested, 32 gave greater than 30 per cent protection, taken as an arbitrary cut-off, of the mice compared with the preimmune control.

9.3.5   Confirmation of protection

In order to confirm the capacity of our antigens to induce protective immunity and to extend the results to a model that reflects better the requirements of an anti-GBS vaccine, we tested the antigens in an active maternal immunization model (Paoletti *et al.*, 1994). Female mice were first immunized with the recombinant antigens, then mated and the resulting offspring were challenged with a lethal dose of GBS within the first 48 h of life. This is quite an exacting model. It requires that the immunization induces a sufficiently high level of antibodies of the IgG1 subclass that can cross the placenta and confer protection to the newborn pups. Using this model, 23 of the recombinant antigens identified in the primary screen induced 30–80 per cent protection of the challenged pups.

While these results were very encouraging, it was clear that the level of protection obtained with a single protein antigen was in most cases significantly lower than the levels of serotype-specific protection obtained using capsular polysaccharide conjugate antigens. The major immunologic correlate with protection in these models and probably in humans is the induction of anti-bodies capable of causing complement-dependent opsonophagocytic killing of the bacteria in the presence of human neutrophils (Baker *et al.*, 1988). We reasoned therefore that the lower levels of the protection obtained with the protein antigens may reflect the lower number of single protein antigens on the surface of the bacteria compared with the number of capsule epitopes available. This suggested that using more than one protein antigen might increase the level of protection obtained. To test this hypothesis, we immunized female mice with all possible triple combinations of five of the single candidate antigens and challenged the offspring as before. Table 9.1 shows the results of these experiments. Five of the 10 combinations gave more than 95 per cent protection and all but one combination gave higher levels of protection than the single antigens alone. These data provide strong indication that an efficacious vaccine for use in man may be developed based on the antigens identified in this genome wide screen.

Table 9.1 Protection against lethal GBS challenge of pups delivered to mice immunized with combinations of three antigens

Antigen combination	Dead/challenged	Survival %
80 + 338 + 330	1/40	97.5
80 + 338 + 104	1/34	97
80 + 338 + 404	0/30	100
80 + 104 + 404	0/24	100
80 + 330 + 104	2/40	95
80 + 330 + 404	3/28	89
338 + 330 + 104	8/30	73
338 + 330 + 404	31/36	14
338 + 104 + 404	13/37	65
330 + 104 + 404	2/10	80
80 alone (best)	10/34	70.5
mock vaccine	23/26	11.5
GBS whole cell	1120	45

9.3.6   Strain variation in antigen sequences

Obviously, for a vaccine to give broad protective coverage against the majority of the circulating disease causing strains, the antigens must be conserved and expressed by these strains. We have taken two approaches to defining antigen variation in GBS strains. These involve on the one hand gene chip technology, and on the other high-throughput PCR and direct gene sequencing.

In the first approach, we prepared a gene chip microarrays corresponding to every gene in the genome of the completely sequenced serotype V strain of GBS. These chips were then hybridized to labelled total DNA extracted from 20 independently isolated strains of GBS reflecting most of the known serotypes (Tettelin *et al.*, 2002). In these experiments, hybridization was performed using the test DNA labelled with a red fluorescent dye mixed with the DNA from the strain used to create the chip labelled with a green fluorescent dye. Genes present in both strains would therefore give a hybridization signal equally intense in both colours, whereas genes lacking in the test DNA or sufficiently diverged to reduce hybridization efficiency would show a biased level of staining in favour of the standard DNA. In these experiments, we considered genes that gave hybridization ratios of between 1.0 and 3.0 to be present in both strains and greater than 10.0 to be absent from the test strain. Ambiguous values of between 3.0 and 10.0 may have been due to highly diverged genes or hybridization to paralogous genes.

The results of these experiments revealed 1698 genes to be conserved in all 20 strains tested. 401 genes were lacking or highly diverged in at least one strain tested. Interestingly, more than 90 per cent of the genes absent from one or more strains were clustered in 15 discrete regions (Figure 9.3). Ten of these regions display an atypical nucleotide composition in the sequenced strain, suggesting that they may have been acquired laterally in this strain. In support of this hypothesis, two of the regions also have characteristics of a prophage and a transposon. These data allowed us to further refine our vaccine candidate shortlist by eliminating antigens whose genes were not present in at least a majority of the strains tested.

Surface-exposed proteins in pathogenic bacteria tend to be highly variable, particularly in obligate pathogens. This is most likely due to the selective pressure of the host immune response. To determine the extent of possible antigenic variation in the remaining candidates, the gene for each was amplified from a subset of 11 strains. In parallel, eight housekeeping genes and a further 11 genes coding for surface-exposed proteins were amplified and sequenced from the same strains. Remarkably, the candidate antigens and the other surface-exposed proteins were found to be not significantly more divergent than the housekeeping genes. All genes sequenced coded for proteins with greater than 96 per cent amino acid identity across strains. A possible explanation of the high conservation of the surface proteins in this bacterium is that it is not an obligate pathogen. In fact, GBS exists as a commensal in the digestive tract of between 30 and 50 per cent of humans, and is an opportunistic pathogen when it colonizes the vagina of women during delivery and can infect unprotected newborn babies. On the other hand, some major antigenic proteins such as the Rib protein and C-protein are both highly repetitive in structure and highly divergent, suggesting that these at least are under selective pressure. Regardless of the reason, the fact that several protective antigens could be found that are highly conserved over a broad range of strains is certainly good news for the development of a protein-based vaccine capable of protecting against all serotypes and at least the majority of circulating strains of GBS.

Currently, experiments are underway to define the best combination among the protective antigens. Once the basic components of the vaccine are defined, the mode of delivery and an appropriate adjuvant for use in man must be selected. For these experiments, the active maternal immunization model will initially be used and at later stages the vaccine may be tested in larger animals. The strategy to be used to demonstrate safety and efficacy is still a major outstanding issue. There may be some reluctance to test a novel vaccine in pregnant women, where efficacy in preventing neonatal disease could in principle be demonstrated. Capsule glycoconjugates have however been shown to be safe in pregnant women in trials overseen by the NIH of the USA. An alternative may be to test the vaccine in elderly people, where the toll of GBS invasive disease is increasing significantly.

9.4   A vaccine against group A streptococcus

Our success in identifying protective protein antigens for GBS prompted us to undertake a similar genome wide search for protective antigens for GAS. The diseases caused by these two closely related pathogens are in some respects similar but in other respects very different. GAS colonizes the throat and causes relatively mild inflammation and soreness. This is the classical strep throat that is the bane of parents of young children. Although, in most cases, the disease is mild, it is nevertheless very costly in terms of health care visits and days of school lost (Cunningham, 2000). Like GBS, GAS can also cause severe invasive disease including scarlet fever and a frequently lethal toxic shock syndrome. Perhaps of more importance are the autoimmune sequelae which frequently follow scarlet fever, rheumatic fever and glomerulonephritis. Rheumatic fever is still a major cause of cardiac pathology in children. Last but not least, GAS is one of the major causes of necrotizing fasciitis, the infamous flesh eating bacteria, which can turn a small wound into massive necrosis and necessitates emergency intervention including extensive surgical intervention and tissue reconstruction. There is no vaccine available for GAS and attempts to identify conserved protective antigens have so far been unsuccessful (Dale, 1999). Although encapsulated, the capsular polysaccharide is hyaluronic acid, which is found in many human tissues and is thus not immunogenic. A major immunodominant antigen, the M-protein has shown type-specific protection in animal models. However there are over 100 known serotypes of this protein. As a minimum, a vaccine against GAS should protect against invasive disease. However, a vaccine capable of preventing colonization of the throat and inflammation in the tonsils would be of tremendous advantage. Hence, vaccination aimed at developing protective mucosal immunity would be of enormous value.

Since our first success in taking a genome wide screening approach to the development of a vaccine against serogroup B meningococcus (Pizza *et al.*, 2000), we have gradually refined our strategy to make the initial selection and screening of potential antigen candidates more efficient. In the GBS project, we eliminated from the *in silico* surface predictions all genes coding for proteins with more than four predicted transmembrane spanning regions, since we knew from the serogroup B meningococcus project that these genes are extremely difficult if not impossible to express in *E. coli*. This led to a reduction of approximately 30 per cent in the number of genes that were subsequently cloned and expressed. The success rate for expression of the selected antigens was thus significantly enhanced. For the GAS project, in addition to eliminating these genes from consideration, we used a second criterion based on microarray analysis of the transcriptome of the bacteria. A retrospective analysis of the protective antigens discovered in the meningococcus B project revealed that all were expressed at levels above the median of the expression of all genes in the genome. In other words, genes that were expressed at very low levels, corresponding to only a few protein molecules per cell, produced poorly protective antigens. This argument is not particularly counterintuitive.

To assess the level of expression of each gene in the genome, DNA microarrays representing each and every open reading frame in the sequenced genome were hybridized with total labelled RNA extracted from exponentially growing GAS bacteria. To eliminate variation due to individual RNA preparations, two independent RNA preparations from two independent cultures were labelled with two different fluorescent dyes and used together in the hybridizations. A graph of the signal obtained on each gene for the two dyes gave a diagonal distribution reflecting the level of expression of the gene. By selecting only genes whose level of expression was above the median level of expression of all genes, we reduced the number of candidates for cloning and expression by approximately 50 per cent.

Obviously, these experiments only assessed the level of expression of the genes in the *in vitro* culture conditions used to prepare the RNA. These conditions may not necessarily reflect the specific gene expression and growth in the course of *in vivo* infection. To address this variable, RNA was extracted from bacteria isolated from the bloodstream of lethally infected mice and used in microchip hybridizations with RNA extracted from *in vitro* grown bacteria. Genes that were predicted to code for surface-exposed proteins and that were found to be up-regulated during *in vivo* growth were thus added to the list of potential vaccine candidates.

In silico prediction algorithms initially identified 684 genes that produced products likely to be secreted or associated with the bacterial surface. Of these, 207 were predicted to contain more than two transmembrane spanning regions and were thus removed from the list of candidates. Microarray analysis of the mRNA extracted from *in vitro* grown bacteria revealed that 267 of the candidate genes were expressed below the median level of expression in the genome and were thus also removed from the list. To these 210 genes were added a further 76 whose expression was up-regulated *in vivo*. Hence the final candidate list for expression in *E. coli* contained 286 genes.

As with the GBS candidates, each GAS gene was cloned in parallel in two vectors containing either sequences coding for six histidine residues or for GST. 285 recombinant proteins were successfully cloned and expressed in *E.coli*; however, about 40 were insoluble and had to be solubilized in urea before use. The success rate of protein expression in this project was significantly higher than in our previous whole-genome screens for vaccine candidates. This is certainly due to the procedures used to eliminate potentially difficult genes.

In order to test the candidates for their capacity in induce protective immunity, it was necessary to develop a mouse model of immunization and challenge of sufficient simplicity to permit the testing of a large number of antigens. For this, we decided to use an adult mouse model of invasive disease. Initially, we compared Balb/c and CD1 mice, and either intraperitoneal challenge or subcutaneous challenge. A strain of GAS that was known to be very virulent in mice was chosen and the LD50 was determined in both strains of mice. Comparison of the results revealed that IP challenge in CD1 mice gave by far the most consistent results and this model was chosen for the experiments. We next established that protective immunity could be achieved in this model. To do this, we immunized mice with recombinant M-protein before challenge. As mentioned above, the abundantly expressed, surface-exposed M-protein was known to confer protection against GAS challenge but only against strains of the same M-serotype. In fact, immunization with the recombinant M-protein cloned from the challenge strain consistently conferred 100 per cent protection. We did not, however, expect other surface proteins, which are expressed at much lower levels on the surface of the bacteria, to be as effective protective antigens. We therefore assessed the capacity of a second M-protein cloned from a different M-serotype in this model. As expected, this heterologous M-protein was much less effective at inducing protective immunity. Finally we defined challenge conditions in which 90 per cent of mock immunized mice were killed, 100 per cent homologous M-protein immunized mice survived and between 60 and 80 per cent of heterologous M-protein immunized mice survived. The model was found to be highly reproducible and the use of the three control groups allowed the elimination of occasional experiments in which the challenge dose was too high or too low.

Using this model, 285 antigens have been tested and over 30 have shown indications of protective efficacy. Currently, combinations of antigens are being tested in order to define vaccine components that will synergize in their capacity to induce protective immunity. In parallel, the conservation of the antigens and their expression in a large panel of clinical isolates are being assayed. Future work will address the efficacy of vaccine candidates in more relevant small–animal models. For example, models of mucosal (intranasal) (Guzman et al., 1999) challenge and of subcutaneous challenge that leads to necrotizing colonization (Ashbaugh et al., 1998) are available. We expect that these experiments will lead to the definition of a candidate vaccine for development and testing in human clinical trials.

9.5   Conclusions

The development of extremely rapid DNA sequencing technology has led to the possibility to obtain complete genome sequences of bacterial pathogens in very short periods of time and at almost negligible cost. The availability of these genome sequences holds the promise of rapid advances in combating infectious disease. The genome wide screening strategy dubbed 'reverse vaccinology' and pioneered in the development of novel vaccine candidates against serogroup B meningococcus (Pizza et al., 2000) is the first practical application of this new wealth of knowledge. Our progress to date in our programmes to use this strategy to develop vaccines against streptococcal pathogens extends its validity to Gram-positive bacteria and together with other ongoing projects demonstrates the advantages of the approach in general.

The bottleneck in identifying protective antigens in a pathogen is now the animal models required for the screening. These must be relatively simple in order to screen a large number of antigens but must also be sufficiently reliable to show protection with small numbers of animals. For the streptococci, these models have bean found and used. For other pathogens, this may be more difficult. Future advances in vaccine design will come from a better understanding of the immunology of protection. In some cases, as for the meningococci, a good antibody-mediated, complement-dependent bactericidal activity is a strong correlate of protection and as such is accepted by regulatory agencies as a surrogate for protection in clinical trials. Unfortunately, this assay does not work with the streptococci and with Gram-positive bacteria generally. A related assay, which measures complement-dependent opsonophagocytic activity with human neutrophils, may be an alternative for Gram positives but the assay is still clumsy and requires the use of human blood.

A better understanding of the microbiological and immunological characteristics that make some antigens better at inducing protective responses may lead to better prediction of the outcome of immunization. One clearly important characteristic is exposure on the surface of the bacteria. This is still difficult to predict precisely, particularly in Gram-positive bacteria, but also in capsulated Gram negatives, where cell wall proteoglycans and capsule can mask surface antigens. This is not, however, the only problem. It is clear from our screens of streptococcal antigens that some may be easily detected on the bacteria by flow cytomeometry using immune sera but still not induce protective immunity. Similar observations have been made in genome wide screens for protective antigens in other bacteria (see chapters on meningococcus and chlamydia in this book). Surprisingly, antibody responses to different exposed epitopes in the same protein can vary dramatically in their capacity to confer protection. A careful study of the antigens that confer protection identified in our genome wide screens may throw light on the qualities of the antigens involved in protection and together with a better understanding of the nature of the protective immune response may ultimately permit the prediction of good vaccine candidates without the need to use large numbers of experimental animals. Who knows, one day in the not too distant future it may be possible to put bacteria in a machine that will determine the complete genome sequence then use sophisticated computer algorithms based on our knowledge of bacterial pathogenesis and immunology to predict a small number of highly efficacious vaccine candidates.

References

Ashbaugh, C. D., Warren, H. B., Carey, V. J., and Wessels, M. R. (1998) Molecular analysis of the role of the group A streptococcal cysteine protease, hyaluronic acid capsule, and M protein in a murine model of human invasive soft-tissue infection. *J. Clin. Invest.* **102**, 550–560.

Baker, C. J., Rench, M. A., Edwards, M. S., Carpenter, R. J., Hays, B. M., and Kasper, D. L. (1988) Immunization of pregnant women with a polysaccharide vaccine of group B streptococcus. *N. Engl. J. Med.* **319**, 1180–1185.

Baltz, R. H., Norris, F. H., Matsushima, P., DeHoff, B. S., Rockey, P., Porter, G., Burgett, S., Peery, R., Hoskins, J., Braverman, L., Jenkins, I., Solenberg, P., Young, M., McHenney, M. A., Skatrud, P. L., and Rosteck, P. R. Jr. (1998) DNA sequence sampling of the Streptococcus pneumoniae genome to identify novel targets for antibiotic development. *Microb. Drug Resist.* **4**, 1–9.

Beres, S. B., Sylva, G. L., Barbian, K. D., Lei, B., Hoff, J. S., Mammarella, N. D., Liu, M. Y., Smoot, J. C., Porcella, S. F., Parkins, L. D., Campbell, D. S., Smith, T. M., McCormick, J. K., Leung, D. Y., Schlievert, P. M., and Musser, J. M. (2002) Genome sequence of a serotype M3 strain of group A Streptococcus: phage-encoded toxins, the high-virulence phenotype, and clone emergence. *Proc. Natl. Acad. Sci. USA* **99**, 10 078–10 083.

Berg, S., Trollfors, B., Lagergard, T., Zackrisson, G., and Claesson, B. A. (2000) Serotypes and clinical manifestations of group B streptococcal infections in western Sweden. *Clin. Microbiol. Infect.* **6**, 9–13.

Cartwright, K. (2002) Pneumococcal disease in western Europe: burden of disease, antibiotic resistance and management. *Eur. J. Pediatr.* **161**, 188–195.

Cunningham, M. W. (2000) Pathogenesis of group A streptococcal infections. *Clin. Microbiol. Rev.* **13**, 470–511.

Dale, J. B. (1999) Group A streptococcal vaccines. *Infect. Dis. Clin. North Am.* **13**, 227–243, viii.

Davies, H. D., Raj, S., Adair, C., Robinson, J., and McGeer, A. (2001) Population-based active surveillance for neonatal group B streptococcal infections in Alberta, Canada: implications for vaccine formulation. *Pediatr. Infect. Dis. J.* **20**, 879–884.

Dopazo, J., Mendoza, A., Herrero, J., Caldara, F., Humbert, Y., Friedli, L., Guerrier, M., Grand-Schenk, E., Gandin, C., de Francesco, M., Polissi, A., Buell, G., Feger, G., Garcia, E., Peitsch, M., and Garcia-Bustos, J. F. (2001) Annotated draft genomic sequence from a Streptococcus pneumoniae type 19F clinical isolate. *Microb. Drug Resist.* **7**, 99–125.

Ferretti, J. J., McShan, W. M., Ajdic, D., Savic, D. J., Savic, G., Lyon, K., Primeaux, C., Sezate, S., Suvorov, A. N., Kenton, S., Lai, H. S., Lin, S. P., Qian, Y., Jia, H. G., Najar, F. Z., Ren, Q., Zhu, H., Song, L., White, J., Yuan, X., Clifton, S. W., Roe, B. A., and McLaughlin, R. (2001) Complete genome sequence of an M1 strain of Streptococcus pyogenes. *Proc. Natl. Acad. Sci. USA* **98**, 4658–4663.

Glaser, P., Rusniok, C., Buchrieser, C., Chevalier, F., Frangeul, L., Msadek, T., Zouine, M., Couve, E., Lalioui, L., Poyart, C., Trieu-Cuot, P., and Kunst, F. (2002) Genome sequence of Streptococcus agalactiae, a pathogen causing invasive neonatal disease. *Mol. Microbiol.* **45**, 1499–1513.

Guzman, C. A., Talay, S. R., Molinari, G., Medina, E., and Chhatwal, G. S. (1999)

Protective immune response against Streptococcus pyogenes in mice after intranasal vaccination with the fibronectin-binding protein SfbI. *J. Infect. Dis.* **179**, 901–906.

Hickman, M. E., Rench, M. A., Ferrieri, P., and Baker, C. J. (1999) Changing epidemiology of group B streptococcal colonization. *Pediatrics* **104**, 203–209.

Lin, F. Y., Clemens, J. D., Azimi, P. H., Regan, J. A., Weisman, L. E., Philips, J. B. III, Rhoads, G. G., Clark, P., Brenner, R. A., and Ferrieri, P. (1998) Capsular polysaccharide types of group B streptococcal isolates from neonates with early-onset systemic infection. *J. Infect. Dis.* **177**, 790–792.

Paoletti, L. C., Kennedy, R. C., Chanh, T. C., and Kasper, D. L. (1996) Immunogenicity of group B Streptococcus type III polysaccharide–tetanus toxoid vaccine in baboons. *Infect. Immun.* **64**, 677–679.

Paoletti, L. C., Wessels, M. R., Rodewald, A. K., Shroff, A. A., Jennings, H. J., and Kasper, D. L. (1994) Neonatal mouse protection against infection with multiple group B streptococcal (GBS) serotypes by maternal immunization with a tetravalent GBS polysaccharide–tetanus toxoid conjugate vaccine. *Infect. Immun.* **62**, 3236–3243.

Pizza, M., Scarlato, V., Masignani, V., Giuliani, M. M., Arico, B., Comanducci, M., Jennings, G. T., Baldi, L., Bartolini, E., Capecchi, B., Galeotti, C. L., Luzzi, E., Manetti, R., Marchetti, E., Mora, M., Nuti, S., Ratti, G., Santini, L., Savino, S., Scarselli, M., Storni, E., Zuo, P., Broeker, M., Hundt, E., Knapp, B., Blair, E., Mason, T., Tettelin, H., Hood, D. W., Jeffries, A. C., Saunders, N. J., Granoff, D. M., Venter, J. C., Moxon, E. R., Grandi, G., and Rappuoli, R. (2000) Identification of vaccine candidates against serogroup B meningococcus by whole-genome sequencing. *Science* **287**, 1816–1820.

Rappuoli, R. (2000) Reverse vaccinology. *Curr. Opin. Microbiol.* **3**, 445–450.

Riley, M. (1993) Functions of the gene products of *Escherichia coli. Microbiol. Rev.* **57**(4), 862–952.

Schuchat, A. (1998) Epidemiology of group B streptococcal disease in the United States: shifting paradigms. *Clin. Microbiol. Rev.* **11**, 497–513.

Schuchat, A. (1999) Group B streptococcus. *Lancet* **353**, 51–56.

Shinefield, H. R., and Black, S. (2000) Efficacy of pneumococcal conjugate vaccines in large scale field trials. *Pediatr. Infect. Dis. J.* **19**, 394–397.

Smoot, J. C., Barbian, K. D., Van Gompel, J. J., Smoot, L. M., Chaussee, M. S., Sylva, G. L., Sturdevant, D. E., Ricklefs, S. M., Porcella, S. F., Parkins, L. D., Beres, S. B., Campbell, D. S., Smith, T. M., Zhang, Q., Kapur, V., Daly, J. A., Veasy, L. G., and Musser, J. M. (2002) Genome sequence and comparative microarray analysis of serotype M18 group A Streptococcus strains associated with acute rheumatic fever outbreaks. *Proc. Natl. Acad. Sci. USA* **99**, 4668–4673.

Suara, R. O., Adegbola, R. A., Mulholland, E. K., Greenwood, B. M., and Baker, C. J. (1998) Seroprevalence of antibodies to group B streptococcal polysaccharides in Gambian mothers and their newborns. *J. Natl. Med. Assoc.* **90**, 109–114.

Tettelin, H., Masignani, V., Cieslewicz, M. J., Eisen, J. A., Peterson, S., Wessels, M. R., Paulsen, I. T., Nelson, K. E., Margarit, I., Read, T. D., Madoff, L. C., Wolf, A. M., Beanan, M. J., Brinkac, L. M., Daugherty, S. C., DeBoy, R. T., Durkin, A. S.,

Kolonay, J. F., Madupu, R., Lewis, M. R., Radune, D., Fedorova, N. B., Scanlan, D., Khouri, H., Mulligan, S., Carty, H. A., Cline, R. T., Van Aken, S. E., Gill, J., Scarselli, M., Mora, M., Iacobini, E. T., Brettoni, C., Galli, G., Mariani, M., Vegni, F., Maione, D., Rinaudo, D., Rappuoli, R., Telford, J. L., Kasper, D. L., Grandi, G., and Fraser, C. M. (2002) Complete genome sequence and comparative genomic analysis of an emerging human pathogen, serotype V Streptococcus agalactiae. *Proc. Natl. Acad. Sci. USA* **99**, 12 391–12 396.

Tettelin, H., Nelson, K. E., Paulsen, I. T., Eisen, J. A., Read, T. D., Peterson, S., Heidelberg, J., DeBoy, R. T., Haft, D. H., Dodson, R. J., Durkin, A. S., Gwinn, M., Kolonay, J. F., Nelson, W. C., Peterson, J. D., Umayam, L. A., White, O., Salzberg, S. L., Lewis, M. R., Radune, D., Holtzapple, E., Khouri, H., Wolf, A. M., Utterback, T. R., Hansen, C. L., McDonald, L. A., Feldblyum, T. V., Angiuoli, S., Dickinson, T., Hickey, E. K., Holt, I. E., Loftus, B. J., Yang, F., Smith, H. O., Venter, J. C., Dougherty, B. A., Morrison, D. A., Hollingshead, S. K., and Fraser, C. M. (2001) Complete genome sequence of a virulent isolate of Streptococcus pneumoniae. *Science* **293**, 498–506.

10

Identification of the 'Antigenome' – a Novel Tool for Design and Development of Subunit Vaccines Against Bacterial Pathogens

Eszter Nagy, Tamás Henics, Alexander von Gabain[1] and **Andreas Meinke**

10.1 Introduction

The adaptive immune system is capable of identifying and eradicating pathogens and pathogen-infected cells, based on a specific repertoire of B-cell-derived antibodies and T-cell-presented receptors that bind to cognate structures of an intruding microbe, defined as antigens. Upon infection, T-cells and circulating antibodies, distinguished by their antigen specificity, are able to defend the host against re-emerging challenges of the original intruder. Vaccines take advantage of this notion such that pathogen-specific antigens formulated in vaccines are able to induce protective B- and T-cell responses. Traditional vaccines contain inactivated/attenuated microbes (e.g. vaccines directed against the bacteria *Bordetella pertussis*, *Vibrio cholera* and *Salmonella typhi* or against the viruses rabies and polio) or microbes with reduced virulence that are related to the pathogen but do not cause severe infections in

[1] Corresponding author: agabain@intercell.com

Genomics, Proteomics and Vaccines edited by Guido Grandi
© 2004 John Wiley & Sons, Ltd   ISBN 0 470 85616 5

humans (e.g. the bovine-specific mycobacterium BCG). In such vaccines whole microbes seem to deliver the set of antigens needed to induce a repertoire of protective antibodies and T-cells. Analysis of traditional vaccines indeed disclosed that a proper set of antibodies mounted in the best case against only a single antigenic structure, such as a surface-displayed carbohydrate or a particular protein, can be sufficient to provide protective immunity (e.g. HBsAg of hepatitis B virus, or OspA of *Borrelia burgdorferi*). Thus, the improved knowledge of the immune system and the progress in molecular biology made it possible to design vaccines with defined antigens. In so-called conjugated vaccines the protective antigenic structure, e.g. a pathogen-derived carbo-hydrate entity, is coupled to a carrier protein and in subunit vaccines antigenic proteins delineated from the pathogen are formulated. The success of vaccines with defined antigens and the advent of recombinant protein technology opened the gate to search systematically for pathogen-specific protein antigens that induce protective immunity in vaccinated individuals.

Sporadic approaches have embarked on the reservoir of previously character-ized pathogen surface proteins and validated their antigenic nature in animal models. However, this relatively narrow selection of available proteins, when formulated in vaccines, only exceptionally contains 'lead' antigens that mediate the degree of immunity needed to protect test animals against the respective pathogen.

A genomic extension of this strategy is to include all proteins encoded or secreted by the pathogen genome, which are forecasted to be surface displayed by gene annotations and predictive programs. Then all candidate genes can be engineered into expression systems and the resulting proteins tested for the protective immunity that they induce in pre-clinical vaccine models (for a review see Adu-Bobie *et al.*, 2003). While this scheme clearly helps to identify a significant pool of pathogen-specific surface proteins, the approach is confined by the limits of known gene annotation methods and by the restrictions to express predicted genes in heterologous hosts. Furthermore, the approach is impeded by the dilemma that it is hard to predict whether antigens shielding animals against a pathogen will also induce protective immunity against the same germ in humans.

Vaccine candidate antigens have also been identified by proteomic tech-niques. For example, proteins of whole-cell bacterial extracts are fractionated by two-dimensional electrophoresis that is followed by western blot analysis, using antibodies from human individuals who have been exposed to the pathogen. Proteins binding to cognate antibodies are then applied to direct peptide sequencing and bioinformatic analysis for gene identification (see e.g. Montigiani *et al.*, 2002; Covert *et al.*, 2001; Sanchez-Campillo *et al.*, 1999). The advantage of this approach lies in the direct interactions between pathogen-expressed proteins and human antibodies. This approach is, however, limited to proteins expressed by the pathogen cultured under laboratory conditions, thus many proteins expressed only during infection will not be identified this way.

Genomic techniques, such as phage display, 'couple' the protein reacting with the cognate antibody to its genotype (see e.g. Santamaria *et al.*, 2001; Hansen, Ostenstad and Sioud, 2001; Mintz *et al.*, 2003). Thus, bacterial and phage-based expression libraries of large polypeptides derived from genomes or small peptides encoded by random synthetic oligonucleotides of identical length have been frequently used to identify antigens. The expression of large polypeptides in bacterial and phage-based systems is hampered by the fact that, due to structural constraints, heterologous expression of many large proteins in host cells is difficult or impossible as part of phage capsids, thereby causing biased display libraries. On the other hand expression of small insert peptides in display libraries is predicted to be less troubled and expected to reduce the bias.

Inspired by the known shot-gun strategy used for sequencing whole genomes, we decided to employ bacterial surface proteins as display platforms that would allow us to present at the surface of *E. coli* cells frame selected, either uniformly small (10–30 amino acids) or uniformly medium-sized (50–100 amino acids) peptide libraries, derived from pathogen genomes. The bacterial display libraries were found to cover a substantial portion of the repertoire of pathogen proteins in form of linear antigens (small-size inserts) and even possibly of sterical epitopes (medium-size inserts). The libraries were exposed to selection with antibodies derived from cohorts of individuals that have to a various degree encountered the pathogen. With this – as we believe – most comprehensive approach, about 100 antigenic proteins per pathogen (including their antibody-binding sites) can be identified (Table 10.1). We refer to this compiled array of antigenic proteins, with all their identified antibody-binding epitopes, as the antigenic fingerprint or the 'antigenome' of a pathogen. Among the antigens, in their majority surface-displayed proteins, those that previously had been shown to protect animals against pathogens in vaccination models were re-discovered. We also describe a validation scheme that filters the best candidate antigens for vaccine development from the entire pool of proteins discovered by our approach.

Table 10.1 Antigenic proteins identified by bacterial surface display. [#]Data from multiple screens are summarized and the number of antigens identified with human sera is shown. [*]The number of protective antigens that have been previously reported is listed. [§]All protective antigens that have been reported in the literature previously and are encoded by the genome applied for the bacterial surface display screens are listed. n.a., not applicable

Pathogen used for library construction	[#]Antigens identified	[*]Reported protective antigens identified	[§]Reported protective antigens for the applied strain		
			Number	Protein	Reference
Staphylococcus aureus *COL*	108	3	3	ClfA	Josefson *et al.*, 2001
				FnbpA and B	Rozalska and Wadstrom, 1993
Staphylococcus epidermidis RP62A	68	n.a.	n.a.		
Streptococcus pyogenes Sf370/M1	95	5	6	[*]SpeB	Kapur *et al.*, 1994
				[*]SpeC	McCormick *et al.*, 2000
				[*]C5A peptidase	Ji *et al.*, 1997
				[*]M1 protein	Beachey *et al.*, 1981
				[*]Spy0843	Reid *et al.*, 2002
				Fbp54	Kawabata *et al.*, 2001
Streptococcus pneumoniae TIGR4	99	11	17	[*]NanA, NanB neuraminidases	Lock, Paton and Hansman,1988
				[*]PspA	McDaniel *et al.*, 1991
				[*]LytA, autolysin	Lock, Hansman and Paton, 1992
				[*]CbpA (PspC, Hic, SpsA)	Rosenow *et al.*, 1997
				[*]LytB glucosaminidase	Wizemann *et al.*, 2001
				[*]LytC muramidase	Wizemann *et al.*, 2001
				[*]PrtA serine protease	Wizemann *et al.*, 2001
				[*]PhtA (histidine triad A)	Wizemann *et al.*, 2001
				[*]HtpB	Adamou *et al.*, 2001
				[*]HtpD	Adamou *et al.*, 2001
				PvaA	Wizemann *et al.*, 2001

Table 10.1 (*continued*)

Pathogen used for library construction	#Antigens identified	*Reported protective antigens identified	§Reported protective antigens for the applied strain		
			Number	Protein	Reference
				PiaA (Pit1)	Brown *et al.*, 2001
				PiuA (Pit2)	Brown *et al.*, 2001
				Pneumolysin	Lock *et al.*, 1988
				PsaA	Talkington *et al.*, 1996
				Hyaluronidase	Lock *et al.*, 1988
Streptococcus agalactiae ATCC12403	140	2	2	Sip	Martin *et al.*, 2002
				C5a peptidase	Cheng *et al.*, 2002
Helicobacter pylori	124	5	9	*GroEL chaperonin	Ferrero *et al.*, 1995
				*UreB (urease beta subunit)	Michetti *et al.*, 1994
KTH Du & Ca1				*Cag26 (CAG-PAI protein)	Marchetti *et al.*, 1995
				*Catalase	Radcliff *et al.*, 1997
				*VacA (vacuol. cytotoxin)	Ghiara *et al.*, 1997
				HP0231 (hypothetical protein)	Sabarth *et al.*, 2002
				HpaA homologue	Sabarth *et al.*, 2002
				NapA	Satin *et al.*, 2000
				UreA (urease alpha subunit)	Michetti *et al.*, 1994

10.2 Small DNA insert libraries – a tool to cover a pathogen's 'antigenome'

Our antigen identification technology is based on the construction of small insert libraries. Ideally they present – in an entirely random manner – the whole arsenal of protein peptides/epitopes covering all coding regions that are defined by the pathogen genome and thereby make them accessible for any subsequent screening and/or selection procedure. It is emphasized here that the availability of whole-genome sequence information of the given pathogen (Fraser and Fleischmann, 1997) supports the evaluation of library quality, but is not a prerequisite for the antigen identification approach.

For the library construction the pathogen genome is randomly sheared into 30–300 bp DNA fragments. Both mechanical (sonication, nebulization) and enzymatic (e.g. DNaseI digestion) fragmentation approaches have their advantages and limitations (Bankier, 1993; Oefner et al., 1996; McKee et al., 1977; Anderson, 1981). The method of choice largely depends on the desired fragment size as well as the nature of subsequent applications. Regardless of the applied fragmentation procedure, the fragments are carefully end repaired and introduced into our novel frame-selection vector system, which had been

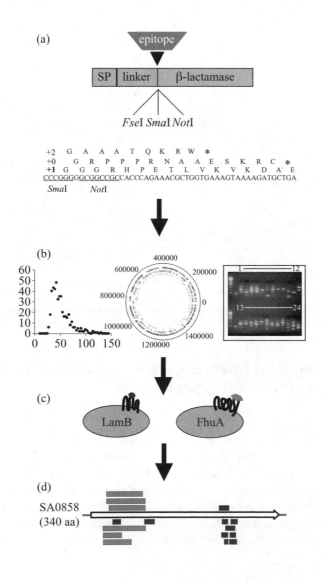

described elsewhere (Etz *et al.*, 2002; Henics *et al.*, 2003). Briefly, the genomic fragments are expressed as fusion chimeras of the β-lactamase protein in the periplasmic space of *E. coli* and this construct is designed in such a way that it allows the elimination of those random sequences that possess a stop codon within the expressed reading frame, leading consequently to ampicillin sensitivity (Figure 10.1(a)). Following antibiotic selection, 500–1000 randomly picked clones are subjected to sequence analysis, in order to monitor the quality of the library. Sequence analysis of individual clones allowed us to assess the quality regarding insertion junctions and to determine the efficiency of the frame-selection system. Generally, more than 95 per cent of the clones with intact insertion junctions follow what we termed the '$3n + 1$ rule'; i.e., only those fragments promote the expression of the downstream β-lactamase gene that possess the $+1$ frame out of the three possibilities as a consequence of random fragmentation (Figure 10.1(a)). After the quality assessment and the trimming step, the clones selected from the libraries are aligned against the corresponding whole genome sequence using a basic local alignment tool (BLASTN, Altschul *et al.*, 1990) that yields a simple measure of how randomly the sampled genomic fragments are distributed (Figure 10.1(b)). Additionally, a comprehensive PCR-based analysis is applied to monitor the presence of arbitrarily chosen sequences in the library (assessment of its representative nature), i.e. search for previously known or randomly *in silico* selected sequences (Figure 10.1(b)). Our analyses indicate that all the targeted sequences are represented in the libraries in at least one orientation, provided there is a sufficient, usually, more than 10-fold coverage of the genome (Henics *et al.*, 2003).

Figure 10.1 Small fragment genomic libraries for bacterial surface display. (a) Genomic DNA of the pathogen is sheared into random fragments (30–300 bp) and inserted into the β-lactamase-based frame selection vector. This procedure allows the elimination of random sequences that possess a stop codon (asterisks) within the expressed reading frame so that only fragments fused in-frame direct the expression of β-lactamase resulting in the gain of ampicillin resistance. (b) 500–1000 randomly picked clones are sequenced and trimmed sequences are analysed to determine fragment size distribution (left-hand panel) and the distribution of sequences over the circular chromosome (middle panel) and to assess the presence of arbitrarily chosen sequences within the library using a PCR-based method (right-hand panel). (c) Libraries are transferred in frame into one of our display vectors, which allow efficient presentation of the library-encoded peptides on the bacterial cell surface in the context of outer membrane proteins, such as LamB or FhuA. (d) After multiple screening of the differently sized displayed libraries, a complex epitope profile is obtained for each identified antigenic protein, as demonstrated with the SA0858 ORF of *Staphylococcus aureus*. Light- and dark-grey bars represent epitopes identified with the medium- and small-size libraries in the platform proteins FhuA and LamB, respectively, and mark the antigenic regions of the protein

Exploiting two rare eight-cutter restriction sites positioned upstream and downstream of the library insertion junctions (see Figure 10.1(a)), the library is directly and in-frame transferred into one of our display vectors, which allow efficient presentation of the fragment-encoded peptides on the bacterial cell surface in the context of outer membrane proteins, such as LamB or FhuA (Etz *et al.*, 2002). The choice of the vector system is dictated primarily by the average fragment size of the library for LamB, allowing the expression of smaller fragments (30–150 bp; i.e. small epitopes), whereas FhuA accepts considerably larger ones (up to 300 bp), possibly including structural epitopes (Figure 10.1(c)). Relying on high cloning efficiencies, routinely, we reach 10- to 20-fold coverage of the genome for each investigated organism.

We have been successful in generating small-fragment genomic libraries using the genomes of a variety of important human pathogens belonging to the genera *Staphylococcus, Streptococcus, Helicobacter, Escherichia, Chlamydia, Enterococcus* and *Campylobacter.*

10.3   Proper display platforms

The identification of antigenic proteins from pathogenic bacteria requires the successful expression and separation of individual proteins in order to determine the corresponding amino acid and gene sequences. We envisioned displaying genome-derived antigenic epitopes of bacterial pathogens in an environment that is related to the natural context where antibodies bind antigens during infection: on the outer surface of bacteria. Therefore we have chosen to use *E. coli* cells as a platform to present peptide epitopes at the surface, as part of proteins that are incorporated into the outer membrane. The outer membrane proteins LamB, BtuB and FhuA are all characterized by flexible loops that allow the insertion of foreign peptides without gross disturbance of their structure and biological function (Etz *et al.*, 2001; Figure 10.1(c)). The foreign peptide sequences are anchored within the outer membrane proteins on both the N- and C-terminus. Since it has been shown that the extra-cellular loops are flexible entities, it can be assumed that the foreign peptide inserted within a loop may potentially adopt a structural configuration that comes close to its natural structure, i.e. the one it maintains as part of the native protein. Therefore, antibodies raised against a native protein during infection are predicted to recognize epitopes derived from the same protein when they are displayed on the surface of *E. coli*. We have characterized the outer membrane proteins LamB, BtuB and FhuA for their ability to display model epitopes on the surface of *E. coli*, demonstrating that they readily accept and display foreign peptides of up to 50, 90 and 200 amino acids, respectively (Etz *et al.*, 2001). The platform proteins were therefore employed to display diverse peptide epitopes from a pathogen's genome ranging from small linear epitopes of six to nine amino acids in size up to larger, potentially conformational epitopes of up to 150 amino acids in size. The libraries were constructed with a sufficient complexity (10^5 to 2×10^6 members) capable of encompassing all possible epitopes encoded by the pathogen's genome (see preceding paragraph). Since the selection of the displayed epitopes is not dependent on the annotation of the genome, but only on the re-generation of the correct reading frame (this problem was solved by the above described β-lactamase selection approach), our libraries will also contain such epitopes, which have not been predicted by the genome annotation process. In addition, peptides not derived from coding regions and not possessing a stop codon will also be contained in our libraries, rendering them to some extent, although genome derived, diversely sized random peptide libraries (Taschner *et al.*, 2002).

The genomic libraries derived from several pathogenic bacteria were subjected to screens with human sera, identifying numerous antigenic proteins (Table 10.1). For our first target, *Staphylococcus aureus*, we applied the LamB and FhuA libraries for selection (Etz *et al.*, 2002). The analysis showed that the two platform proteins selected a distinct set of antigens with a considerable overlap, especially among the frequently identified antigens. Therefore we were interested to see whether a third platform protein would further increase the coverage of the screens. For the human pathogen *Streptococcus pyogenes* the screens were performed subsequently with all three outer membrane proteins in order to evaluate the selection bias of the individual platforms. This analysis showed (unpublished data) that LamB and BtuB libraries selected a very similar repertoire of antigens despite the different insert size of the two libraries (35 and 100 bp on average, respectively). In contrast, the set of antigens again differed significantly from the one found in the FhuA library (300 bp), but many of the antigens were identified in all three libraries. Our data confirm that a comprehensive selection of antigens can be obtained applying the two platform vectors encoding LamB and FhuA.

10.4   Selected human sera provide imprints of pathogen encounters

It is generally accepted that antibodies are crucial in protecting individuals against extra-cellular bacteria (Kaufmann, Sher and Ahmed, 2002). In recent years, serological identification of antigens has proven to be an effective strategy for the identification of antigens (*e.g.* Utt *et al.*, 2002; Lei *et al.*, 2000; for a review see Klade, 2002). The antibodies induced by pathogens and present in serum and other body fluids, although not all of them protective, constitute a molecular imprint of the *in vivo* expression of the corresponding antigens and have therefore been chosen in our approach as a tool for the identification of a comprehensive set of antigens. It is our preference to use human instead of animal sera as antibody source, for several reasons. Animals and humans induce different immune responses towards a pathogen and show different disease manifestations regarding the same pathogen, and infectious disease models in animals are often artificial (see e.g. Josefsson *et al.*, 2001).

In order to gain access to relevant antibody repertoires for the identification of bacterial antigens, a comprehensive set of samples from different donors needs to be collected. Patients with invasive diseases represent the most 'visible' group of individuals with the highest susceptibility for infections. However, a large proportion of the community has multiple encounters with the most common human pathogens, which result in mild diseases and colonization, or most importantly in elimination or prevention of carriage. For the identification of protective antigens aimed to prevent colonization, the specific and functional antibodies present in highly exposed healthy non-carriers are the most valuable. In contrast, candidate antigens for vaccines against invasive diseases should be selected with the antibodies present in the sera of convalescent phase patients overcoming serious infections. The susceptibility of the latter patients is reflected by the low level or lack of protective antibodies in their acute phase sera.

For selection of the most appropriate screening reagents, we characterized sera obtained from different donor groups by several immune assays. First, as a global assessment of the levels of anti-bacterial antibodies of high affinity (IgG and IgA), we performed ELISA using total bacterial extracts or whole cells, and secreted components isolated from culture supernatants of pathogens (Figure 10.2). It is advisable to use capsule negative mutant strains as well as to measure pathogen specific antibodies against proteinaceous components, since many human pathogens produce highly immunogenic, serotype-specific polysaccharide capsules. The abundantly induced antibodies, in spite of their lower affinity, may otherwise dominate the reactivity as detected by ELISA. Second, a small set of already known recombinant proteins derived from the pathogen (such as virulence factors, cell surface proteins and/or described antigens) can be used as coating reagents for benchmarking. Because bactericidal antibodies are the most critical class of antibodies to eliminate extra-cellular pathogens (see e.g. Johnson *et al.*, 1999; Cunningham, 2000), we also measure the level of antibody-induced, complement-dependent phagocytosis or *in vitro* killing using phagocytic cells (e.g. isolated human PMNs or monocytic cell lines). Further-

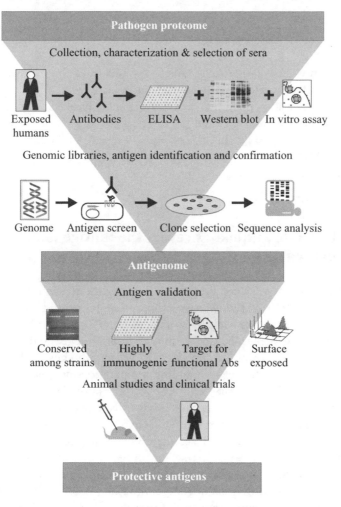

Figure 10.2 Schematic scheme of the antigen identification and validation approach. Human antibodies are characterized by ELISA and Western blot and *in vitro* assays and purified to generate pathogen-specific screening reagents. These antibodies are used to screen genomic peptide libraries presented on the surface of *E. coli* cells for immunogenic epitopes applying magnetic bead sorting technology. Sequencing and bioinformatic analysis determines the encoded peptide sequence and thereby identifies the corresponding antigenic protein. The antigen identification approach thus reduces the complex pathogen proteome to an antigenome consisting of a small subset of proteins. The validation procedure determines the presence of the corresponding genes in a large array of clinical isolates, the reactivity of antigenic peptides with multiple human sera, the exposure of the antigenic proteins on the surface of the pathogen and the potential of the antigenic proteins being targeted by functional antibodies. Among the most promising candidates the protective antigens are identified in animals models before entering clinical studies

more, levels of antibodies with the highest opsonic and complement-fixing activity can be determined with subclass-specific detection reagents (IgG1 and 3; Figure 10.2).

This comprehensive approach leads to the identification of human sera with high levels of functional anti-bacterial antibodies, which are then used for antigen identification. To further enrich for prevalent antibodies, pools of sera (five in a pool) are prepared from the different donor groups. After removal of the antibodies directed against *E. coli*, IgG and IgA antibodies are purified by standard affinity chromatography with protein G and anti-human IgA antibodies, respectively.

10.5   Cognate antibodies reveal the 'antigenome' of a pathogen

The construction of very comprehensive peptide libraries and the characterization and purification of pathogen-specific antibodies provide the basis for our antigen identification approach. We incubate *E. coli* cells displaying a genomic peptide library with purified and biotinylated antibodies, and those cells that display peptides recognized by pathogen-specific antibodies bind to streptavidin coupled to paramagnetic beads, enabling a separation by magnetic force. From a single screen (one library and one serum pool), we typically analyse approximately 1000 individual clones by DNA sequencing. Bioinformatic analysis (assignment to the genome by BLASTN) and confirmation of immunogenicity by Western blot analyses identifies all peptide epitopes selected by pathogen-specific antibodies. The set of antigens that are determined by this approach, and which we refer to as the 'antigenome', comprises not only all proteins encompassed by a pathogen genome that react with an antibody, but also the antibody-binding domains (epitopes) within the reactive protein (Figure 10.1(d); Figure 10.2).

It is evident that the repertoire of antigens identified from genomic peptide libraries is largely determined by the human sera applied for selection, assuming that all potential antigenic peptides are presented in the libraries. As described above, human sera from patients who have successfully recovered from an infection, or from healthy individuals, who are less prone to infection, most likely contain antibodies contributing to protection against disease. The validity of our approach became immediately evident when, not only in the case of *Staphylococcus aureus* (Etz *et al.*, 2002), but also in subsequent screens with several human streptococcal pathogens, most of the published protective antigens were among those identified by the genomic screens (Table 10.1), such as PspA (McDaniel *et al.*, 1991) or several of the choline binding proteins (Wizemann *et al.*, 2001) of *Streptococcus pneumoniae* and the M1 protein (Hu *et al.*, 2002) and C5a peptidase (Ji *et al.*, 1997) of *Streptococcus pyogenes*. It was therefore of great interest to analyse the selection of antigens in relation to the applied serum. A comparison of the antigens recovered from the screens provided a first qualitative distinction, showing that a number of antigens was identified only by human sera obtained from healthy individuals, while others were mainly selected by patient sera or by both types of serum. A carefully performed comparative study with sera that were known to provide protection and those leaving a patient unprotected should thus be very valuable in the identification of the most potent vaccine candidates.

Among the antigens that have not been described before, we interestingly identified several antigenic peptides that could not be assigned to any annotated ORF, but were reactive with human sera when tested individually. These peptides could potentially be encoded by novel, previously not annotated ORFs, or the antibodies in human sera may have recognized this peptide epitope because it mimics an epitope from an unrelated protein that has elicited the antibody response during infection. Although we have not yet shown that these potentially novel or alternative ORFs give rise to the expression of the predicted proteins in the bacterium, the selection of these epitopes clearly confirms the potential of our libraries to encompass most if not all potential epitopes encoded by the respective genome.

In summary, the 'antigenome' allows us to compare and to catalogue subsets of antigens that are specific for the way an individual has encountered the pathogen, providing thereby a basis to retrieve from the 'antigenome' the most promising and potent candidates for the development of a vaccine against the respective human pathogen (Figure 10.2).

10.6 How to retrieve from the 'antigenome' the candidate antigens for vaccine development

It is an important aspect of our approach that we employ several independent selection criteria based on a variety of straightforward *in vitro* assays in order to retrieve from the about 100 antigens defining the 'antigenome', the most promising candidates for vaccine development. All these assays are geared up such that the identification of the most promising candidate antigens will be obtained by relatively rapid and simple screening methods. The key selection criteria are based on the notion that the most valuable protective antigens are expected to be conserved among various clinical strains, surface exposed and widely immunogenic in humans (Figure 10.2).

An important criterion of antigen selection is the occurrence of the candidate antigens in the majority of clinical isolates causing the targeted disease; low gene distribution among clinical isolates unequivocally will preclude antigens from further consideration. In order to address this question, we perform PCR analysis of genomic DNA (typically from 50 characterized strains) using gene-specific primers designed by available genomic information (negative results are re-confirmed by Southern blotting). DNA sequence analysis of the widely distributed genes provides further important information about antigenic variation in the target proteins. Applying this analysis, we exclude up to 20 per cent of the antigens confined by an 'antigenome'.

An equally important criterion in the selection for vaccine candidates is their *in vivo* expression under disease conditions against which the vaccine is targeted, as well as their high immunogenicity, i.e. the induction of antibodies against a given antigen in the majority of individuals. Serological analysis can address both aspects simultaneously given that appropriate samples – such as sera obtained from patients with different disease manifestations and clinical outcome, healthy adults with or without carriage and 'naïve' infants – are tested and compared for antibody levels. In our approach, antigens are tested by ELISA using synthetic peptides corresponding to epitopes selected in the context of the platform proteins. In a series of automated ELISAs we are thus able to identify the most immunogenic epitopes inducing high levels of antibodies in many individuals. Having analysed epitopes identified from four different pathogens, we can conclude that there is a considerable overlap in anti-bacterial antibody levels between high-titre patient and high-titre healthy sera, suggesting an overlapping pattern of expression of the corresponding antigens during invasive disease and colonization/interaction without infection. The application of this analysis allows us to rank the selected antigens for further examination.

Finally, regarding the criterion to select surface-displayed antigens, our cumulative data obtained from the analysis of six 'antigenomes' disclosed that a large fraction of the pathogen antigens identified by our method represents cell-surface-localized and secreted proteins based on bioinformatics (i.e. prediction programs), homologies to other bacterial proteins and published data. Additionally, in order to determine the surface localization of our antigens experimentally, we perform at least one of three different assays most applicable for a given pathogen: first, staining of living bacteria in a FACS-based assay employing antibodies purified from human high-titre sera or from sera obtained by vaccination of mice with *E. coli* extracts carrying the epitopes embedded in the platform proteins (Etz *et al.*, 2001); second, more sensitive methods, such as bactericidal killing or opsonophagocytosis assays with fluorescently labelled pathogens, are performed when low *in vitro* expression of the antigen or high background reactivity of the antibody reagents prevents surface detection.

The application of the major selection criteria reduces the number of candidate antigens derived from the 'antigenome' to a small number of candidate proteins that can be rapidly expressed in recombinant form for subsequent studies. Having performed and completed this comprehensive antigen validation procedure with several of the important human pathogens, we learned that – in addition to numerous novel antigens – many of the previously identified protective antigens of for instance *Staphylococci* and *Streptococci* are among the selected potential vaccine candidates (Table 10.1). Furthermore, evaluation of predicted vaccine candidate that had fulfilled the selection criteria revealed that they are indeed able to protect test animals against the underpinning disease.

10.7   Summary and discussion

We describe here a novel approach for the identification of protein antigens from bacterial pathogens that are promising candidates for the development of subunit vaccines against bacterial diseases. The underlying technology confronts genome-derived peptide libraries comprising the nearly complete protein repertoire of a pathogen with comprehensive collections of antibodies derived from individuals exposed to the respective germ. As a result, we establish what we refer here to as the 'antigenome' of a pathogen, i.e. the catalogue of the most relevant antigenic proteins, including their antibody-binding sites that are targeted by the human immune system. At this stage we have identified more than 700 antigenic proteins and their antibody-binding sites by applying this approach to eight different human pathogens.

The particular profile of the pathogen 'antigenome' obtained will depend on the antibody repertoire used for epitope selection. Thus, the 'antigenome' will differ when antibodies are derived from individuals who have encountered the pathogen in dissimilar fashion, e.g. when non-symptomatic carriers are compared with cohorts that have gone through an acute infection. Such a notion is supported by our data obtained when establishing the 'antigenome' of cohorts that have encountered *Staphylococcus aureus* under different circumstances (Etz *et al.*, 2002). Similarly, the composition of the 'antigenome' of a pathogen is expected to diverge when different disease manifestations of the same pathogen are compared. It is, however, also anticipated that the 'antigenome' derived when applying sera from different individuals will potentially vary, even if they have gone through a nearly identical course of the disease, mainly for the following reasons: (1) genetic differences that may govern the human immune response, (2) differences in the exact encounter of the microbe and disease course and (3) strain differences among the pathogens that infect the compared individuals. It is therefore instrumental to our approach to derive sera from different cohorts of humans and to pool the highest-titre sera within these cohorts for antigen identification in order to obtain a most comprehensive 'antigenome'. 'Antigenomes' therefore not only provide a sound resource for design and development of vaccines, as discussed below, but also allow us to monitor antigenic fingerprints that reflect the specific conditions of a pathogen encounter and the underpinning disease. For example, the 'antigenome' of a pathogen that is obtained when humans and animals are compared may well answer the question whether an animal disease model has the potential to be predictive for human disease studies.

The 'antigenome' specific for a pathogen is of course also dependent on the nature and the composition of the genomic peptide libraries employed for the selection procedure. Although our libraries are derived from relevant strains that are isolated from diseased individuals or from diagnosed carriers, they may be incomplete in that certain antigens are not encoded by the genome of the strain used for construction. On the other hand, antigens only occurring in a subset of strains are not favourable candidates for the development of subunit vaccines. Our approach to display antigens in the form of genome-derived peptides as part of surface proteins has proven to establish comprehensive libraries that cover major parts of all ORFs confined by the pathogen genome. We could also demonstrate that the coverage of the antigen repertoire is definitely increased when libraries with different insert size displayed in suitable platform proteins are used in the selection procedure (Etz *et al.*, 2002). The FhuA display platform is capable to tolerate inserts up to 150 amino acids in length, a size that would include epitopes recognized by antibodies through their sterical configuration. This hypothesis is supported by the observation that – in contrast to many epitopes found in small- and large-insert libraries – some are exclusively detected in large-insert libraries. Further analysis of such epitopes by synthetic peptides supports the conclusion that their binding to cognate antibodies requires their 3D structure (Etz, 2002).

Our studies aim to establish 'antigenomes' from pathogens for the purpose of designing and developing subunit vaccines directed against diseases that cannot be properly mastered by present vaccination strategies. In order to increase the likelihood of selecting antigens that – formulated in vaccines – induce protective immunity, we carefully choose sera from patients and pathogen carriers that contain high-titre and protective antibodies. The 'antigenomes' we retrieve with such sera in fact contain numerous antigens that have previously been shown to mediate protective immunity in animals challenged by the respective pathogen. In order to pinpoint further candidate antigens for vaccine development, we implemented a coherent and rapid strategy to retrieve the most promising antigens from the 'antigenome' of a pathogen. We believe that the most promising antigen candidates for vaccine development must be conserved among clinical isolates of the same pathogen, reactive with the sera of many individuals who have successfully overcome the disease through their own immunity, predicted to be surface displayed and/or involved in vital virulence functions and targeted by functional antibodies. Normally, only a small number of antigens remain in the race as strong candidates. Interestingly, we obtained results suggesting that related pathogens taking part in mixed infections 'share' identical antigens; i.e., antibodies may be able to fight more than one pathogen by recognition of epitopes conserved between bacterial species. This discovery may provide an additional clue for picking antigen candidates. The potency of candidate antigens to protect against a pathogen can be tested in animal models by vaccination with the recombinant form. We also were able to validate candidate antigens in animal models as they are integrated in their platform proteins. At this stage, data emerge from our studies showing that among the closest choice of antigens retrieved from an 'antigenome' are candidates inducing protective immunity in animal vaccine models.

Since antibodies of the human immune system recognize the antigenic proteins that are selected by the genomic screens, these proteins must have been expressed *in vivo* and most of them are very likely accessible to the immune response without lysis of the pathogen. This perception does not only indicate that these proteins are localized outside the bacterium, but also suggests that they serve a function for the establishment of the infection, the propagation in the host or potentially for the distraction of the host's immune response by the pathogen. All these functions are attractive targets for the development of vaccines, but they also provide a tool to identify novel anti-microbial compounds (monoclonal antibodies, antibiotics small-molecule drugs) that interfere with these processes and are thereby capable of counteracting pathogens causing infections.

References

Adamou, J. E., Heinrichs, J. H., Erwin, A. L., Walsh, W., Gayle, T., Dormitzer, M., Dagan, R., Brewah, Y. A., Barren, P., Lathigra, R., Langermann, S., Koenig, S., and Johnson, S. (2001) Identification and characterization of a novel family of pneumococcal proteins that are protective against sepsis. *Infect. Immun.* **69**, 949–958.

Adu-Bobie, J., Capecchi, B., Serruto, D., Rappuoli, R., and Pizza, M. (2003) Two years into reverse vaccinology. *Vaccine* **21**, 605–610.

Altschul, S. F., Gish, W., Miller, W., Myers, E. W., and Lipman, D. J. (1990) Basic local alignment search tool. *J. Mol. Biol.* **215**, 403–410.

Anderson, S. (1981) Shotgun DNA sequencing using cloned DNase I-generated fragments. *Nucl. Acids Res.* **9**, 3015–3027.

Bankier, A. T. (1993) Generation of random fragments by sonication. *Methods Mol. Biol.* **23**, 47–50.

Beachey, E. H., Seyer, J. M., Dale, J. B., Simpson, W. A., and Kang, A. H. (1981) Type-specific protective immunity evoked by synthetic peptide of *Streptococcus pyogenes* M protein. *Nature* **292**, 457–459.

Brown, J. S., Ogunniyi, A. D., Woodrow, M. C., Holden, D. W., and Paton, J. C. (2001) Immunization with components of two iron uptake ABC transporters protects mice against systemic *Streptococcus pneumoniae* infection. *Infect. Immun.* **69**, 6702–6706.

Cheng, Q., Debol, S., Lam, H., Eby, R., Edwards, L., Matsuka, Y., Olmsted, S. B., and Cleary, P. P. (2002) Immunization with C5a peptidase or peptidase-type III poly-saccharide conjugate vaccines enhances clearance of group B Streptococci from lungs of infected mice. *Infect. Immun.* **70**, 6409–6415.

Covert, B. A., Spencer, J. S., Orme, I. M., and Belisle, J. T. (2001) The application of proteomics in defining the T cell antigens of *Mycobacterium tuberculosis*. *Proteomics* **1**, 574–586.

Cunningham, M. W. (2000) Pathogenesis of group A streptococcal infections. *Clin. Microbiol. Rev.* **13**, 470—511.

Etz, H. (2002) Identification and evaluation of *Staphylococcus aureus* antigens from a genomic bacterial surface display screen using FhuA as a novel platform protein. *Ph.D. thesis*. University of Vienna.

Etz, H., Bui, D. M., Schellack, C., Nagy, E., and Meinke, A. (2001) Bacterial phage receptors – versatile tools for the display of polypeptides on the cell surface. *J. Bacteriol.* **183**, 6924–6935.

Etz, H., Minh, D. B., Henics, T., Dryla, A., Winkler, B., Triska, C., Boyd, A. P., Söllner, J., Schmidt, W., von Ahsen, U., Buschle, M., Gill, S. R., Kolonay, J., Khalak, H., Fraser, C. M., von Gabain, A., Nagy, E., and Meinke, A. (2002) Identification of *in vivo* expressed vaccine candidate antigens from *Staphylococcus aureus*. *Proc. Natl. Acad. Sci. USA* **99**, 6573–6578.

Ferrero, R. L., Thiberge, J. M., Kansau, I., Wuscher, N., Huerre, M., and Labigne, A. (1995) The GroES homolog of *Helicobacter pylori* confers protective immunity against mucosal infection in mice. *Proc. Natl. Acad. Sci. USA* **92**, 6499–6503.

Fraser, C. M., and Fleischmann, R. D. (1997) Strategies for whole microbial genome sequencing and analysis. *Electrophoresis* **18**, 1207–1216.

Ghiara, P., Rossi, M., Marchetti, M., Di Tomasso, A., and Vindigni, C. (1997) Therapeutic intragastric vaccination against *Helicobacter pylori* in mice eradicates an otherwise chronic infection and confers protection against reinfection. *Infect. Immun.* **65**, 4996–5002.

Hansen, M. H., Ostenstad, B., and Sioud, M. (2001) Identification of immunogenic antigens using a phage-displayed cDNA library from an invasive ductal breast carcinoma tumour. *Int. J. Oncol.* **19**, 1303–1309.

Henics, T., Winkler, B., Pfeifer, U., Gill, S. R., Buschle, M., von Gabain, A., and Meinke, A. (2003) Small-fragment genomic libraries for the display of putative epitopes of pathogens with high medical importance. *BioTechniques* **35**, 196–202.

Hu, M. C., Walls, M. A., Stroop, S. D., Reddish, M. A., Beall, B., and Dale, J. B. (2002) Immunogenicity of a 26–valent group A streptococcal vaccine. *Infect. Immun.* **70**, 2171–2177.

Ji, Y., Carlson, B., Kondagunta, A., and Cleary, P. P. (1997) Intranasal immunization with C5a peptidase prevents nasopharyngeal colonization of mice by the group A Streptococcus. *Infect. Immun.* **65**, 2080–2087.

Johnson, S. E., Rubin, L., Romero-Steiner, S., Dykes, J. K., Pais, L. B., Rizvi, A., Ades, E., and Carlone, G. M. (1999) Correlation of opsonophagocytosis and passive protection assays using human anticapsular antibodies in an infant mouse model of bacteremia for *Streptococcus pneumoniae*. *J. Infect. Dis.* **180**, 133–140.

Josefsson, E., Hartford, O., O'Brien, L., Patti, J. M., and Foster, T. (2001) Protection against experimental *Staphylococcus aureus* arthritis by vaccination with clumping factor A, a novel virulence determinant. *J. Infect. Dis.* **184**, 1572–1580.

Kapur, V., Maffei, J. T., Greer, R. S., Li, L. L., Adams, G. J., and Musser, J. M. (1994) Vaccination with streptococcal extracellular cysteine protease (interleukin-1 beta convertase) protects mice against challenge with heterologous group A streptococci. *Microb. Pathog.* **16**, 443–450.

Kaufmann, S., Sher, A., and Ahmed, R. (Eds.). (2002) *Immunology of Infectious Diseases*. ASM Press, Washington, DC.

Kawabata, S., Kunitomo, E., Terao, Y., Nakagawa, I., Kikuchi, K., Totsuka, K., and Hamada, S. (2001) Systemic and mucosal immunizations with fibronectin-binding protein FBP54 induce protective immune responses against *Streptococcus pyogenes* challenge in mice. *Infect. Immun.* **69**, 924–930.

Klade, C. S. (2002) Proteomics approaches towards antigen discovery and vaccine development. *Curr. Opin. Mol. Ther.* **3**, 216–223.

Lei, B., Mackie, S., Lukomski, S., and Musser, J. M. (2000) Identification and immunogenicity of group A Streptococcus culture supernatant proteins. *Infect. Immun.* **68**, 6807–6818.

Lock, R. A., Hansman, D., and Paton, J. C. (1992) Comparative efficacy of autolysin and pneumolysin as immunogens protecting mice against infection by *Streptococcus pneumoniae*. *Microb. Pathog.* **12**, 137–143.

Lock, R. A., Paton, J. C., and Hansman, D. (1988) Comparative efficacy of pneumococcal neuraminidase and pneumolysin as immunogens protective against *Streptococcus pneumoniae*. *Microb. Pathog.* **5**, 461–467.

Marchetti, M., Arico, B., Burroni, D., Figure, N., Rappouli, R., and Ghiara, P. (1995) Development of a mouse model of *Helicobacter pylori* infection that mimics human disease. *Science* **267**, 1655–1658.

Martin, D., Rioux, S., Gagnon, E., Boyer, M., Hamel, J., Charland, N., and Brodeur,

B. R. (2002) Protection from group B streptococcal infection in neonatal mice by maternal immunization with recombinant Sip protein. *Infect. Immun.* **70**, 4897–4901.

McCormick, J. K., Tripp, T. J., Olmsted, S. B., Matsuka, Y. V., Gahr, P. J., Ohlendorf, D. H., and Schlievert, P. M. (2000) Development of streptococcal pyrogenic exotoxin C vaccine toxoids that are protective in the rabbit model of toxic shock syndrome. *J. Immunol.* **165**, 2306–2312.

McDaniel, L. S., Sheffield, J. S., Delucchi, P., and Briles, D. E. (1991) PspA, a surface protein of *Streptococcus pneumoniae*, is capable of eliciting protection against pneumococci of more than one capsular type. *Infect. Immun.* **59**, 222–228.

McKee, J. R., Christman, C. L., O'Brien, W. D. Jr., and Wang, S. Y. (1977) Effects of ultrasound on nucleic acid bases. *Biochemistry* **16**, 4651–4654.

Michetti, P., Corthesy-Thuelaz, C., Davin, C., Haas, R., Vaney, A. C., Heitz, M., Bille, J., Kraehenbuhl, J. P., Saraga, E., and Blum, A. L. (1994) Immunization of BALB/c mice against *Helicobacter felis* infection with *Helicobacter pylori* urease. *Gastroenterology* **107**, 1002–1011.

Mintz, P. J., Kim, J., Do, K. A., Wang, X., Zinner, R. G., Cristofanilli, M., Arap, M. A., Hong, W. K., Troncoso, P., Logothetis, C. J., Pasqualini, R., and Arap, W. (2003) Fingerprinting the circulating repertoire of antibodies from cancer patients. *Nat. Biotechnol.* **21**, 57–63.

Montigiani, S., Falugi, F., Scarselli, M., Finco, O., Petracca, R., Galli, G., Mariani, M., Manetti, R., Agnusdei, M., Cevenini, R., Donati, M., Nogarotto, R., Norais, N., Garaguso, I., Nuti, S., Saletti, G., Rosa, D., Ratti, G., and Grandi, G. (2002) Genomic approach for analysis of surface proteins in *Chlamydia pneumoniae*. *Infect. Immun.* **70**, 368–379.

Oefner, P. J., Hunicke-Smith, S. P., Chiang, L., Dietrich, F., Mulligan, J., and Davis, R. W. (1996) Efficient random subcloning of DNA sheared in a recirculating point-sink flow system. *Nucleic Acids Res.* **24**, 3879–3886.

Radcliff, F. J., Hazell, S. L., Kolesnikow, T., Doidge, C., and Lee, A. (1997) Catalase, a novel antigen for *Helicobacter pylori* vaccination. *Infect. Immun.* **65**, 4668–4766.

Reid, S. D., Green, N. M., Sylva, G. L., Voyich, J. M., Stenseth, E. T., DeLeo, F. R., Palzkill, T., Low, D. E., Hill, H. R., and Musser, J. M. (2002) Postgenomic analysis of four novel antigens of group a streptococcus: growth phase-dependent gene transcription and human serologic response. *J. Bacteriol.* **184**, 6316–6324.

Rosenow, C., Ryan, P., Weiser, J. N., Johnson, S., Fontan, P., Ortqvist, A., and Masure, H. R. (1997) Contribution of novel choline-binding proteins to adherence, colonization and immunogenicity of *Streptococcus pneumoniae*. *Mol. Microbiol.* **25**, 819–829.

Rozalska, B., and Wadstrom, T. (1993) Protective opsonic activity of antibodies against fibronectin-binding proteins (FnBPs) of *Staphylococcus aureus*. *Scand. J. Immunol.* **37**, 575–580.

Sabarth, N., Hurwitz, R., Meyer, T. F., and Bumann, D. (2002) Multiparameter selection of *Helicobacter pylori* antigens identifies two novel antigens with high protective efficacy. *Infect. Immun.* **70**, 6499–6503.

Sanchez-Campillo, M., Bini, L., Comanducci, M., Raggiaschi, R., Marzocchi, B., Pallini, V., and Ratti, G. (1999) Identification of immunoreactive proteins of *Chlamydia trachomatis* by Western blot analysis of a two-dimensional electrophoresis map with patient sera. *Electrophoresis* **20**, 2269–2279.

Santamaria, H., Manoutcharian, K., Rocha, L., Gonzalez, E., Acero, G., Govezensky, T., Uribe, L. I., Olguin, A., Paniagua, J., and Gevorkian, G. (2001) Identification of peptide sequences specific for serum antibodies from human papillomavirus-infected patients using phage display libraries. *Clin. Immunol.* **101**, 296–302.

Satin, B., Del Guidice, G., Della Bianca, V., Dusi, S., Laudanna, C., Tonello, F., Kelleher, D., Rappouli, R., Montecucco, C., and Rossi, F. (2000) The neutrophyl-activating protein (HP-NAP) of *Helicobacter pylori* is a protective antigen and a major virulence factor. *J. Exp. Med.* **191**, 1467–1476.

Talkington, D. F., Brown, B. G., Tharpe, J. A., Koenig, A., and Russell, H. (1996) Protection of mice against fatal pneumococcal challenge by immunization with pneumococcal surface adhesin A (PsaA). *Microb. Pathog.* **21**, 17–22.

Taschner, S., Meinke, A., von Gabain, A., and Boyd, A. P. (2002) Selection of peptide entry motifs by bacterial surface display. *Biochem. J.* **367**, 393–402.

Utt, M., Nilsson, I., Ljungh, A., and Wadstrom, T. J. (2002) Identification of novel immunogenic proteins of *Helicobacter pylori* by proteome technology. *Immunol. Methods* **259**, 1–10.

Wizemann, T. M., Heinrichs, J. H., Adamou, J. E., Erwin, A. L., Kunsch, C., Choi, G. H., Barash, S. C., Rosen, C. A., Masure, H. R., Tuomanen, E., Gayle, A., Brewah, Y. A., Walsh, W., Barren, P., Lathigra, R., Hanson, M., Langermann, S., Johnson, S., and Koenig, S. (2001) Use of a whole genome approach to identify vaccine molecules affording protection against *Streptococcus pneumoniae* infection. *Infect. Immun.* **69**, 1593–1598.

11

Searching the Chlamydia Genomes for New Vaccine Candidates

Giulio Ratti, Oretta Finco and Guido Grandi

11.1   Old problems and new perspectives for chlamydial vaccines

The obligate intracellular bacteria, *Chlamydia trachomatis* and *Chlamydia pneumoniae* (*Chlamydophila pneumoniae*[1]) are widespread human pathogens, with a unique life cycle organization (see Chapter 12) that renders them very successful in avoiding host immune responses and establishing a chronic infection, often leading to serious disease. Both species can infect a wide range of eukaryotic cells; however, human chlamydial infection generally occurs in mucosal tissues and causes inflammatory disease leading to diverse pathological consequences, according to the site of infection and the type of immune response from the infected individual (Stephens, 1999; Peeling and Brunham, 1996). *C. trachomatis* chronic infection of the ocular mucosa can lead to blindness, or, in the female, infection of the upper genital tract can lead to pelvic inflammatory disease, ectopic pregnancy and sterility. Sexually transmitted disease (STD) induced by *C. trachomatis* has been also implicated as a

---

[1] The taxonomy of chlamydia has been recently revised with the institution of a new genus and species based on host specificity, and in particular the renaming of *Chlamydia pneumoniae* as *Chlamydophila pneumoniae*. While a discussion on the opportunity of such re-classification is still open among research scientists in the field (Schacter *et al.*, 2001; Stephens, 2003), we maintain in this chapter the old classification.

Genomics, Proteomics and Vaccines edited by Guido Grandi
© 2004 John Wiley & Sons, Ltd   ISBN 0 470 85616 5

risk factor for the sexual transmission of the human immunodeficiency virus (HIV). Since *in vitro* studies indicate that the presence of co-infecting chlamydiae increases several-fold the efficiency of monocyte infection by HIV (Ho *et al.*, 1995), chronic chlamydial infection of the genital tract could increase the risk of becoming infected, or infecting others. Worldwide prevalence rates of *C. trachomatis* infections according to the World Health Organization are shown in Box 11.1 (Datamonitor, 2001; http://www.datamonitor.com).

In contrast, *C. pneumoniae* appears to infect preferentially the respiratory tract (Grayston, 1992; Grayston *et al.*, 1990; Marrie *et al.*, 1987). In the US it is reported to be a significant cause of pneumonia, since each year an estimated 50 000 adults are hospitalized with pneumonia caused by *C. pneumoniae* and chlamydial infection is reported to account for approximately 7–10 per cent of cases of community-acquired pneumonia among adults. Unlike *Streptococcus pneumoniae*, which gives peak rates of infection in the winter months, rates of *C. pneumoniae* infection do not vary significantly by season (Jackson and Grayston, 2000).

Box 11.1 WHO estimates that approximately 89 million *C. trachomatis* infections occurred worldwide in 1997 alone. Moreover, *C. trachomatis* infection of the genital tract is considered as the leading sexually transmitted disease in the US: in 2000, 702 093 chlamydial infections were reported to CDC; 4 million cases/year were estimated at a rate of 254 cases/100 000.

Incidence of *Chlamydia trachomatis* in 15–49 year olds (millions)

Region	Male	Female
North America	1.64	2.34
Western Europe	2.30	3.20
Australasia	0.12	0.17
Latin America and the Caribbean	5.01	5.12
Sub-Saharan Africa	6.96	8.44
North Africa and Middle East	1.67	1.28
Eastern Europe and Central Asia	2.15	2.92
East Asia and Pacific	2.70	2.63
South and South-East Asia	20.20	20.28
Total	42.75	46.38

Source: Datamonitor *Drugs of tomorrow 2000 Vaccines* (Lancet)   D A T A M O N I T O R

According to US reports (The Jordan Report, 2002), the highest rates of pneumonia due to *C. pneumoniae* were among the elderly, with a case fatality of 6–23 per cent. However, *C. pneumoniae* has attracted a considerable amount of interest in recent years, mainly because chronic, and presumably asymptomatic, infection by this pathogen has been associated with the development of atherosclerosis and cardiovascular disease. This was initially suggested by seroepidemiologic studies, showing a very high prevalence of seropositivity in apparently healthy individuals (up to 50–80 per cent in some studies), essentially proportional to the mean age of the population, and by the detection of this organism in atherosclerotic plaques (Siscovick *et al.*, 2000; Kuo *et al.*, 1995, 1993). Following the initial statistical and clinical findings, further support was provided by experimental data from *in vitro* cell culture studies, and animal model studies, which actually point to a number of molecular mechanisms by which a persistent *C. pneumoniae* infection could favour the development of cardiovascular disease (Gaydos, 2000; Gupta and Camm, 1997; Krull *et al.*, 1999; Saikku *et al.*, 1998; Coombes, Johnson and Mahony, 2002; Coombes and Mahony, 2001; Fong *et al.*, 1999; Summersgill *et al.*, 2000; Blessing *et al.*, 2001).

Animal model studies have shown that *C. pneumoniae* can disseminate systemically through the bloodstream via infected monocytes and macrophages, and *in vitro* data suggest that chlamydia-infected macrophages can transmit the infection to endothelial cells (Moazed *et al.*, 1998). Therefore, after the initial infection of the respiratory tract, the real target of chronic *C. pneumoniae* infection may well be the endothelial cells of the blood vessels. In this scenario, reports showing that events caused by the peculiar life cycle of chlamydiae can generate oxidized LDL (Netea *et al.*, 2000; Kalayoglu *et al.*, 1999) are consistent with a possible involvement of *C. pneumoniae*, together with other co-factors, in the development or progress of atherosclerosis and related pathology. Other arguments in support of the proposed association could be mentioned; however, atherosclerosis being clearly a complex and multifactorial disease, several scientists are skeptical about the actual relevance of *C. pneumoniae* chronic infection for human cardiovascular disease. Therefore, this aspect of chlamydial pathogenicity can be considered so far an intriguing but still open question, certainly deserving further study.

Although chlamydial infections are susceptible to antibiotic treatment, and new effective treatment schedules have been introduced in clinical practice in recent years, vaccination still appears to be the most desirable approach for controlling chlamydial infections. This opinion stems essentially from the observed high rates of reinfection in some groups of patients, and from the fact that initial infection can be often scarcely symptomatic, or even completely asymptomatic, and therefore it can easily escape early diagnosis and potentially effective treatment. Furthermore, natural infection does not appear to confer an adequate protection and in some groups of patients reinfection rates can be high. Therefore, an effective anti-chlamydia vaccine would be the best solution for a control of chlamydial infections, both from the patient point of view and for the expected reduction in medical expenses.

Anti-chlamydia vaccination against sexually transmitted disease, administered to 12-year-old subjects, fell into the 'more favourable' category (category II) of recommended vaccines according to a committee convened by the US Institute of Medicine to study priorities for vaccine development (*Vaccines for the 21st Century*, Academy National Press). A further obvious consideration is the potential value of a vaccine directed at preventing chlamydial ocular infection in those developing countries where trachoma is still an important cause of blindness and of consequent social costs. However, in spite of years of efforts by several research groups around the world, a vaccine against human chlamydial infection is so far still unavailable.

The drawbacks in this field have been, besides the lack of a genetic transformation procedure for chlamydiae, the availability of only very few potential vaccine candidates (de la Maza and Peterson, 2002; Igietseme, Black and Caldwell, 2002). Essentially, only one of them,[2] the major outer membrane protein (MOMP) of *C. trachomatis* (an immunodominant antigen, with a probable trimeric porin structure partially exposed on the bacterial surface), attracted most of the interest as a vaccine. This happened because this protein for a long time was considered the only known antigen capable of inducing antibodies able to prevent initial host cell infection and subsequent intracellular replication and spread of infection. Unfortunately, MOMP presents in its native state a large epitopic variability among clinical isolates, and MOMP's neutralizing epitopes are composed of discontinuous peptide segments and are conformation dependent. Also, by its very nature as a complex membrane-interacting structure, MOMP (like other similar bacterial outer membrane antigens) is very difficult to prepare as a recombinant vaccine, in a native or quasi-native conformation. Therefore, for various reasons, the protective activity of MOMP has been so far only partial, and ultimately disappointing. Even attempts at delivering it by DNA vaccination did not quite meet initial expectations. DNA vaccination with the *ompA* gene encoding MOMP seemed to induce acquired immunity to *C. trachomatis* lung infection (Zhang *et al.*, 1997) in a murine model of pneumonia; however, vaccination of mice with

---

[2] A few other candidates have been proposed, but so far with minor impact. Excluding some preliminary studies that used DNA immunization, an approach that still presents important problems for a possible future human use, the most prominent candidate in the pre-genomic era is perhaps the anti-idiotypic antibodies to *C. trachomatis* exoglycolipid antigen (Whittum-Hudson *et al.*, 1996). Also, there is currently much hope in the newly described family of surface molecules named PMPs (polymorphic membrane proteins).

DNA plasmids expressing MOMP failed to protect against genital challenge (Pal *et al.*, 1999).

While tackling the problem of finding a good source of antibodies capable of neutralizing EB infectivity, it also became progressively clear that complete elimination of intracellular chlamydial infection could not rely solely on humoral responses but required an effective and adequate cell-mediated immune response. Several, sometime contradictory, studies indicated that both CD4 and CD8 positive T cells have a role in chlamydial clearance (Loomis and Starnbach, 2002); however, on the basis of extensive animal model studies over the last 15 years and immunological observations in patients, there is now a prevailing consensus on the fact that specific $CD4^+$ Th1 cells and B cells are critical for the complete clearance of intracellular chlamydiae and for mediating recall immunity to chlamydial infection (Igietseme, Black and Caldwell, 2002; Igietseme *et al.*, 1999). This is a fundamental observation from the point of view of vaccine development.

The protective role of $CD8^+$ T cells appears still questionable. A current opinion is that $CD8^+$ T cells could exert an *in vivo* anti-chlamydial activity as a result of their ability to produce IFN-γ and not based on their cytotoxic activity (Lampe *et al.*, 1998). However, it can be also noted that some of the apparent experimental discrepancies may be accounted for by the fact that optimal $CD8^+$ T cell function requires $CD4^+$ T cells (Yu *et al.*, 2003; Sun and Bevan, 2003; Shedlock and Shen, 2003). A constantly updated review of this and other topics of chlamydial biology and pathogenicity can be obtained by consulting the excellent web site coordinated by M. E. Ward (Southampton University, UK) at http://www.chlamydiae.com.

Quoting a recent review on vaccine development (de la Maza and Peterson, 2002), 'based on our current knowledge, an ideal vaccine should induce long lasting (neutralizing) antibodies, and a cell mediated immunity that can quickly respond with the production of IFNγ upon exposure to Chlamydia. As of today, this represents a phenomenal challenge'. Considering all the (microbiological, immunological and biotechnological) hurdles of developing an effective, therapeutic or prophylactic vaccination against a well adapted intracellular parasite such as Chlamydia, one cannot but agree with the above statement. However, the large amount of new information on chlamydia made available by recent genomic, proteomic and transcriptomic studies (Kalman *et al.*, 1999; Stephens *et al.*, 1998; Read *et al.*, 2000; Shirai *et al.*, 2000b; Read *et al.*, 2003; Shirai *et al.*, 2000a; Rockey, Stephens and Lenart, 2000; Montigiani *et al.*, 2002; Vandahl *et al.*, 2002a,b; Shaw *et al.*, 2002a,b,c; Vandahl *et al.*, 2001a,b, Shaw *et al.*, 2000a; Shaw, Christiansen and Birkelund, 1999; Shaw *et al.*, 2000b; Bini *et al.*, 1996; Nicholson *et al.*, 2003) has now opened, as in the case of other 'difficult' pathogens such as *Neisseria meningitidis* type B (Pizza *et al.*, 2000) a completely new perspective for the development of an effective anti-chlamydia vaccine. The positive new outlook essentially relies on the reasonable bet that if one can find, by exhaustive whole-genome searches and parallel proteome characterization, a high enough number of viable vaccine candidates, eventually some chlamydial antigen, or better a combination of two or more chlamydial antigens, will eventually prove to have all the right properties required for anti-chlamydia protection in humans.

11.2 Post-genomic approaches

The ways to exploit the new wealth of information provided by genome sequencing data for the development of new anti-bacterial vaccines could be essentially grouped into two categories:

1. *the analytical approach* which aims first at understanding the structure and physiology of the bacterial pathogen and then at selecting possible vaccine candidates on the basis of gene function, and

2. *the screening approach*, which starts from an initial pool of genes of unknown function (which could even coincide with all the ORFs in a genome, but more conveniently a subset of ORFs selected on the basis of *in silico* predictions) whose encoded products are then expressed as recombinant proteins and progressively 'filtered' on the basis of a chosen set of physical–chemical and biological parameters.

Systematic analysis of genome usage by a pathogenic bacterium – using proteomic or microarray hybridization technologies – may well lead eventually to the identification of potential vaccine candidates for further assessment; however, the alternative approach of expressing a large number of selected ORFs in one or more genome variants (high-throughput cloning and expression), followed by multiple experimental filtering, has proven to date a preferable choice for an industrial R&D vaccine project (see Chapter 8).

One advantage of the multiple-screening approach is that it takes a short-cut by trying to obtain the highest number of molecules suitable for vaccine development, without necessarily knowing the function of the molecules or their role in microbial pathogenicity. Obviously, in any one multiple-screening project the choice of the selection procedures (i.e. type and number of screenings and their relative read-outs) is critical (see Chapter 2). Screening strategies need to start from an initial definition of the pathology one wants to counteract with a vaccine, and proceed by taking into account the parameters that basic research studies have found to correlate in model systems with a successful neutralization of infectivity and elimination of the pathogen from the infected host.

11.3 Genomic screening results

11.3.1 *C. pneumoniae* vaccine candidates

Our rationale for vaccine candidate selection stems from the already mentioned observation that for an anti-chlamydia vaccine to be effective both neutralizing antibodies and CD4$^+$ T cells should be induced. Antibodies are important to prevent EBs from invading human tissues, CD4$^+$ T cells have the role of eliminating infected cells, most likely through the action of INF-γ. Considering that chlamydial mechanisms mediating host cell adhesion, invasion and colonization, and possibly also the ability to cope with host immune responses, presumably rely in large part on EB surface organization, we reasoned that by identifying a large number of membrane-associated surface proteins we could eventually identify antigens capable of inducing both antibodies and cell-mediated immunity. Therefore, our strategy for vaccine discovery was based on the following experimental steps (Figure 11.1): (a) analysis of Chlamydia genome sequence to select putative membrane-associated antigens, (b) cloning, expression and purification selected antigens, (c) preparation of antigen-specific sera by mouse immunization with the purified antigens, (d) FACS analysis of chlamydia EBs using the mouse sera to identified surface-exposed antigens, (e) '*in vitro* neutralization' assay to test whether antibodies elicited by a given antigen can interfere with the process of eukaryotic cell infection and (f) use of an appropriate animal model to test the capacity of selected antigens to confer protection against chlamydia challenge.

As recently described by Montigiani *et al.* (2002), from the initial screening of the *C. pneumoniae* genome, a panel of mouse sera was prepared against over 170 recombinant His-tagged or GST-fusion proteins encoded by genes or 'open reading frames' somehow predicted to be peripherally located in the chlamydial cell. When these antibodies were tested in a FACS assay (Figure 11.2) for their ability to bind the surface of purified *C. pneumoniae* EBs, a list of 53 'FACS-positive' sera was obtained. The corresponding putative surface antigens (see Table 11.1) were then further assessed for their capability of inducing neutralizing antibodies. This part of the work (Finco *et al.*, unpublished results) involved testing which of the sera contained antibodies capable of interfering with the

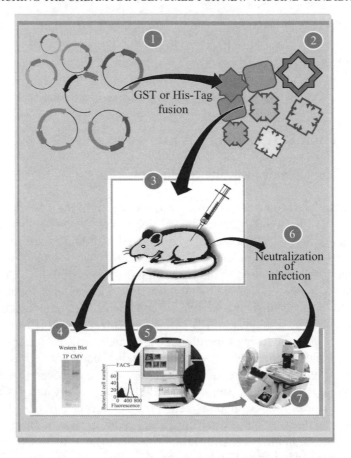

Figure 11.1 Experimental approach for the definition of a short list of potential chlamydial vaccine molecules, starting from a set of genomic ORFs selected *in silico*. The selected ORFs are cloned in *E.coli* plasmid expression vectors (**1**) to obtain a set of purified, preferably water soluble, recombinant fusion proteins (**2**). These are used to immunize mice and obtain antigen-specific antibodies (**3**). The antibody panel is then used for the detection of the native protein in purified EBs by immunoblotting techniques (**4**), for probing the accessibility of the corresponding antigen on the surface of chlamydial cells in a FACS based assay (**5**) and for testing which of these antibodies can inhibit the infectivity of purified chlamydial EBs for *in vitro* cell cultures (**6**). In the neutralization assay cell monolayers are infected in the presence and absence of antibodies and after 48 h the chlamydial cytoplasmic inclusions are stained with chlamydia-specific monoclonal antibodies and counted with a microscope (**7**)

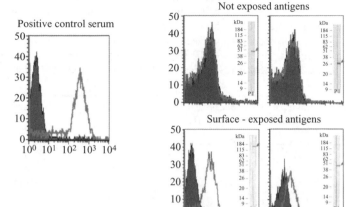

Figure 11.2 Identification of surface accessible epitopes on chlamydial cells by a FACS assay. Antibodies to be tested are labelled with a fluorescent dye and allowed to bind to chlamydial cells. Chlamydiae that have antibodies bound on their surface are separated by FACS associated software from the unreacted cells and the extent of labelling is evaluated. The shift between the histograms shown in the figure was taken as a measure of antibody binding to the EB cell surface. We routinely adopted as a read-out the Kolmorov–Smirnov (K–S) two-sample test and the $D/s(n)$ values (an index of dissimilarity between the two curves) are reported in Table 10.1 as 'K–S score'

process of *in vitro* infection of epithelial cell cultures. In the *in vitro* 'neutralization' assay purified infectious EBs are incubated with progressive dilutions of the immune sera and, in parallel, dilutions of the corresponding pre-immune sera, and of sera against non-chlamydial control antigens. Cell cultures are infected in the presence of cycloheximide, which inhibits host cell protein synthesis and favours chlamydial intracellular growth with the consequent formation of typical cytoplasmic inclusions, which can be stained with chlamydia-specific fluorescence-labelled monoclonal antibodies and counted with an UV light microscope. Working with appropriate pathogen-to-host cell ratios, it can be reasonably assumed that the number of detected cytoplasmic inclusion is proportional to the number of infectious chlamydiae in the original sample, so a reduction in inclusion numbers caused by the presence of an antigen-specific antiserum, as compared with the numbers obtained with control sera, gives a measure of the capability of a given antigen to elicit antibodies that can inhibit some stage of the chlamydial infection process. According to common

Table 11.1 The FACS-positive set of *C. pneumoniae* proteins. Corresponding gene ID in the CWL029 genome, according to Kalman *et al.* (1999), and PID accession numbers of the corresponding protein are reported in the first two columns. The third column shows the type of recombinant fusion protein to which the reported data refer. Positive FACS assay results are reported with the K–S score (index of curve dissimilarity as compared with controls) as explained in the main text. The names of the genes of 28 proteins that have been shown to be present in *C. pneumoniae* purified EB extracts by 2DE mapping and MALDI/TOF analysis are shaded

Gene/ORF ID in CWL029	Encoded protein ID (PID)	Recombinant fusion type	FACS screen (K–S score)	Annotations
CPn0005	4376260	GST	17.55	Pmp_1
CPn0013	4376270	GST	14.70	Pmp_2
CPn0017	–	GST	18.51	Pmp_4.1 (C-terminal half of frameshifted pmp_4)
CPn0444	4376727	His	36.32	Pmp_6
CPn0446	4376729	GST	30.88	Pmp_8
CPn0447	4376731	GST	41.94	Pmp_9
CPn0449&450	(AJ133034)	GST	40.24	Pmp_10
CPn0451	4376733	His	38.82	Pmp_11
CPn0453	4376736	GST	32.38	Pmp_13
CPn0454	4376737	GST	10.77	Pmp_14
CPn0466	4376751	GST	26.76	Pmp_15
CPn0467	4376752	His	26.06	Pmp_16
CPn0540	4376830	GST	38.16	Pmp_20
CPn0963	4377287	GST	40.44	Pmp_21
CPn0020	4376272	GST	29.27	Predicted OMP
CPn0021	4376273	GST	14.11	Predicted OMP
CPn0195	4376466	His	8.92	Oligopeptide binding protein oppA_1; lipoprotein?
CPn0196	4376467	GST	35.75	Oligopeptide binding protein oppA_2; lipoprotein?
CPn0278	4376552	GST	27.98	Conserved outer membrane lipoprotein
CPn0300	4376576	GST	7.36	yaeT (omp85 homologue)
CPn0301	4376577	GST	10.05	Similarity to OmpH-like outer membrane proteins
CPn0324	4376602	GST	21.12	lcrE (low calcium response E)
CPn0385	4376664	GST	20.02	pepA (leucyl aminopeptidase A)
CPn0482	4376767	GST	20.16	art J (arginine binding protein?)
CPn0503	4376790	His	44.58	dnaK (hsp70 heat shock protein)
CPn0557	4376849	GST	31.02	omcB (60 kDa cysteine-rich OMP)
CPn0558	4376850	GST	20.45	omcA (9 kDa cystein-rich lipoprotein)
CPn0562	4376854	GST	12.52	CHLPS 43 kDa protein homologue-1
CPn0584	4376878	GST	14.61	atoS (2-Component sensor histidine kinase)

Table 11.1 (*continued*)

Gene/ORF ID in CWL029	Encoded protein ID (PID)	Recombinant fusion type	FACS screen (K–S score)	Annotations
CPn0604	4376900	GST	14.80	fliY (glutamine binding protein)
CPn0661	4376960	GST	16.26	mip (FKBP peptidyl–prolyl *cis–trans* isomerase)
Pn0695	4376998	GST	17.01	ompA (major outer membrane protein)
CPn0728	4377033	GST	32.21	76 kDa protein homologue_1 (CT622)
CPn0800	4377111	GST	11.64	eno (enolase)
CPn0828	4377140	GST	11.60	yscJ (YOP translocation J protein)
CPn0854	4377170	GST	9.86	porB (outer membrane protein analogue)
CPn0904	4377224	GST	21.17	murG (peptidoglycan transferase)
CPn0979	4377306	GST	28.67	htrA (DO serine protease)
CPn0186	4376456	GST	36.68	Conserved hypothetical, weak similarity to CT768
CPn0415	4376696	GST	29.95	CT266 hypothetical protein
CPn0514	4376802	GST	9.99	CT 427 hypothetical protein
CPn0525	4376814	GST	12.43	CT 398 hypothetical protein
CPn0668	4376968	GST	10.13	CT 547 hypothetical protein
CPn0791	4377101	GST	25.95	CT590 hypothetical protein
CPn0792	4377102	GST	10.05	CT 589 hypothetical protein
CPn0820	4377132	His	12.20	CT 567 hypothetical protein
CPn0042	4376296	GST	22.77	hypothetical protein
CPn0126	4376390	GST	12.95	hypothetical protein
CPn0498	4376784	GST	17.08	hypothetical protein
CPn0794	4377105	GST	11.49	hypothetical protein
CPn0795	4377106	GST	14.84	hypothetical protein
CPn0796	4377107	His	11.14	hypothetical protein
Cpn0797	4377108	GST	22.17	hypothetical protein

convention, an anti-serum is labelled as 'neutralizing' when the reduction of infectivity is equal or greater than 50 per cent, and the serum dilution yielding a 50 per cent reduction in infectivity is referred to as the 50 per cent end-point neutralization titre.

Some of the results obtained by screening the panel of recombinant antigens with the *C. pneumoniae in vitro* neutralization assay are shown in Table 11.2. Just by a cursory look at the 'current annotation' column it can be seen that in both Tables 11.1 and 11.2 are listed antigens, such as the members of the family of heterogeneous polymorphic membrane proteins (PMPs) that, on the basis of published literature data, could be reasonably expected to be surface exposed and possibly induce neutralizing antibodies, but there are also proteins that could be considered so far to be only hypothetical, and proteins that just on the

Table 11.2   Example of results obtained by screening a genomic selection of water-soluble recombinant *C. pneumoniae* antigens for their capacity of inducing antibodies able to bind the chlamydial cell surface (FACS assay) and to inhibit *C. pneumoniae* infectivity *in vitro*. 50 per cent reciprocal neutralization end-point titres are reported. In the rightmost column are shown also the results obtained with the same antigens *in vivo* using the hamster spleen protection assay described in the text. Gene names in the first column refer to the first *C. pneumoniae* genome sequencing project (http://chlamydia www.berkeley.edu:42311)

Gene/ORF ID in CWL029	Protein ID	Recombinant fusion type	Annotation	Reciprocal of 50% neutralization titre	% protection in the hamster spleen test
CPn0449&Cpn0450	AJ133034	GST	pmp 10	640	n.a.
CPn0042	4376296	GST	hypothetical protein	120	80
CPn0186	4376465	GST	hypothetical protein	80	93
CPn0301	4376577	GST	hypothetical (similarity to CT242 'Omp-H-like' outer membrane protein)	1000	83
CPn0482	4376767	GST	artJ–arginine binding protein?	160	n.a.
CPn0558	4376850	GST	omcA, 9 kDa cysteine-rich lipoprotein	380	63
CPn0584	4276878	GST	atoS, 2-component sensor	520	51
CPn0795	4377106	GST	hypothetical protein	640	86
CPn0800	4377111	GST	eno, enolase	640	80
CPn0904	4377224	GST	murG, peptidoglycan transferase	300	56
CPn0979	4377306	GST	htrA, serine protease	380	n.a.
CPn0498	4376784	GST	hypothetical protein	0	94
CPn0525	4376814	GST	hypothetical protein (similarity to CT398)	0	97

n.a, not assessed.

basis of their current functional annotation could not be expected at all to be found on the bacterial surface.

11.3.2   *C. trachomatis* vaccine candidates

The same screening protocols as described for the *C. pneumoniae* vaccine search were subsequently applied to the *C. trachomatis* serovar D, D/UW-3/Cx genome (Stephens *et al.*, 1998). This time however, before engaging in a whole-genome analysis for a *de novo* selection of ORFs to be entered in a cloning-expression list, we preferentially examined *C. trachomatis* proteins having some degree of homology to *C. pneumoniae* antigens already found to elicit either FACS-positive (surface binding) or infectivity neutralizing antibodies (Bonci *et al.*, unpublished results). In Table 11.3 are shown the results of FACS and selected '*in vitro* neutralization' assays obtained from sera raised against a set of 93 *C. trachomatis* recombinant fusion proteins. Although the screening for neutralizing antigens has not been completed yet, we have so far identified 11 'neutralizing' antigens. With the exception of MOMP (Caldwell and Perry, 1982; Peterson *et al.*, 1991; Su and Caldwell, 1991; Zhong, Berry and Brunham, 1994; Fan and Stephens, 1997), none of them had been previously reported as neutralizing. Previous literature also describes PorB as a second neutralizing protein (Kawa and Stephens, 2002; Kubo and Stephens, 2000). However, as shown in Table 10.3, the serum against our recombinant form of PorB failed to neutralize chlamydia infection *in vitro*. This discrepancy could be explained considering that PorB is a membrane-interacting protein and our recombinant antigen might have lost the correct conformation required to induce neutralizing antibodies. The possibility of a similar situation should also be kept in mind in the interpretation of data relative to the other 'insoluble' antigens. It is interesting to note that, besides MOMP, other proteins in this selection, including PepA, DnaK, HtrA and PorB, have been reported as proteins that are immunogenic in the course of genital tract infection in humans (Sanchez-Campillo *et al.*, 1999). This is an interesting 'tag' for the selection of a candidate human vaccine, which anticipates a critical property for a vaccine candidate, i.e. that of being immunogenic in humans.

A comparison of the proteins selected in Tables 11.2 and 11.3 from *C. pneumoniae* and *C. trachomatis* genomic screenings shows that, excluding members of the PMP family, eight antigens encoded by homologous genes are present in both pathogens' lists as candidates for further assessment. All of them have *in vitro* neutralizing properties for *C. pneumoniae* and five are also *in vitro* 'neutralizing' antigens for *C. trachomatis*. Interestingly, the 'hypothetical' protein encoded by CT398, 80 per cent similar to the product of Cpn0525, yielded an antiserum with an *in vitro* neutralization titre of 1/800 for *C. trachomatis* while the antiserum obtained against the corresponding *C. pneumoniae* product gave negative *in vitro* results; however, in a hamster immunization assay (see below) the Cpn0525 product gave 97 per cent protection from spleen infection.

Relevant to the focus of this review is the observation that a positive signal obtained in the FACS assay does not guarantee a corresponding neutralization activity and conversely a negative FACS assay result for a given antibody does

Table 11.3 Example of results which can be obtained by screening a genomic selection of *C. trachomatis* recombinant fusion proteins for their capacity to induce antibodies able to bind the chlamydial cell surface (FACS assay) and inhibit *C. trachomatis* infectivity *in vitro*. This group of proteins was selected for their homology to *C. pneumoniae* gene products already shown to give positive results in at least one of previous screening tests. *C. trachomatis* gene names in the first column refer to the first *C. trachomatis* genome sequencing project (http://chlamydia-www.berkeley.edu:4231/) and are ordered according to decreasing homology levels to *C. pneumoniae* gene products. (*) In these cases the antisera were raised against solubilized forms of antigens originally obtained as insoluble inclusion bodies (ib), so some conformation-dependent epitopes may have been lost

GENE ID in Serovar D (D/UW-3/Cx)	% homology with Cpn-encoded protein	Annotation	Fusion type	FACS K–S score	Reciprocal of 50% neutralization titre
CT 316	87	L7/L12 ribosomal protein	His	9.08	n.a.
CT 396	84	DnaK, HSP-70	His	34.5	300
CT 396	84	"	GST	32.76	0
CT 398	82	hypothetical protein	GST	31.24	800
CT 398	82	"	His	26.1	640
CT 587	73	Eno, enolase	His	20.85	1200
CT 559	72	YscJ, Yop proteins translocation lipoprotein J	His	23.21	0
CT 443	71	OmcB, 60kDa cysteine-rich OMP	His	21.28	0
CT 045	69	PepA, Leucyl aminopeptidase A	His	16.81	160
CT 823	69	HtrA, DO serine protease	His	26.62	n.a.
CT 600	68	Pal, peptidoglycan-associated lipoprotein	His	10.46	n.a.
CT 541	62	Mip, FKBP-type peptidyl–prolyl *cis-trans* isomerase	GST	9.94	n.a.
CT 681	62	OmpA, major outer membrane protein	(*) His-ib	34.66	160
CT 467	61	AtoS; 2-component regulatory system–sensor Histidine kinase	GST	0	1100

CT number	Description	Tag	Value	
CT 381	ArtJ, arginine binding protein	His	32.54	450
CT 381		GST	35.98	0
CT 242	('OmpH-like'/ outer membrane protein)	His	0	100
CT 623	CHLPN 76 kDa homologue	(*) His-ib	20.27	n.a.
CT 623		(*) GST-ib	15,89	n.a.
CT 713	PorB, outer membrane protein analogue	(*) His-ib	25.82	0
CT 761	MurG, peptidoglycan transferase	(*) GST-ib	11.45	n.a.
CT 700	hypothetical protein	(*) GST-ib	8.72	n.a.
CT 547	hypothetical protein	His	28.21	40
CT 547		GST	15,57	n.a.
CT 444	OmcA, 9 kDa-cysteine-rich lipoprotein	GST	15	n.a.
CT 444		His	13,28	>40
CT 266	hypothetical protein	His	21.29	n.a.
CT 077	hypothetical protein	His	9.17	0
CT 414	PmpC, putative outer membrane protein C	(*) GST-ib	8.03	n.a.
CT 456	hypothetical protein	GST	10.9	n.a.
CT 089	LcrE, low calcium response protein E	His	12.59	n.a.
CT 812	PmpD, putative outer membrane protein D	GST	10.43	n.a.
CT 871	PmpG, putative outer membrane protein G	(*) His-ib	8.32	n.a.
CT 869	PmpE, putative outer membrane protein E	(*) His-ib	15.28	n.a.
CT 165	hypothetical protein	GST	10.46	n.a.

n.a., not assessed.

not mean that a neutralizing activity can be excluded. It is clear that several variables (such as conformation and immunogenic properties of the recombinant antigens, level of expression of each antigen *in vivo* etc., chlamydial isolate and type of cell line used in the assays, etc.) can contribute to the appearance of false positive or negative results when adopting screening approaches where there is a trading off between the collection of a large number of data and the benefit of a more detailed scientific investigation. It is therefore recommendable that having obtained the best possible panel of soluble recombinant antigens, all chosen screening procedures should be carried out in parallel independently from the results of each type of screening test. The final short listing for the selection of antigens to carry over into further individual and more extensive evaluation should be done at the end by comparing the performance of each antigen in all screening tests.

11.3.3 *In vivo* evaluations

A further step in the vaccine candidate list refinement was the establishment of an *in vivo* screening assay in laboratory animals in order to probe also the contributions to infectivity neutralization properties that may come from cell-mediated (cytokine-mediated) immune responses. As in all screening procedures there is a need to have a test with high efficiency, simple enough to be carried out on large numbers of antigens, and with a fairly straightforward and robust read-out. In the case of *C. pneumoniae*, since previously described animal models of infection seemed to lack, at least in our hands, one or more of these characteristics, we adopted a new model of infection recently developed in the hamster by Prof. R. Cevenini and collaborators at the Department of Microbiology, Bologna University, Italy (Sambri *et al.*, 2004). This group found that intraperitoneal injection of *C. pneumoniae* EBs in adult Syrian hamsters causes a systemic infection allowing cell-culture isolation of viable chlamydiae from several organs for several days post infection (p.i.). In particular, spleen infection occurs in 100 per cent of injected animals, who eventually all recover from day 20 p.i. onwards. Previous immunization of two groups of nine hamsters with heat-inactivated purified EB completely protected the spleens of 16 of the 18 animals, and substantially reduced infection levels in the remaining two, so this model appeared to provide a suitable screening tool for the assessment of the protective activity of potential vaccine candidates. Intraperitoneal infection in the hamster was therefore used to evaluate the ability of the recombinant antigens to prevent hamster spleen infection. Animals were immunized subcutaneously with three doses of 20 µg of protein, and challenged with 2×10^8 CFU (colony forming units) of *C. pneumoniae*. Spleen infection

infection was assessed 7 d after challenge by tissue culture isolation of chlamydiae from spleen tissue homogenates. *In vivo* protection was estimated by counting the numbers of chlamydia inclusions obtained from samples of immunized groups of hamsters as compared with non-immunized or mock-immunized hamsters. In Table 11.2 some of the results obtained from this assay for this group of 13 antigens are reported in the rightmost column.

With reference to the above comment on the advisability of performing parallel screenings rather that sequential selections, it can be noted from Tables 10.1 and 10.2 how the two 'hypothetical' proteins 6784 and 6814 (encoded by the ORFs Cpn0498 and Cpn0525) yielded FACS-positive sera, which, however, were not able to neutralize host cell infection *in vitro*. According to a sequential strategy these antigens might have been excluded from more expensive and time-consuming *in vivo* evaluations, but in fact these antigens performed remarkably well in the hamster-spleen test and were not excluded from further assessment. When a more detailed study on the immunogenicity of the protein encoded by Cpn0498 was carried out in a mouse model of respiratory infection, immunization with this antigen proved to be able to induce, besides the expected antibody response, also specific, IFNγ-secreting CD4$^+$ cells, which expand and infiltrate the lungs upon respiratory challenge with *C. pneumoniae* (Tracy Hussell, Imperial College London, private communication). Also, in immunized C57BL/6 mice the level of chlamydial infection in the lung was reduced, as compared with non-immunized controls, to a level similar to the degree of protection conferred upon re-challenge by a primary chlamydial infection. In this case, therefore, the triple parallel-screening evaluation, with two positive and one negative result, brought about the identification of a previously unknown antigen having, according to current views, exactly the desirable basic properties of a vaccine candidate, at least in this particular strain of mice.

While for *C. trachomatis* the primary screening work is, at this date, still in progress in our laboratory, we would predict that when this project reaches the stage of *in vivo* immunological and protection studies in animal models of infection and disease, similarly promising results will also be obtained for this pathogen.

As it is a well known observation that different strains of mice, or in general hosts with different genetic backgrounds, have different susceptibilities to chlamydial infection and consequent disease, it can be expected that potentially protective antigens may have a widely different efficacy in different hosts. One of the strong points in favour of the genomic screening approach is that it is expected to yield not one but a consistent number of potential vaccine candidates – with different score profiles in the panel of screening tests – to be used in complementary combinations.

11.4 Concluding considerations

Extensive work throughout the years on various animal models of chlamydial infection provided a fundamental, and in fact essential, contribution to the definition of the pathological mechanisms that may underlie chlamydia induced disease in humans. This comment extends from the early pioneering work on mouse salpingitis induced by *C. trachomatis* (Tuffrey, Falden and Taylor-Robinson, 1982; Tuffrey *et al.*, 1986a,b), to the above-mentioned recent studies aiming at explaining how *C. pneumoniae* chronic infection might promote the development of atherosclerosis and cardiovascular disease, in support of the still controversial epidemiological findings on human populations. Also, animal model work, at least in the case of *C. trachomatis* genital infection, has the merit of having eventually established a common consensus on the type of immune response that is essential for an efficient and successful clearance of the infection from the organism. Such an agreement within the scientific community is very important because it provides health authorities with criteria for granting permission for human clinical trials.

Again stemming from animal model work, it is now becoming clearer that the type of cell-mediated immune response mounted by the infected individual, and its modulation by the cytokine/chemokine regulatory network, is crucial for deciding whether the outcome of a given infection will be a quick resolution with no serious consequences, or rather persistence and reinfection with consequent development of disease due to the chronic localized inflammation. This view of chlamydia-induced disease (which would be therefore caused by the infection process itself and not by antigenic mimicry mechanisms) is in fact favourable to the task of developing a chlamydial vaccine, since it minimizes the prospect of having, at the stage of human clinical trials, unwanted antigen-dependent delayed-type hypersensitivity or adverse autoimmunity effects induced by a chlamydial vaccine component.

However, in spite of a great recent progress in human immunology parallel to the microbial genomics era, we do not know yet how to safely translate immune responses from animal models to humans, nor do we well know how to artificially modulate the type of immune response to a given antigen (vaccine), an outstanding challenge to human chlamydial vaccine development is the poor knowledge we have of correlates of protection and/or quick clearance in humans. In this perspective the larger the list of potential human vaccine candidates the better the chance of obtaining satisfactory results in human populations, so the genomic screening approach providing a wide repertoire of potential vaccines, to be tested in different combinations, may well be the winning way to anti-chlamydia vaccine development.

11.4.1 Summary outline

Systematic analysis of proteome usage by a pathogenic bacterium may well eventually lead to the identification of potential vaccine candidates for further assessment; however, the alternative approach of expressing a large number of selected open reading frames in one or more genome variants (high-throughput cloning and expression), followed by multiple experimental filtering, appears a preferable choice for a vaccine R&D project.

This is particularly true for chlamydial vaccine development, which still presents with a number of unresolved problems concerning both the molecular pathogenicity of chlamydiae and the immunology of host responses to chlamydial infection and to anti-chlamydial vaccination. Such an industrial 'screening' approach applied to the *C. pneumoniae* genome is already showing promising results and is currently being extended to *C. trachomatis*.

The final challenge facing anyone wishing to reach the final goal of an effective human vaccine is to succeed in obtaining a number of positive preclinical findings adequate for entering human trials.

References

Bini, L., Sanchez-Campillo, M., Santucci, A., Magi, B., Marzocchi, B., Comanducci, M., Christiansen, G., Birkelund, S., Cevenini, R., Vretou, E., Ratti, G., and Pallini, V. (1996) *Electrophoresis* **17**, 185–190.

Blessing, E., Campbell, L. A., Rosenfeld, M. E., Chough, N., and Kuo, C. C. (2001) *Atherosclerosis* **158**, 13–17.

Caldwell, H. D., and Perry, L. J. (1982) *Infect. Immun.* **38**, 745–754.

Coombes, B. K., Johnson, D. L., and Mahony, J. B. (2002) *Curr. Drug Targets Infect. Disord.* **2**, 201–216.

Coombes, B. K., and Mahony, J. B. (2001) *Infect. Immun.* **69**, 1420–1427.

Datamonitor (2001) *Drugs of Tomorrow 2000 Vaccines: A New Era of Disease Targets.*

de la Maza, L. M., and Peterson, E. M. (2002) *Curr. Opin. Investig. Drugs* **3**, 980–986.

Fan, J., and Stephens, R. S. (1997) *J. Infect. Dis.* **176**, 713–721.

Fong, I. W., Chiu, B., Viira, E., Jang, D., and Mahony, J. B. (1999) *Infect. Immun.* **67**, 6048–6055.

Gaydos, C. A. (2000) *J. Infect. Dis.* **181**, S473–S478.

Grayston, J. T. (1992) *Annu. Rev. Med.* **43**, 317–323.

Grayston, J. T., Campbell, L. A., Kuo, C. C., Mordhorst, C. H., Saikku, P., Thom, D. H., and Wang, S. P. (1990) *J. Infect. Dis.* **161**, 618–625.

Gupta, S., and Camm, A. J. (1997) *Clin. Cardiol.* **20**, 829–836.

Ho, J. L., He, S., Hu, A., Geng, J., Basile, F. G., Almeida, M. G., Saito, A. Y., Laurence, J., and Johnson, W. D. Jr. (1995) *J. Exp. Med.* **181**, 1493–1505.

Igietseme, J. U., Ananaba, G. A., Bolier, J., Bowers, S., Moore, T., Belay, T., Lyn, D., and Black, C. M. (1999) *Immunology* **98**, 510–519.

Igietseme, J. U., Black, C. M., and Caldwell, H. D. (2002) *BioDrugs* **16**, 19–35.

Jackson, L. A., and Grayston, J. T. (2000) In Mandell, G. L., Bennett, J. E., and Dolin, R. (Eds.). *Mandell Douglas & Bennett's Principles and Practice of Infectious Diseases 5th edn.* Churchill Livingstone, 2007–2014.

The Jordan Report, 20th Anniversary, Accelerated Development of Vaccines 2002. U.S. Department of Health and Human Services, National Institutes of Health, National Institute of Allergy and Infectious Diseases.

Kalayoglu, M. V., Hoerneman, B., LaVerda, D., Morrison, S. G., Morrison, R. P., and Byrne, G. I. (1999) *J. Infect. Dis.* **180**, 780–790.

Kalman, S., Mitchell, W., Marathe, R., Lammel, C., Fan, J., Hyman, R. W., Olinger, L., Grimwood, J., Davis, R. W., and Stephens, R. S. (1999) *Nat. Genet.* **21**, 385–389.

Kawa, D. E., and Stephens, R. S. (2002) *J. Immunol.* **168**, 5184–5191.

Krull, M., Klucken, A. C., Wuppermann, F. N., Fuhrmann, O., Magerl, C., Seybold, J., Hippenstiel, S., Hegemann, J. H., Jantos, C. A., and Suttorp, N. (1999) *J. Immunol.* **162**, 4834–4841.

Kubo, A., and Stephens, R. S. (2000) *Mol. Microbiol.* **38**, 772–780.

Kuo, C. C., Grayston, J. T., Campbell, L. A., Goo, Y. A., Wissler, R. W., and Benditt, E. P. (1995) *Proc. Natl. Acad. Sci. USA* **92**, 6911–6914.

Kuo, C. C., Shor, A., Campbell, L. A., Fukushi, H., Patton, D. L., and Grayston, J. T. (1993) *J. Infect. Dis.* **167**, 841–849.

Lampe, M. F., Wilson, C. B., Bevan, M. J., and Starnbach, M. N. (1998) *Infect. Immun.* **66**, 5457–5461.

Loomis, W. P., and Starnbach, M. N. (2002) *Curr. Opin. Microbiol.* **5**, 87–91.

Marrie, T. J., Grayston, J. T., Wang, S. P., and Kuo, C. C. (1987) *Ann. Intern. Med.* **106**, 507–511.

Moazed, T. C., Kuo, C. C., Grayston, J. T., and Campbell, L. A. (1998) *J. Infect. Dis.* **177**, 1322–1325.

Montigiani, S., Falugi, F., Scarselli, M., Finco, O., Petracca, R., Galli, G., Mariani, M., Manetti, R., Agnusdei, M., Cevenini, R., Donati, M., Nogarotto, R., Norais, N., Garaguso, I., Nuti, S., Saletti, G., Rosa, D., Ratti, G., and Grandi, G. (2002) *Infect. Immun.* **70**, 368–379.

Netea, M. G., Dinarello, C. A., Kullberg, B. J., Jansen, T., Jacobs, L., Stalenhoef, A. F., and Van Der Meer, J. W. (2000) *J. Infect. Dis.* **181**, 1868–1870.

Nicholson, T. L., Olinger, L., Chong, K., Schoolnik, G., and Stephens, R. S. (2003) *J. Bacteriol.* **185**, 3179–3189.

Pal, S., Barnhart, K. M., Wei, Q., Abai, A. M., Peterson, E. M., and de la Maza, L. M. (1999) *Vaccine* **17**, 459–465.

Peeling, R. W., and Brunham, R. C. (1996) *Emerg. Infect. Dis.* **2**, 307–319.

Peterson, E. M., Cheng, X., Markoff, B. A., Fielder, T. J., and de la Maza, L. M. (1991) *Infect. Immun.* **59**, 4147–4153.

Pizza, M., Scarlato, V., Masignani, V., Giuliani, M. M., Arico, B., Comanducci, M., Jennings, G. T., Baldi, L., Bartolini, E., Capecchi, B., Galeotti, C. L., Luzzi, E.,

Manetti, R., Marchetti, E., Mora, M., Nuti, S., Ratti, G., Santini, L., Savino, S., Scarselli, M., Storni, E., Zuo, P., Broeker, M., Hundt, E., Knapp, B., Blair, E., Mason, T., Tettelin, H., Hood, D. W., Jeffries, A. C., Saunders, N. J., Granoff, D. M., Venter, J. C., Moxon, E. R., Grandi, G., and Rappuoli, R. (2000) *Science* **287**, 1816– 1820.

Read, T. D., Brunham, R. C., Shen, C., Gill, S. R., Heidelberg, J. F., White, O., Hickey, E. K., Peterson, J., Utterback, T., Berry, K., Bass, S., Linher, K., Weidman, J., Khouri, H., Craven, B., Bowman, C., Dodson, R., Gwinn, M., Nelson, W., DeBoy, R., Kolonay, J., McClarty, G., Salzberg, S. L., Eisen, J., and Fraser, C. M. (2000) *Nucleic Acids Res.* **28**, 1397–1406.

Read, T. D., Myers, G. S., Brunham, R. C., Nelson, W. C., Paulsen, I. T., Heidelberg, J., Holtzapple, E., Khouri, H., Federova, N. B., Carty, H. A., Umayam, L. A., Haft, D. H., Peterson, J., Beanan, M. J., White, O., Salzberg, S. L., Hsia, R. C., McClarty, G., Rank, R. G., Bavoil, P. M., and Fraser, C. M. (2003) *Nucleic Acids Res.* **31**, 2134–3147.

Rockey, D. D., Lenart, J., and Stephens, R. S. (2000) *Infect. Immun.* **68**, 5473–5479.

Saikku, P., Laitinen, K., and Leinonen, M. (1998) *Atherosclerosis* **140**, S17–S19.

Sambri, V., Donati, M., Storni, E., Di Leo, K. D., Agnusdei, M., Petracca, R., Finco, O., Grandi, G., Ratti, G. and Cevenini, R. (2004) *Vaccine*, **22**, 1131–7.

Sanchez–Campillo, M., Bini, L., Comanducci, M., Raggiaschi, R., Marzocchi, B., Pallini, V., and Ratti, G. (1999) *Electrophoresis* **20**, 2269–2279.

Schachter, J., Stephens, R. S., Timms, P., Kuo, C., Bavoil, P. M., Birkelund, S., Boman, J., Caldwell, H., Campbell, L. A., Chernesky, M., Christiansen, G., Clarke, I. N., Gaydos, C., Grayston, J. T., Hackstadt, T., Hsia, R., Kaltenboeck, B., Leinonnen, M., Ocjius, D., McClarty, G., Orfila, J., Peeling, R., Puolakkainen, M., Quinn, T. C., Rank, R. G., Raulston, J., Ridgeway, G. L., Saikku, P., Stamm, W. E., Taylor– Robinson, D. T., Wang, S. P., and Wyrick, P. B. (2001) *Int. J. Syst. Evol. Microbiol.* **51**, 249, 251–253.

Shaw, A. C., Christiansen, G., and Birkelund, S. (1999) *Electrophoresis* **20**, 775–780.

Shaw, A. C., Christiansen, G., Roepstorff, P., and Birkelund, S. (2000a) *Microbes Infect.* **2**, 581–592.

Shaw, A. C., Gevaert, K., Demol, H., Hoorelbeke, B., Vandekerckhove, J., Larsen, M. R., Roepstorff, P., Holm, A., Christiansen, G., and Birkelund, S. (2002a) *Proteomics* **2**, 164–186.

Shaw, A. C., Larsen, M. R., Roepstorff, P., Christiansen, G., and Birkelund, S. (2002b) *FEMS Microbiol. Lett.* **212**, 193–202.

Shaw, A. C., Vandahl, B. B., Larsen, M. R., Roepstorff, P., Gevaert, K., Vandekerckhove, J., Christiansen, G., and Birkelund, S. (2002c) *Cell Microbiol.* **4**, 411–424.

Shaw, E. I., Dooley, C. A., Fischer, E. R., Scidmore, M. A., Fields, K. A., and Hackstadt, T. (2000b) *Mol. Microbiol.* **37**, 913–925.

Shedlock, D. J., and Shen, H. (2003) *Science* **300**, 337–339.

Shirai, M., Hirakawa, H., Kimoto, M., Tabuchi, M., Kishi, F., Ouchi, K., Shiba, T., Ishii, K., Hattori, M., Kuhara, S., and Nakazawa, T. (2000a) *Nucleic Acids Res.* **28**, 2311– 2314.

Shirai, M., Hirakawa, H., Ouchi, K., Tabuchi, M., Kishi, F., Kimoto, M., Takeuchi, H., Nishida, J., Shibata, K., Fujinaga, R., Yoneda, H., Matsushima, H., Tanaka, C., Furukawa, S., Miura, K., Nakazawa, A., Ishii, K., Shiba, T., Hattori, M., Kuhara, S., and Nakazawa, T. (2000b) *J. Infect. Dis.* **181**, S524–S527.

Siscovick, D. S., Schwartz, S. M., Caps, M., Wang, S. P., and Grayston, J. T. (2000) *J. Infect. Dis.* **181**, S417–S420.

Stephens, R. S. (Ed.). (1999) *Chlamydia: Intracellular Biology, Pathogenesis and Immunology.* American Society for Microbology Press, Washington, DC.

Stephens, R. S. (2003) *Trends Microbiol.* **11**, 44–51.

Stephens, R. S., Kalman, S., Lammel, C., Fan, J., Marathe, R., Aravind, L., Mitchell, W., Olinger, L., Tatusov, R. L., Zhao, Q., Koonin, E. V., and Davis, R. W. (1998) *Science* **282**, 754–759.

Su, H., and Caldwell, H. D. (1991) *Infect. Immun.* **59**, 2843–2845.

Summersgill, J. T., Molestina, R. E., Miller, R. D., and Ramirez, J. A. (2000) *J. Infect. Dis.* **181**, S479–S482.

Sun, J. C., and Bevan, M. J. (2003) *Science* **300**, 339–342.

Tuffrey, M., Falder, P., Gale, J., Quinn, R., and Taylor–Robinson, D. (1986a) *J. Reprod. Fertil.* **78**, 251–260.

Tuffrey, M., Falder, P., Gale, J., and Taylor–Robinson, D. (1986b) *Br. J. Exp. Pathol.* **67**, 605–616.

Tuffrey, M., Falder, P., and Taylor–Robinson, D. (1982) *Br. J. Exp. Pathol.* **63**, 539–546.

Vandahl, B. B., Birkelund, S., and Christiansen, G. (2002a) *Methods Enzymol.* **358**, 277–288.

Vandahl, B. B., Birkelund, S., Demol, H., Hoorelbeke, B., Christiansen, G., Vandekerckhove, J., and Gevaert, K. (2001a) *Electrophoresis* **22**, 1204–1223.

Vandahl, B. B., Gevaert, K., Demol, H., Hoorelbeke, B., Holm, A., Vandekerckhove, J., Christiansen, G., and Birkelund, S. (2001b) *Electrophoresis* **22**, 1697–1704.

Vandahl, B. B., Pedersen, A. S., Gevaert, K., Holm, A., Vandekerckhove, J., Christiansen, G., and Birkelund, S. (2002b) *BMC Microbiol.* **2**, 36.

Whittum–Hudson, J. A., An, L. L., Saltzman, W. M., Prendergast, R. A., and MacDonald, A. B. (1996) *Nat. Med.* **2**, 1116–1121.

Young, I. T. (1977) *J. Histochem. Cytochem.* **25**, 935–941.

Yu, P., Spiotto, M. T., Lee, Y., Schreiber, H., and Fu, Y. X. (2003) *J. Exp. Med.* **197**, 985–995.

Zhang, D., Yang, X., Berry, J., Shen, C., McClarty, G., and Brunham, R. C. (1997) *J. Infect. Dis.* **176**, 1035–1040.

Zhong, G., Berry, J., and Brunham, R. C. (1994) *Infect. Immun.* **62**, 1576–1583.

12

Proteomics and Anti-Chlamydia Vaccine Discovery

Gunna Christiansen*, Svend Birkelund,
Brian B. Vandahl and Allan C Shaw

12.1 Introduction

The severe consequences of *Chlamydia trachomatis* infections justify the development of a vaccine. The obligate intracellular Gram-negative bacteria *C. trachomatis* are important human pathogens causing ocular and genital infections. *C. trachomatis* is divided into three biovars comprising 15 serological variants (Wang and Grayston, 1970). The trachoma biovar comprises the serovars A–K. *C. trachomatis* A–C are responsible for the disease trachoma, which is endemic in many developing countries. Infections of epithelial cells in the conjunctiva are followed by inflammatory responses, which lead to scarring of the cornea, with blindness as a result (Schachter and Caldwell, 1980). Serovars D–K are agents of sexually transmitted diseases and cause infections in the epithelial cells of the genital or urinary tract that can spread to the fallopian tubes and cause salpingitis, resulting in sterility or increased risk for ectopic pregnancy (Schachter and Caldwell, 1980). The *C. trachomatis* lympho-granuloma venereum (LGV) biovar comprises the serovars L1–L3, which can spread to lymph nodes, causing systemic infections (Schachter and Caldwell, 1980). The molecular background for serovar variation is found in the variation in the major outer membrane protein (MOMP). In *C. trachomatis* MOMP is the

* Corrresponding author. Telephone: +45 8942 1749. Fax: +45 8619 6128. E-mail address: gunna@medmicro.au.dk

Genomics, Proteomics and Vaccines edited by Guido Grandi
© 2004 John Wiley & Sons, Ltd   ISBN 0 470 85616 5

predominant outer membrane protein covering the surface of reticulate bodies (RBs) and elementary bodies (EBs). Genetically MOMP has four variable surface-exposed domains alternating with four constant sequences. MOMP is highly immunogenic and contains both species- and serovar-specific epitopes. When a panel of EBs from *C. trachomatis* serovars is used as antigens in microimmunofluorescence microscopy, sera from patients can be analysed for the presence of antibodies and for serovar typing.

All *Chlamydia* share a unique biphasic developmental cycle in which they alternate between the extracellular infectious EBs of ~0.3 μm in diameter, and the intracellular non-infectious and replicating RBs of ~1 μm in diameter. EBs attach to the host cell and induce their own uptake into a specialized vacuole termed 'the chlamydial inclusion'. Internalization of EBs is followed by reorganization into RBs, which multiply by binary fission. Towards the end of the intracellular stage, RBs are transformed into EBs and ultimately a new generation of infectious EBs is released upon disruption of the host cell. Transformation from RBs to EBs is accompanied by the synthesis of four late-phase proteins, the cysteine-rich outer membrane proteins Omp2 and Omp3 and the histone H1-like proteins, Hc1 and Hc2. These proteins bind to the chlamydial chromosome and RNA, thereby mediating the condensation of the chlamydial nucleoid that inhibits transcription and translation. Omp2 and 3 are cross-linked by disulphide bonds. Together with MOMP they form a supramolecular complex structure that provides the EBs envelope with a rigidity and osmotic stability that is not seen in RBs (for a review see Christiansen and Birkelund, 2002). Alteration of the developmental cycle can be induced by interferon-gamma (IFN-γ) (Beatty *et al.*, 1993) or amino acid starvation (Allan and Pearce, 1983), and it is speculated that such alteration may lead to a chronic or persistent infection, resulting in continued tissue damage.

Chlamydiae have a genome size of only ~1 Mb (Birkelund and Stephens, 1992; Kalman *et al.*, 1999), which is among the smallest known in bacteria. Valuable data were obtained by sequencing the chlamydial genomes (Stephens *et al.*, 1998b; Kalman *et al.*, 1999), but the lack of tools for genetic manipulation makes alternative strategies to exploit this information necessary. The development in separation of proteins by immobilized pH-gradient two-dimensional polyacrylamide gel electrophoresis (IPG 2D-PAGE) and identification of proteins by rapid and sensitive mass spectrometry methods have made it possible to study the protein expression in *Chlamydia* on a large scale. This will identify expressed proteins. However, proteins separated by IPG 2D-PAGE can also be transferred to PVDF membranes and used in immunoblotting to identify immunogenic proteins. We here review how proteomics can be exploited to identify potential vaccine candidates.

12.2   Proteome analysis

Due to the high resolving power of 2D PAGE, this technique currently provides the best solution for global visualization of proteins from micro-organisms. The resulting gel image is a pattern of spots, each representing a protein, for which pI and M_w can be read as in a coordinate system.

12.2.1   EBs proteins

The first attempt to characterize the *C. trachomatis* proteome by 2D PAGE (IPG) was performed by Bini *et al.* (1996), who made a preliminary reference map of silver-stained gels of purified *C. trachomatis* L2 EB. By using immunoblot analysis with specific antibodies and/or N-terminal amino acid sequencing they established the map positions of a number of described chlamydial proteins encoded both by the genome and the plasmid. Antibodies from patients with genital *C. trachomatis* infections or sequelae were used to identify antigenic proteins (Sanchez-Campillo *et al.*, 1999). In addition to silver staining, autoradiography of 2D-PAGE (IPG) gels with [35S]-labelled *Chlamydia* proteins has been successfully used for visualizing the chlamydial proteomes (Shaw *et al.*, 2002a; Vandahl *et al.*, 2001a). Cycloheximide, which efficiently blocks eukaryotic protein synthesis, was added to the labelling medium in order to restrict the labelling to chlamydial proteins. An advantage of [35S]-labelling is that the expression of chlamydial proteins at different time points in the developmental cycle can be visualized on autoradiographs of 2D gels, and the turnover or processing of specific proteins can be determined in pulse chase studies (Shaw *et al.*, 2002b,c; Vandahl *et al.*, 2001b).

   The chlamydial genome sequences allow the majority of proteins separated by 2D PAGE to be identified using fast and sensitive mass-spectrometry-based protein identification strategies (Karas and Hillenkamp, 1988; Shaw *et al.*, 2002a; Vandahl *et al.*, 2001a), but unambiguous identifications by peptide mass mapping may be difficult to obtain when proteins with low molecular weight or low abundancy or several proteins are present in the sample. In these cases tandem MS sequencing of the peptides derived from the protein can be used. Classical methods for protein identification from 2D gels also include N-terminal amino acid sequencing by Edman degradation.

   Genome sequences provide an important understanding of the biological potential of a microorganism, but the genomes are static and do not provide information concerning expression of a particular gene, post-translational modifications of proteins or how proteins are regulated in specific biological situations. Currently, complete genomes exist for *C. trachomatis* D (UW-3/Cx) (Stephens *et al.*, 1998b), *C. trachomatis* MoPn (Read *et al.*, 2000), *C. pneumoniae* VR1310 (Kalman *et al.*, 1999), *C. pneumoniae* J138 (Shirai *et al.*, 2001) and *C. pneumoniae* AR39 (Read *et al.*, 2000). A partly finished and not annotated low-resolution version of *C. trachomatis* L2 (434/Bu) is available on the internet (Stephens *et al.*, 1998a). 894 and 1073 open reading frames (ORFs) were predicted to encode proteins in the *C. trachomatis* and *C. pneumoniae* genomes, respectively (Stephens *et al.*, 1998b, Kalman *et al.*, 1999).

The genome of *C. trachomatis* D showed that 28 per cent of the predicted ORFs did not have any significant similarity to ORFs from other organisms outside the genus *Chlamydia* (Stephens *et al.*, 1998b). An important aspect of future research is to characterize the function and the expression pattern of such proteins, but although *C. trachomatis* and some *C. psittaci* possess plasmids (Palmer and Falkow, 1986; Joseph *et al.*, 1986) or bacteriophages (Storey, Luchen and Richmond, 1989; Hsia *et al.*, 2000) it has not been possible to perform stable transformation of chlamydiae with these potential vectors. Therefore, as a basis for protein expression studies 2D-PAGE reference maps for *C. pneumoniae* (Vandahl *et al.*, 2001a) and *C. trachomatis* serovars A, D and L2 (Shaw *et al.*, 2002a) were generated using both silver staining of purified EBs and autoradiography of gels with [35S]-labelled proteins at various time points in the chlamydial developmental cycle. Approximately 700 protein spots could be visualized using the IPG system, proteins from more than 200 protein spots were identified and 144 different protein species corresponding to 16 per cent of the proteins predicted from the *C. trachomatis* D genome were identified and annotated. 50 per cent of the proteins were experimentally identified in more than one serovar. The 2D-PAGE reference maps of *C. trachomatis* serovars (Shaw *et al.*, 2002a) and *C. pneumoniae* (Vandahl *et al.*, 2001a) are integrated in the parasite host cell interaction database (PHCI) 2D-PAGE database (Shaw *et al.*, 1999) available at http://www.gram.au.dk for interlaboratory comparisons.

The investigation of the *C. trachomatis* A, D and L2 proteomes showed expression of 25 unique *C. trachomatis* ORFs. Among the most abundant were CT577 and CT579 (Figure 12.1), which are encoded by genes located in a type III secretion subcluster (Shaw *et al.*, 2002a; Figure 12.1). Proteins from the type III secretion apparatus were identified in both proteomes (Shaw *et al.*, 2002a; Vandahl *et al.*, 2001a). In *C. trachomatis* YscL and YscN (Figure 12.1) were expressed at 24 and 36 h.p.i. and were also present in *C. trachomatis* EBs (Shaw *et al.*, 2002a). This suggests that chlamydiae may be able to secrete proteins via type III secretion throughout most of the developmental cycle. Secreted proteins may serve as contact molecules between *Chlamydia* and the surroundings inside the inclusion or in the host cell cytosol. If transported to

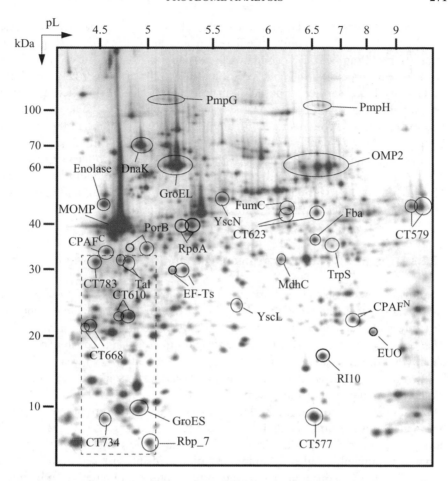

Figure 12.1 Autoradiograph of 2D-PAGE (3-10NL IPG) separations of *C. trachomatis* D proteins from infected HeLa cells [^{35}S] labelled from 34–36 h.p.i. The dashed box represents the enlargement shown in Figure 12.2. CPAF^{N}, N-terminal fragment of CPAF; CPAF^{C}, C-terminal fragment of CPAF

the host cell cytosol, the moleculae may undergo processing and end up at the surface of infected cells within the MHC class I molecule, thereby being potential vaccine candidates.

Genome sequencing revealed the presence of a large gene family encoding polymorphic membrane proteins (Pmps). These proteins have the structure of autotransporter proteins with a C-terminal part forming a β-barrel in the outer membrane and a N-terminal passenger domain with the structure of parallel

β-helices (Henderson and Lam, 2001; Birkelund *et al.*, 2002). Of the nine members of the Pmp family predicted from the *C. trachomatis* genome, translation of Pmp B, D, F, G and H was revealed and it was observed that these proteins were processed or modified (Shaw *et al.*, 2002a). It is likely that Pmp E, G and H are surface exposed (Tanzer and Hatch, 2001), thereby being exposed to contact with the immune system and thus being vaccine candidates. The expression of 31 *C. pneumoniae* unique ORFs was verified in the study of *C. pneumoniae* (Vandahl *et al.*, 2001a). In *C. pneumoniae* Pmp 2, 6, 7, 8, 10, 11, 13, 14, 20 and 21 were expressed, Pmp 6, 20 and 21 were cleaved and it is likely that Pmp 6, 8, 10, 11 and 21 are surface exposed (Vandahl *et al.*, 2002). The function of these proteins is unknown, but the expression of a high number of the genes indicates that the proteins may serve important functions.

12.2.2   Comparison of EBs proteomes from different serovars

Chlamydia species show remarkable differences in terms of biology and pathogenesis even though *C. trachomatis* D and L2 have the same genomic structure and gene complement (Stephens, 1999). A comparative analysis of *C. trachomatis* serovars A, D and L2 representing organisms causing different chlamydial diseases (i.e. trachoma, genital infections and LGV) was performed to identify differences in protein migration and expression, which may be linked to differences in pathogenesis.

Of 144 different proteins identified, 26 proteins showed migrational variation between *C. trachomatis* A and D, 55 between serovars D and L2 and 52 between serovars A and L2 (Shaw *et al.*, 2002a). The majority of migration variants presumably resulted from minor amino acid substitutions in the protein sequence leading to different net charge of the protein. Estimated from both the identified migrational variants and the degree of differences in protein abundance, the *C. trachomatis* L2 proteome profile differed from the *C. trachomatis* A and D profiles. Since *C. trachomatis* A and D are more similar in tissue tropism and virulence compared with serovar L2, the variability in proteome profiles therefore agrees with differences in biology and pathogenesis.

A truncation of the *C. trachomatis* L2 fumarate hydratase (FumC) was observed. FumC comigrated on gels of *C. trachomatis* A and D with a molecular weight of $\sim$50 kDa (Figure 12.1), but was not detected on *C. trachomatis* L2 gels. SDS-PAGE separation of [35S]-labelled *in vitro* translation products of fumC showed that *C. trachomatis* L2 had a $\sim$30 kDa truncation, which may impair the enzyme activity (Shaw *et al.*, 2002a). Malate dehydrogenase (MdhC) was remarkably more abundant in *C. trachomatis* L2 than in *C. trachomatis* A and D (Figure 12.1). MdhC is metabolically downstream of FumC in the tricarboxylic acid cycle and the two observations may therefore be linked, so that *C. trachomatis* L2 increases the MdhC activity for a more efficient utilization of malate obtained from the host cell to compensate for the FumC impairment (Shaw *et al.*, 2002a). Also tryptophanyl-tRNA synthetase (TrpS) was remarkably more abundant in *C. trachomatis* L2 than in A and D (Figure 12.1).

12.2.3   IFN-γ dependent inhibition of Chlamydia

Dynamic use of proteomics was exploited in the studies of IFN-γ-dependent inhibition of *Chlamydia*. IFN-γ is a potent immunoregulator involved in the control of chlamydial infections and may also be implicated in the development of chronic infections. To elucidate how IFN-γ affects protein synthesis in *Chlamydia*, IFN-γ-mediated inhibition of *C. trachomatis* A, D and L2 was studied. Immunofluorescence microscopy of *C. trachomatis* A-infected HeLa cells cultivated for 24 h in the presence or absence of human IFN-γ confirmed that RBs of *C. trachomatis* A were atypically large and stained weakly with anti-MOMP antibodies (Beatty *et al.*, 1993; Shaw, Christiansen and Birkelund, 1999; Shaw *et al.*, 2000a), but although *C. trachomatis* D and L2 displayed smaller inclusions in the presence of IFN-γ, no aberrant RBs morphology or reduction in MOMP staining was observed (Shaw, Christiansen and Birkelund, 1999; Shaw *et al.*, 2000a). To visualize differences in proteome profiles caused by IFN-γ treatment, HeLa cells infected with *C. trachomatis* A, D or L2 were cultivated in the presence or absence of IFN-γ and [35S]-labelled 22–24 h.p.i. To reveal proteins that are up- or down-regulated in response to IFN-γ, autoradiographs of 2D-PAGE separations of these samples were compared.

12.2.4   IFN-γ-down-regulated proteins in *C. trachomatis*

2D-PAGE analysis of *C. trachomatis* A revealed that several proteins besides MOMP are down-regulated due to IFN-γ treatment compared with steady levels of GroEL. These include ClpC protease and fructose-bisphosphate aldolase class I (Fba) (Shaw, Christiansen and Birkelund, 1999; Shaw *et al.*, 2002a). The observed down-regulation of proteins compared to GroEL in *C. trachomatis* A corresponds well with previous observations (Beatty *et al.*, 1993, 1994, 1995), but seems to be more general and not restricted to the important immunogens. Significant IFN-γ dependent-down-regulation of proteins was not observed on 2D gels of *C. trachomatis* D and *C. trachomatis* L2 (Shaw, Christiansen and Birkelund, 1999; Shaw *et al.*, 2000a) elucidating a difference in the response towards IFN-γ between these three serovars, and no down-regulation was observed in *C. pneumoniae*-infected cells due to IFN-γ treatment (Molestina *et al.*, 2002).

12.2.5   Induction of *C. trachomatis* tryptophan synthase by IFN-γ

A tryptophan operon (trp operon) containing orthologues of tryptophan synthase A (trpA) and B (trpB) subunit and the tryptophan repressor (trpR) is present in the *C. trachomatis* D genome (Stephens *et al.*, 1998b). This suggests that *C. trachomatis* has the potential to perform at least the last step in tryptophan synthesis (Stephens *et al.*, 1998b). 2D-PAGE studies revealed that *C. trachomatis* A, D and L2 all respond to IFN-γ by a strong induction of TrpA and TrpB (Shaw *et al.*, 2000a). This indicates that *C. trachomatis* TrpR is functional and can regulate the expression of tryptophan synthase (TrpBA) in response to tryptophan limitation.

TrpB comigrated on autoradiographs of both *C. trachomatis* A, D and L2 gels, and the *C. trachomatis* D and L2 TrpB sequences were identical, indicating high conservation of this protein between serovars. *C. trachomatis* A TrpA migrated with a remarkably lower molecular weight than the *C. trachomatis* D and L2 TrpA (Shaw, Christiansen and Birkelund, 1999; Shaw *et al.*, 2000a) due to a 1 bp deletion, which resulted in a frameshift and a premature stop codon (Shaw *et al.*, 2000a) truncating 70 amino acids ($\sim$7.7 kDa). Amplification of the trp operon from *C. trachomatis* B and C by PCR showed that *C. trachomatis* C had the same truncation as observed for *C. trachomatis* A, but the tryptophan synthesizing genes could not be amplified from *C. trachomatis* B. This is in agreement with an observation by Stephens and Zhao, who, during the selection of a suitable *C. trachomatis* serovar for genome sequencing, observed that *C. trachomatis* B had a chromosomal deletion of a genomic region containing the trp operon (Stephens, 1999). These findings correlate with earlier studies showing that *C. trachomatis* A–C in contrast to other serovars have a distinctive requirement for tryptophan for growth (Allan and Pearce, 1983). The increased IFN-γ susceptibility among trachoma-causing serovars was recently verified by Morrison (2000), who showed that infectious *C. trachomatis* A, B and C recovered with lower yields from IFN-γ-treated HeLa cells compared with other human *C. trachomatis* serovars. The truncation or absence of TrpBA in serovars A–C may render these serovars more sensitive to IFN-γ-mediated tryptophan degradation, increasing the ability of these serovars to persistently infect their host cells and cause chronic infections (Shaw *et al.*, 2000a). The *C. pneumoniae* genome also has a chromosomal deletion of the trp operon, but upon treatment with IFN-γ an almost general up-regulation of proteins was observed (Molestina *et al.*, 2002).

12.2.6 Protein compartmentalization

Secreted proteins are of special interest in chlamydial research as they may modify the host cell environment during the intracellular development and thus be responsible for host cell/parasite interactions. In addition, secreted effector proteins are natural candidates for major histocompatibility (MHC) class I antigen presentation and thus possible targets for development of a vaccine against chlamydiae. To elucidate whether chlamydiae secrete proteins into the host cell cytoplasm, a proteomics-based strategy was developed for comparing [35S]-labelled chlamydial proteins from whole lysates of infected cells (WLICs) with [35S]-labelled chlamydial proteins from purified RBs and EBs. Secreted proteins should be present in WLICs but absent from purified chlamydiae.

The only identified chlamydial protein that is known to be secreted into the host cell cytoplasm is the chlamydial protease- or proteasome-like activity factor (CPAF) (Zhong *et al.*, 2001) corresponding to CT858 in *the C. trachomatis* D genome (Stephens *et al.*, 1998a) and CPN1016 in the *C. pneumoniae* genome (Kalman *et al.*, 1999). CPAF is secreted by *C. trachomatis* L2 (Zhong *et al.*, 2001) and by *C. pneumoniae* (Fan *et al.*, 2002) into the host cell cytoplasm, where it degrades the host cell transcription factors RFX5 (Zhong *et al.*, 2000) and USF-1 (Zhong, Fan and Liu, 1999). RFX5 and USF-1 are required for MHC class I and II antigen presentation, respectively (Gobin *et al.*, 1998; Muhlethaler-Mottet *et al.*, 1998). N-terminal and C-terminal fragments of *C. trachomatis* CPAF were isolated from column chromatography fractions of infected cells showing proteolytic activity (Zhong *et al.*, 2001).

The results of Zhong *et al.* (2001) were verified and extended in studies by Shaw *et al.* (2002c). Using proteomics the N-terminal and C-terminal fragments of *C. trachomatis* CPAF (Figure 12.1) and *C. pneumoniae* CPAF were observed on 2D gels of WLICs, but were absent from RBs and EBs. It was shown that both *C. trachomatis* A, D and L2 and *C. pneumoniae* secrete CPAF from the middle of the developmental cycle and towards the end of the cycle. The CPAF sequence is highly conserved between *C. trachomatis, C. pneumoniae* and *C. psittaci* and it is thus likely that all *Chlamydia* species secrete CPAF during the developmental cycle.

The two fragments of *C. trachomatis* D CPAF showed high stability in the host cell cytoplasm based on pulse-chase studies in combination with 2D PAGE. The very slow turnover of the CPAF fragment may reflect limited degradation of the secreted protein by the host cell proteasome. During evolution of host cell/chlamydiae interactions CPAF may have evolved an amino acid sequence that is not readily processed in the host cell proteasome. In this way *Chlamydia* may secrete a protein that inhibits MHC class I and II antigen presentation without itself being a target for proteasome degradation and MHC class I antigen presentation (Shaw *et al.*, 2002c). The *C. trachomatis* and *C. pneumoniae* CPAF fragments (or full length CPAF) were not observed on autoradiographs when protease inhibitors such as N-tosyl-L-phenylalanyl-chloromethane (TPCK) were added during the labelling and chase periods, suggesting that either cleavage or secretion of CPAF is prevented by inhibition of proteases (Shaw *et al.*, 2002c). CPAF is presumably not secreted via a typical type III secretion system, because a strong prediction of a cleavable signal peptide could be obtained using Signal P for both *C. trachomatis* and *C. pneumoniae* CPAF. The development of a rapid proteomics-based method to screen for secreted proteins facilitates identification of other proteins secreted from *Chlamydia*.

12.2.7   RB-specific proteins

Due to the biphasic developmental cycle of chlamydiae it is of biological interest to determine whether a protein is associated primarily with the intracellular stage of chlamydial development. With the chlamydial genome information as a basis, studies based on reverse transcriptase PCR assays have been used to determine the developmental stage specific transcription of genes during the developmental cycle (Shaw *et al.*, 2000b). This technique does, however, not provide answers to whether a given early transcript is specific for the RBs stage, as the corresponding protein may have long turnover and thus still be present in EBs.

Comparison of proteome profiles of WLICs, purified RBs/EBs and estimations of turnover for proteins include discovery of proteins that are specific for the intracellular stage of chlamydial development. An example is CT783 (Figures 12.1 and 12.2), a predicted protein disulphide isomerase (Stephens *et al.*, 1998a), which was present in very low amounts in labelled EBs, but present in high amounts in [35S]-labelled RBs (Shaw *et al.*, 2002c). This protein displayed a very fast turnover as it was almost completely degraded after chase periods of 4–6 h (Figure 12.2). Examples of other identified proteins with similar characteristics include CT668, CT610 and the 7 kDa reticulate body protein (Rbp_7) (Figures 12.1 and 12.2). Analysis of silver-stained gels of purified EBs confirmed that these proteins are either absent or significantly reduced in EBs. The unifying fast turnover time (Figure 12.2)

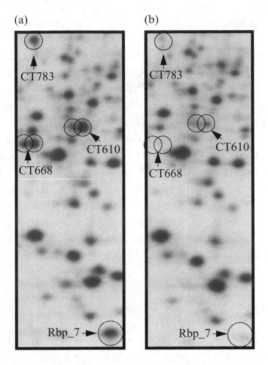

Figure 12.2   Pulse-chase study showing the rapid turnover of CT610, CT668, CT783 and Rbp_7. *C. trachomatis* D proteins from infected HeLa cells were [^{35}S] labelled from 22–24 h.p.i. Cells were either harvested directly after labelling (a) or the labelling medium was changed to normal infection medium and the proteins were chased for 6 h (b). Labelled proteins were separated by 2D PAGE and visualized by autoradiography. Enlargements (a) and (b) correspond to the dashed section in Figure 12.1

suggests that the proteins are presumably degraded before or during conversion to the EBs stage, where no new protein expression is apparent.

12.3   Proteomics as a complement for genomics

12.3.1   Correcting computer-annotated ORFs

As a complement for genomics, proteomics is valuable for identification of the exact start point of proteins, as an indication of modification and processing and for determination of un-annotated proteins. A computer-predicted translation start site in mRNA does not always correspond to the first amino acid in proteins expressed by bacteria. Post-translational processing may result in a mature protein very different from the one that can be predicted. Thus, in many cases, the correct N-terminal sequence can only be determined by experimental approaches. A study of low-molecular-weight proteins from *E. coli* revealed proteins that migrated differently than predicted, and corrections of the predicted N-terminal amino acid sequence could be performed on the basis of Edman degradation of excised spots (Wasinger and Humphery-Smith, 1998).

Experimental determination of the N-terminus of *C. trachomatis* CT610 (Figure 12.1) showed that the initiator methionine is presumably located five amino acids downstream of the initiator methionine predicted from genome analysis (Shaw *et al.*, 2002a). It is likely that some of the proteins that migrate below their estimated molecular weights, such as CT734 (Figure 12.1), (Shaw *et al.*, 2002b), may also be incorrectly predicted in the *C. trachomatis* D genome.

12.3.2   Small unrecognized ORFs

In large genome sequencing projects, a molecular weight threshold for protein annotation is often used to avoid annotation of many false positive ORFs. Annotation of very small proteins that are below the selected threshold may therefore be neglected (Rudd *et al.*, 1998). Identification of such small proteins visualized on 2D gels can be obtained by Edman degradation or tandem mass spectrometry. This was done on a low-molecular-weight protein from *C. trachomatis* D using tandem mass spectrometry. A novel ORF (Rbp_7) not annotated in the *C. trachomatis* D genome was identified (Figure 12.1), (Shaw *et al.*, 2002b). Rbp_7 was significantly expressed from the middle of the *C. trachomatis* A, D and L2 developmental cycle and was predominantly expressed in RBs or intermediate bodies as determined by immunofluorescence microscopy. Based on analysis of translated DNA sequences, *C. trachomatis* A and D Rbp_7 were found to be identical. Only one amino acid substitution, resulting in a p*I* shift on the 2D gels, distinguished the *C. trachomatis* L2 sequence from *C. trachomatis* A/D sequence. The *C. trachomatis* A/D Rbp_7 sequence showed 75 per cent identity to an ORF in the *C. trachomatis* MoPn, which was also not annotated in the *C. trachomatis* MoPn genome (Read *et al.*, 2000). Rbp_7 has a genomic location between CT804 and CT805 in the *C. trachomatis* D genome (Stephens *et al.*, 1998a) with an overlap between Rbp_7 and CT804. No significant similarity was observed to other proteins present in databases and no unambiguous motifs were discovered in the protein sequence, thus rendering the function of Rbp_7, if any, unknown (Shaw *et al.*, 2002b).

12.4   Benefits that proteomics provide for vaccine development

Proteomics as a complement for genomics has proven important to determine the correct ORFs, identify secreted proteins and determine regulation of proteins due to changes in growth conditions. Furthermore, comparison of the proteomes of closely related serovars provides insight into delicate variations that may have severe consequences for the life of the microorganisms. For chlamydiae, for which no genetic tools for transformation exist, proteomics has provided valuable new knowledge which may be of importance for vaccine development.

Whole-cell chlamydial vaccines have in animal models shown development of delayed-type hypersensitivity reactions. Therefore, a subunit vaccine should be developed. Selection of relevant components is a major challenge. Proteomics may help to identify suitable components by limiting the number of candidate proteins as it can identify proteins synthesized under various growth conditions. In the comparative analysis, several of the *C. trachomatis* A and L2 proteins were observed to migrate differently from *C. trachomatis* D proteins. Some of these may be responsible for biological differences. It was reported that at least *C. trachomatis* D and L2 have the same genomic structure and gene complement (Stephens, 1999). This may indicate that minor mutations within homologues or in promoter regions may have a large impact on biology and/or pathogenesis. The discovery of serovar-specific truncations (e.g. FumC and TrpA) or up-regulations (e.g. TrpS and MdhC) demonstrate that proteomics can visualize subtle differences in genes.

Selection of proteins as vaccine candidates can be facilitated by identification of immunogenic proteins. Using immunoblotting of *C. trachomatis* proteins separated by 2D PAGE, patient serum samples were shown to react with a high number of proteins (Sanchez-Campillo *et al.*, 1999). Due to lack of the *C. trachomatis* genome sequence at the time when this study was performed it was only possible to identify a few of the reacting proteins. Sanchez-Campillo *et al.*, (1999) identified the known immunogenic proteins as MOMP, Omp2, Hsp60/GroEL and DnaK but also OmpB and five conserved bacterial proteins: ribosomal protein S1, elongation factor EF-Tu, a putative stress-induced protease of the HtrA family and the ribosomal protein L7/L12. Thus, more house-keeping proteins than expected were immunogenic. With the genome data now available it is possible to identify the chlamydial proteins to which antibodies are generated under different infections, and such data should help in identifying candidate proteins for a subunit vaccine.

The development of a strategy for discovering secreted chlamydial proteins that are currently not predictable from the gene or protein sequence alone is a major advantage for prediction of proteins harbouring CD8+ T-cell epitopes. All *Chlamydia* strains investigated so far possess the gene encoding CPAF and are presumably capable of secreting this protease into the host cell cytoplasm and evade the cellular immune response. Blocking of the function of this protein may increase killing of infected cells.

A vaccine should be developed to provide full protection against infection but this is not always possible to obtain. Sequelae caused by genital *C. trachomatis* infections are expected to be caused by a delayed-type hypersensitivity reaction. It is therefore crucial that a vaccine does not mediate such reactions. Selected vaccine candidates should therefore be analysed not only for their capacity to mediate protection but also for the direction in which they trigger the immune response.

References

Allan, I., and Pearce, J. H. (1983) Amino acid requirements of strains of *Chlamydia trachomatis* and *C. psittaci* growing in McCoy cells: relationship with clinical syndrome and host origin *J. Gen. Microbiol.* **129**(7), 2001–2007.

Beatty, W. L., Belanger, T. A., Desai, A. A., Morrison, R. P., and Byrne, G. I. (1994) Tryptophan depletion as mechanism for gamma-interferin-mediated chlamydial persistence. *Infect. Immun.* **62**(9), 3705–3711.

Beatty, W. L., Byrne, G. I., and Morrison, R. P. (1993) Morphologic and antigenic characterization of interferon gamma-mediated persistent *Chlamydia trachomatis* infection in vitro, *Proc. Natl. Acad. Sci. USA* **90**(9), 3998–4002.

Beatty, W. L, Morrison, R. P., and Byrne, G. I. (1995) Reactivation of persistent *Chlamydia trachomatis* infection in cell culture. *Infect. Immun.* **63**(1), 199–205.

Bini, L., Sanchez-Campillo, M., Santucci, A., Magi, B., Marzocchi, B., Comanducci, M., Christiansen, G., Birkelund, S., Cevenini, R., Vretou, E., Ratti, G., and Pallini, V. (1996) Mapping of *Chlamydia trachomatis* proteins by immobiline polyacrylamide two-dimensional electrophoresis: spot identification by N-terminal sequencing and immunoblotting. *Electrophoresis* **17**(1), 185–190.

Birkelund, S., Christiansen, G., Vandahl, B. B., and Pedersen, A. S. (2002) Are the Pmp proteins parallel beta-helices? In: Schachter, J., Christiansen, G., Clarke, I. N., Hammerschlag, M. R., Kaltenboeck, B., Kuo, C.-C., Rank, R. G., Ridgway, G. L., Saikku, P., Stamm, W. E., Stephens, R. S., Summersgill, J. T., Timms, P., and Wyrick, P. B. (Ed.). *Proceedings of the 10th Symposium on Human Chlamydial Infections*, 16–21 June, *International Chlamydia Symposium*. San Francisco, CA. 551–554.

Birkelund, S., and Stephens, R. S. (1992) Construction of physical and genetic maps of *Chlamydia trachomatis* serovar L2 by pulsed-field gel electrophoresis. *J. Bacteriol.* **174**(9), 2742–2747.

Christiansen, G., and Birkelund, S. (2002) *Chlamydia* structure – a molecular approach to understand structure of *Chlamydia*. In Schachter, J., Christiansen, G., Clarke, I. N., Hammerschlag, M. R., Kaltenboeck, B., Kuo, C.-C., Rank, R. G., Ridgway, G. L., Saikku, P., Stamm, W. E., Stephens, R. S., Summersgill, J. T., Timms, P., and Wyrick, P. B. (Ed.). *Proceedings of the 10th Symposium on Human Chlamydial Infections*, 16–21 June, *International Chlamydia Symposium*. San Francisco, CA. 537–546.

Fan, P., Dong, F., Huang, Y., and Zhong, G. (2002) *Chlamydia pneumoniae* secretion of a protease-like activity factor for degrading host cell transcription factors is required for major histocompatibility complex antigen expression. *Infect. Immun.* **70**(1), 345–349.

Gobin, S. J., Peijnenburg, A., van Eggermond, M., van Zutphen, M., van den Berg, R., and van den Elsen, P. J. (1998) The RFX complex is crucial for the constitutive and CIITA-mediated transactivation of MHC class I and beta2-microglobulin genes. *Immunity* **9**(4), 531–541.

Henderson, I. R., and Lam, A. C. (2001) Polymorphic proteins of *Chlamydia* spp. – autotransporters beyond the Proteobacteria. *Trends Microbiol.* **9**(2), 573–578.

Hsia, R. C., Ohayon, H., Gounon, P., Dautry-Varsat, A., and Bavoil, P. M. (2000) Phage infection of the obligate intracellular bacterium, *Chlamydia psittaci* strain guinea pig inclusion conjunctivitis. *Microbes Infect.* **2**(7), 761–772.

Joseph, T., Nano, F. E., Garon, C. F., and Caldwell, H. D. (1986) Molecular characterization of *Chlamydia trachomatis* and *Chlamydia psittaci* plasmids. *Infect. Immun.* **51**(2), 699–703.

Kalman, S., Mitchell, W., Marathe, R., Lammel, C., Fan, J., Hyman, R. W., Olinger, L., Grimwood, J., Davis, R. W., and Stephens, R. S. (1999) Comparative genomes of *Chlamydia pneumoniae* and *C. trachomatis*. *Nat. Genet.* **21**(4), 385–389.

Karas, M., and Hillenkamp, F. (1988) Laser desorption ionization of proteins with molecular masses exceeding 10,000 daltons. *Anal. Chem.* **60**(20), 2299–2301.

Molestina, R. E., Klein, J. B., Miller, R. D., Pierce, W. H., Ramirez, J. A., and Summersgill, J. T. (2002) Proteomic analysis of differentially expressed *Chlamydia pneumoniae* genes during persistent infection of HEp-2 cells. *Infect. Immun.* **70**(6), 2976–2981.

Morrison, R. P. (2000) Differential sensitivities of *Chlamydia trachomatis* strains to inhibitory effects of gamma interferon. *Infect. Immun.* **68**(10), 6038–6040.

Muhlethaler-Mottet, A., Di Berardino, W., Otten, L. A., and Mach, B. (1998) Activation of the MHC class II transactivator CIITA by interferon-gamma requires cooperative interaction between Stat1 and USF-1. *Immunity* **8**(2), 157–166.

Palmer, L., and Falkow, S. A. (1986) A common plasmid of *Chlamydia trachomatis*. *Plasmid* **16**(1), 52–62.

Read, T. D., Brunham, R. C., Shen, C., Gill, S. R., Heidelberg, J. F., White, O., Hickey, E. K., Peterson, J., Utterback, T., Berry, K., Bass, S., Linher, K., Weidman, J., Khouri, H., Craven, B., Bowman, C., Dodson, R., Gwinn, M., Nelson, W., DeBoy, R., Kolonay, J., McClarty, G., Salzberg, S. L., Eisen, J., and Fraser, C. M. (2000) Genome sequences of *Chlamydia trachomatis* MoPn and *Chlamydia pneumoniae* AR39. *Nucleic Acids Res.* **28**(6), 1397–1406.

Rudd, K. E., Humphery-Smith, I., Wasinger, V. C., and Bairoch, A. (1998) Low molecular weight proteins: a challenge for post-genomic research. *Electrophoresis* **19**(4), 536–544.

Sanchez-Campillo, M., Bini, L., Comanducci, M., Raggiaschi, R., Marzocchi, B., Pallini, V., and Ratti, G. (1999) Identification of immunoreactive proteins of *Chlamydia trachomatis* by Western blot analysis of a two-dimensional electrophoresis map with patient sera *Electrophoresis* **20**(11), 2269–2279.

Schachter, J., and Caldwell, H. D. (1980) "Chlamydiae". *Annu. Rev. Microbiol.* **34**, 285–309.

Shaw, A. C., Christiansen, G., and Birkelund, S. (1999) Effects of interferon gamma on *Chlamydia trachomatis* serovar A and L2 protein expression investigated by two-dimensional gel electrophoresis. *Electrophoresis* **20**(4–5), 775–780.

Shaw, A. C., Larsen, M. R., Roepstorff, P., Holm, A., Christiansen, G., and Birkelund, S. (1999) Mapping and identification of HeLa cell proteins separated by immobilized pH-gradient two-dimensional gel electrophoresis and construction of a two-dimensional polyacrylamide gel electrophoresis database. *Electrophoresis* **20**, 977–983.

Shaw, A. C., Christiansen, G., Roepstorff, P., and Birkelund, S. (2000a) Genetic differences in the *Chlamydia trachomatis* tryptophan synthase a-subunit can explain variations in serovar pathogenesis. *Microbes. Infect.* **2**(6), 581–592.

Shaw, A. C., Gevaert, K., Demol, H., Hoorelbeke, B., Vandekerckhove, J., Larsen, M. R., Roepstorff, P., Holm, A., Christiansen, G., and Birkelund, S. (2002a) Comparative analysis of *Chlamydia trachomatis* serovar A, D and L2. *Proteomics* **2**(2), 164–186.

Shaw, A. C., Larsen, M. R., Roepstorff, P., Christiansen, G., and Birkelund, S. (2002b) Identification and characterization of a novel *Chlamydia trachomatis* reticulate body protein. *FEMS Microbiol. Lett.* **212**(2), 193–202.

Shaw, A. C., Vandahl, B. B., Larsen, M. R., Roepstorff, P., Gevaert, K., Vanderkerckhove, J., Christiansen, G., and Birkelund, S. (2002c) Characterization of a secreted *Chlamydia* protease. *Cell Microbiol.* **4**(7), 411–424.

Shaw, E. I., Dooley, C. A., Fischer, E. R., Scidmore, M. A., Fields, K. A., and Hackstadt, T. (2000b) Three temporal classes of gene expression during the *Chlamydia trachomatis* developmental cycle. *Mol. Microbiol.* **37**(4), 913–925.

Shirai, M., Hirakawa, H., Kimoto, M., Tabuchi, M., Kishi, F., Ouchi, K., Shiba, T., Ishii, K., Hattori, M., Kuhara, S., and Nakazawa, T. (2001) Comparison of whole genome sequences of *Chlamydia pneumoniae* J138 from Japan and CWL029 from USA. *Nucleic Acids Res.* **28**(12), 2311–2314.

Stephens, R. S. (Ed.). (1999) *Chlamydia. Intracellular Biology, Pathogenesis, and Immunity.* ASM Press, Washington, DC, 9–27.

Stephens, R. S., Kalman, S., Fenner, C., and Davis, R. (1998a) The *Chlamydia* genome project. http://socrates.berkeley.edu:4231/.

Stephens, R. S., Kalman, S., Lammel, C., Fan, J., Marathe, R., Aravind, L., Mitchell, W., Olinger, L., Tatusov, R. L., Zhao, Q., Koonin, E. V., and Davis, R. W. (1998b) Genome sequence of an obligate intracellular pathogen of humans: *Chlamydia trachomatis. Science* **282**(5389), 754–759.

Storey, C. C., Lusher, M., and Richmond, S. J. (1989) Analysis of the complete nucleotide sequence of Chp 1, a phage which infects *Chlamydia psittaci. J. Gen. Virol.* **70**(12), 3381–3390.

Tanzer, R. J., and Hatch, T. P. (2001) Characterization of outer membrane proteins in *Chlamydia trachomatis* LGV serovar L2. *J. Bacteriol.* **183**(8), 2686–2690.

Vandahl, B. B., Birkelund, S., Demol, H., Hoorelbeke, B., Christiansen, G., Vandekerckhove, J., and Gevaert, K. (2001a) Proteome analysis of the *Chlamydia pneumoniae* elementary body. *Electrophoresis* **22**(6), 1204–1223.

Vandahl, B. B., Gevaert, K., Demol, H., Hoorelbeke, B., Holm, A., Vandekerckhove, J., Christiansen, G., and Birkelund, S. (2001b) Time-dependent expression and processing of a hypothetical protein of possible importance for regulation of the *Chlamydia pneumoniae* developmental cycle. *Electrophoresis* **22**(9), 1697–1704.

Vandahl, B., Pedersen, A., Gevaert, K., Holm, A., Vandekerckhove, J., Christiansen, G., and Birkelund, S. (2002) The expression, processing and localization of polymorphic membrane proteins in *Chlamydia pneumoniae* strain CWL029. *BMC Microbiol.* **2**(1), 36.

Wang, S. P., and Grayston, J. T. (1970) Immunologic relationship between genital TRIC, lymphogranuloma venereum, and related organisms in a new microtiter indirect immunofluorescence test. *Am. J. Ophthalmol.* **70**(3), 367–374.

Wasinger V. C., and Humphery-Smith, I. (1998) Small genes/gene-products in *Escherichia coli* K-12. *FEMS Microbiol. Lett.* **169**(2), 375–382.

Zhong, G., Fan, P., Ji, H., Dong, F., and Huang, Y. (2001) Identification of a chlamydial protease-like activity factor responsible for the degradation of host transcription factors. *J. Exp. Med.* **193**(8), 935–942.

Zhong, G., Fan, T., and Liu, L. (1999) Chlamydia inhibits Interferon-g inducible major histocompability complex class II expression by degradation of upstream stimulatory factor 1. *J. Exp. Med.* **189**(12), 1931–1938.

Zhong, G., Liu, L., Fan, T., Fan, P., and Ji, H. (2000) Degradation of transcription factor RFX5 during the inhibition of both constitutive and interferon-g inducible major histocompability complex class I expression in Chlamydia infected cells. *J. Exp. Med.* **191**(9), 1525–1534.

13

Proteome Analysis of Outer Membrane and Extracellular Proteins from *Pseudomonas aeruginosa* for Vaccine Discovery

Stuart J. Cordwell and **Amanda S. Nouwens**

13.1 Introduction

The capacity of an organism to be pathogenic is dependent on an ability to (1) enter the host, (2) acquire the necessary nutrients to grow and proliferate in the host, (3) inhibit and defend itself against host responses, (4) attack (damage) the host and (5) infiltrate and spread through the host. In addition to the production of pathogenic factors, the degree of infection is also dependent on the immune status of the host, the particular cell type invaded and the nutrients available in the host environment (reviewed by Casadevall and Pirofski, 1999; Finlay and Falkow, 1997; Smith, 2000). As an opportunistic pathogen, *P. aeruginosa* infections are of particular concern in immuno-compromised individuals. Healthy individuals can resist pseudomonad infections through normal defence mechanisms, including physical barriers such as skin, and clearance mechanisms including macrophages in the blood and cilia in the lungs (Lyczak, Cannon and Pier, 2000). *P. aeruginosa* is able to colonize lung epithelia in people with cystic fibrosis (CF), as the CF patient has a defect in the ciliated epithelial cells preventing clearance of the lungs (Govan and

Genomics, Proteomics and Vaccines edited by Guido Grandi
© 2004 John Wiley & Sons, Ltd   ISBN 0 470 85616 5

Deretic, 1996). Once established, the pseudomonad infection becomes chronic, and the extreme damage caused to the lungs by *P. aeruginosa* normally leads to death in these patients. The organism is also a severe problem in burns patients since damage to the skin provides the perfect combination of site colonization and nutrients to disseminate from the initial point of infection (Pavlovskis and Wretlind, 1979).

The difficulty in controlling and defeating pseudomonad infections is compounded by the vast array of virulence factors it produces, and the resistance *P. aeruginosa* displays to many classes of antibiotics, including beta-lactams and aminoglycosides (Nakae *et al.*, 1999; Westbrock-Wadman *et al.*, 1999; Zhao *et al.*, 1998). Pathogenicity factors produced by *P. aeruginosa* are diverse in both the structure of the molecule (proteins, pigments and ions such as cyanide) and the targets in the host, improving the ability of the bacterium to protect itself and proliferate in the host environment. Not surprisingly, virulence factors can be either localized on the cell surface or released into the external environment. Surface-associated proteins, such as flagellin and pilin subunits, play a role in motility and adherence to host cells and are also highly immunogenic (Arora *et al.*, 1998; Comolli, *et al.*, 1999; Ramphal *et al.*, 1991; Wu, Gupta and Hazlett, 1995). Many other virulence factors are secreted into the external environment, while some, such as exoenzyme S, appear to be specifically 'injected' into host cells (Vallis *et al.*, 1999). Extracellular proteins, in particular, are primarily produced at or near stationary phase, with many virulence factors under the control of one or both quorum-sensing regulatory systems identified in *P. aeruginosa* (Nouwens *et al.*, 2003).

13.2   Membrane proteins in *P. aeruginosa*

P. aeruginosa is a major cause of hospital-acquired infections and a number of strains have evolved a high resistance to one or more antibiotic classes (Hancock, 1998). The resistance of *P. aeruginosa* to antibiotics (and other xenobiotic compounds) can in part be accredited to its highly impermeable membrane, as well as through the role of membrane proteins that form efflux pumps capable of removing antibiotics from the cell (Lee *et al.*, 2000; Westbrock-Wadman *et al.*, 1999). Pseudomonal membrane proteins function to promote adherence to surfaces, transport solutes and nutrients into the cell, export proteins and macromolecules destined for the external environment, allow cell–cell signalling, sense changes in the external environment, provide membrane stabilization and maintain a high osmotic pressure within the cell relative to the external environment. Membrane proteins, particularly those localized on the exterior surface, often also elicit an immune response in the host, and, as such, represent potential targets for vaccine development (Jang *et al.*, 1999; Mansouri *et al.*, 1999; Staczek *et al.*, 1998). A database of outer membrane proteins (OMPs) from *P. aeruginosa* has been established (http://cmdr.ubc.ca/bobh/omps) and in May 2003 lists 30 functionally verified OMPs, with more than 100 open reading frames (ORFs) predicted to encode OMPs.

Proteins in the *P. aeruginosa* outer membrane include porins, lipoproteins and receptor proteins. Porins are barrel-shaped proteins that function as channels, allowing the uptake and removal of solutes and small molecules into and out of the cell. Some porins are non-specific in the molecules they transport, such as porin F (*oprF*), a major structural porin constitutively expressed in *P. aeruginosa* (Woodruff and Hancock, 1989). Others are specific for the uptake of biochemical compounds, for example porin D (*oprD*) channels basic amino acids (glutamate, histidine, arginine, alanine) and the antibiotic imipenem. OprD is also constitutively expressed in *P. aeruginosa* at low levels and is upregulated in the presence of basic amino acids. Many other porins are not constitutively expressed and require specific nutrients to induce expression; for example, OprB is induced by glucose, and OprJ, part of the MexCD–OprJ efflux pump, is induced by a range of compounds including ethidium bromide and acriflavine (Morita *et al.*, 2001). Other membrane proteins, such as OprM, the outer membrane porin forming part of the MexAB–OprM efflux system, is growth-phase regulated (Evans and Poole, 1999), but also subject to the *mexR* regulatory gene (Evans *et al.*, 2001; Srikumar, Paul and Poole, 2000). Optimal expression of OprM occurs during exponential to late exponential phase, with OprM functioning to remove antibiotics, and export other compounds, such as secondary metabolites (Evans and Poole, 1999; Zhao *et al.*, 1998).

Although porins are embedded in the outer membrane, they are surprisingly hydrophilic. Bacterial porins do not have transmembrane spanning regions such as those found in typical membrane proteins from eukaryotes or the inner membrane of Gram-negative bacteria (Santoni, Molloy and Rabilloud, 2000; Schirmer, 1998). Instead, porins consist of anti-parallel β-sheets in monomeric or trimeric arrangements (Koebnik, Locher and Van Gelder, 2000) and, therefore, do not have the series of hydrophobic amino acids that constitute typical transmembrane domains. Instead, the barrel arrangement is such that the interior of the barrel is hydrophilic, while the exterior surface, embedded in the membrane, is hydrophobic. Santoni, Molloy and Rabilloud (2000) examined over 70 outer membrane proteins, only two of which were hydrophobic, as defined by grand average hydropathy or GRAVY values (Kyte and Doolittle, 1982). Lipoproteins consist of a lipid moiety covalently attached to the protein. The lipid moiety forms part of the membrane and thus the proteinaceous part of the molecule is not necessarily embedded in the membrane. The lipid moiety is similar across bacterial lipoproteins as it serves as an anchor for the (generally hydrophilic) protein to the membrane (Wu, 1996). Lipoproteins are identifiable by a hydrophobic signal sequence with a terminal cysteine residue, which serves as the lipid attachment site (Wu, 1996). Lipoproteins in *P. aeruginosa* include a major outer-membrane lipoprotein OprI (Cornelis *et al.*, 1989) and OmlA, a constitutively expressed lipoprotein believed to function in maintaining cell envelope integrity (Ochsner *et al.*, 1999).

The periplasmic space is a gel-like matrix sandwiched between the inner and outer membranes (Hobot *et al.*, 1984). It contains a large range of proteins, including hydrolytic enzymes, chemoreceptors and binding proteins. Periplasmic proteins process and direct the transport of nutrients, proteins and other compounds into and out of the cell (reviewed by Oliver, 1996). In addition to those proteins located entirely within the periplasm, some proteins embedded in the outer or inner membranes may have domains localized in the periplasmic space (Oliver, 1996). Unlike the cytoplasmic environment of the cell, which is reducing, the periplasm is an oxidizing environment, to allow formation of disulphide bonds in proteins destined for the external environment (Oliver, 1996). A number of proteins, including DsbA, DsbB and DsbC, are located in the periplasm and are important in disulphide bond formation (Urban *et al.*, 2001).

13.2.1   Proteomics of *P. aeruginosa* outer membrane proteins

Proteomics encompasses many useful tools for examining the expression of bacterial proteins under a variety of biological conditions (Figure 13.1). Currently, the most common experimental approach to examining global protein expression is separation of complex mixtures via two-dimensional gel electrophoresis (2DGE). Other methods based on multi-dimensional liquid chromatography (Washburn, Wolters and Yates, 2001), or with quantitation provided by isotope-coded affinity tags (ICAT; Gygi, Rist and Aebersold, 2000; Smolka *et al.*, 2001) also provide excellent alternatives for high-throughput mapping and may also reveal significant subsets of proteins that complement those separated by 2DGE. Advances in 2DGE in the past ten years include reproducible immobilized pH gradients (IPGs), improved solubilizing and reducing agents and novel synthetic zwitterionic detergents, including amidosulphobetaine 14 (ASB-14; Chevallet *et al.*, 1998), and improved detection methods, such as fluorescent dyes (e.g. Sypro Ruby; Lopez *et al.*, 2000). These improvements, coupled with the rapid advances in peptide mass spectrometry, now allow for the rapid identification of many expressed, gel-purified proteins (Chalmers and Gaskell, 2000; Gygi and Aebersold, 2000) and are the basis for

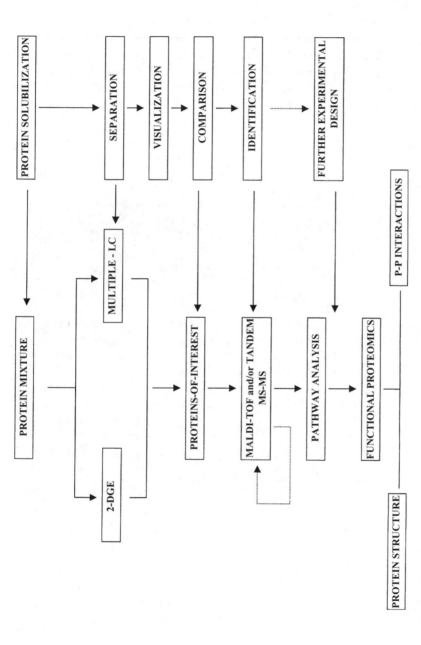

Figure 13.1 Schematic representation of the proteomics approach. Comparative analyses determine a set of 'proteins of interest' from amongst a complex mixture. This is performed experimentally in five steps: protein solubilization, separation (2-DGE or LC), visualization, comparison (spots or peaks of significance) and identification (via MS). The 'proteins of interest' are then functionally characterized

renewed interest in 2DGE and associated proteomic tools for the analysis of *P. aeruginosa* proteins.

Fractionation based on sub-cellular location is an approach particularly useful for aiding in the characterization of membrane proteins from *P. aeruginosa* (Nouwens *et al.*, 2000). Outer membranes are precipitated following cell disruption using a high-pH sodium carbonate buffer, and the membrane proteins extracted in a strong solubilizing buffer containing ASB-14 (Figure 13.2). Approximately 300 protein spots were separated using this approach and

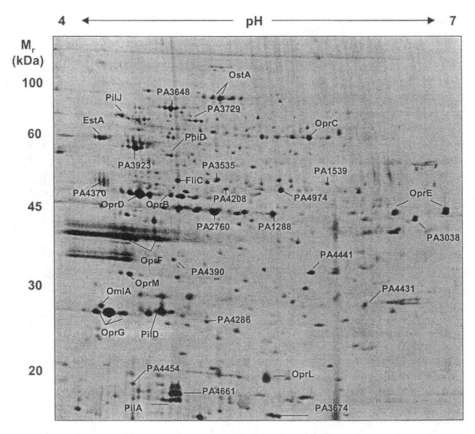

Figure 13.2 Reference map of *P. aeruginosa* OMPs. Membrane proteins were enriched with a sodium carbonate treatment and separated on pH 4–7 2D gels. Protein spots were characterized by mass spectrometry, and protein identities assigned by comparison with the translated PA01 genome sequence. Protein spots are labelled with gene names or *P. aeruginosa* (PA) accession numbers

189 of these were subsequently identified using mass spectrometry. The predicted roles of the identified proteins could be divided into three categories – (1) porins, (2) receptors and (3) unknown function. Seventeen proteins with significant sequence similarity to porins (porins C, D, E1, F, H1, OprM, H.8 and OprF) were identified, and five of these (OprD, E, F, G and potential OmpE3) accounted for over 50 per cent of the total protein visible. As an illustration OprF is defined as the major pseudomonal OMP and appears to have multiple functions, including the maintenance of cell shape and association with peptidoglycan (Nicas and Hancock, 1983; Rawling, Brinkman and Hancock, 1998). OprF has also been the basis for both DNA and protein vaccines (Mansouri *et al.*, 1999; Price *et al.*, 2001). In the second category, receptors for several siderophores, including pyochelin, pyoverdine and aerobactin, were identified. The expression of such proteins may be constitutive, or indeed the cells may have been iron starved even in the complex media used to perform the study.

The *P. aeruginosa* OMP data set revealed that 16 per cent of the abundant, 2DGE-compatible *P. aeruginosa* membrane proteins had no previously known function, and a further 46 per cent had function defined solely on sequence similarity to genes/proteins from other organisms. The identifications also showed that this methodology resulted in minimal contamination from abundant cytosolic proteins, at least in *P. aeruginosa*, since all but one protein was predicted to have either an outer or inner membrane location, or to reside in the periplasmic space. The 28 proteins of unknown function determined in that previous study are therefore likely to be novel OMPs not previously characterized in other microorganisms and thus an excellent set of proteins suitable for further study, especially in the search for new drug targets and vaccines. The proteomics technologies defined in this preliminary study were used to examine OMP expression across *P. aeruginosa* strains with different pathogenic phenotypes (Nouwens *et al.*, 2002). The majority of strain differences visible at the membrane protein level appeared to be the result of minor amino acid sequence variations resulting in isoelectric point shifts visible on 2D gels. However, some unique protein spots were visible in each of the strains, highlighting the need for such comparisons when defining subsets of proteins involved in pathogenic processes. Other researchers have also examined the outer-membrane and cell surface proteins from *P. aeruginosa*, in particular to determine differential gene expression in response to biological conditions. Multiple differences were detected in *P. aeruginosa* membrane protein composition either in response to the presence of antibiotics (Michea-Hamzehpour *et al.*, 1993) or due to nutrient limitation (Cowell *et al.*, 1999).

13.3   Extracellular proteins in *P. aeruginosa*

Extracellular products produced by strains of *P. aeruginosa* include a variety of proteins, signalling molecules and pigments. The functions of these extracellular products include cell–cell signalling, defence against competitors or counter-attacks by the host and scavenging of nutrients, as well as initiation and establishment of infection. For *P. aeruginosa*, specific extracellular compounds include pigments (e.g. pyocyanin and pyoverdine; Meyer, 2000), signal molecules (e.g. quinolones, acylated homoserine lactones and cyclic dipeptides; Holden *et al.*, 1999; Pesci *et al.*, 1999; Van Delden and Iglewski, 1998), as well as proteins, such as proteases (reviewed by Steadman, Heck and Abrahamson, 1993) and exotoxins (reviewed by Galloway, 1993). Extracellular products can be differentiated into those products that are transported through the membrane, but remain attached to the external cell surface (exported products), and those transported and released as free molecules into the external environment (secreted products). Not surprisingly, it is the exported and secreted factors produced by pathogenic micro-organisms that have the greatest involvement in pathogenic processes. As *P. aeruginosa* is pathogenic to humans, there has been a concerted effort to characterize the extracellular factors involved in virulence and pathogenicity, and the secretory systems by which they are transported.

There are several well characterized virulence factors produced by *P. aeruginosa*. These factors work synergistically to establish infections in the host and include the abundant extracellular proteases elastase (*lasB*) and LasA protease (*lasA*) that degrade connective tissues (e.g. elastin and collagen), with LasA acting as an endoprotease and LasB as an exoprotease (reviewed by Galloway, 1993; Steadman *et al.*, 1993).

This degradation most likely serves to disseminate the bacterium further into the host tissue. In some circumstances, loss of one or more extracellular factors can have a substantial effect on the ability of the bacterium to initiate and maintain an infection in the host (Tang *et al.*, 1996). However, a number of strains have been found to lack major virulence factors, such as elastase, and still establish an infection in the host (Preston *et al.*, 1997). Additionally, the necessity of particular virulence factors is also dependent in part on the cell type invaded. For example, although LasB is well characterized as a virulence factor, and contributes significantly to the destruction of connective tissues in the host, it has no role in corneal infections. The infection and tissue damage in the cornea is caused not only by bacterial factors, but also by host proteases that exacerbate corneal damage (Gray and Kreger, 1975). However, LasA mutants are avirulent in corneal infections, indicating different virulence factors are required in different cell types (Preston *et al.*, 1997).

P. aeruginosa causes severe infections in immuno-compromised individuals and thus much interest surrounds the disease process and damage caused to host cells. Two main infective approaches by *P. aeruginosa* have been described and these are commonly referred to as (1) cytotoxic and (2) invasive, with most strains believed to be cytotoxic (Fleiszig *et al.*, 1996, 1997). The main feature of cytotoxicity is the secretion or release of molecules, including proteins, that are toxic towards host cells. Cytotoxic molecules are necrotizing and kill the host tissue. In the alternative, invasive approach, bacteria invade the host cells (i.e. are internalized), which potentially represents a mechanism by which the bacterium can avoid host defences, as well as antibiotics such as gentamicin that do not easily penetrate mammalian cells (Fleiszig *et al.*, 1996).

A number of genes in *P. aeruginosa* are involved in determining cytotoxicity and invasiveness, in particular those genes under the control of the transcriptional regulatory gene *ExsA*. *ExsA* regulates a number of proteins including *exoS*, which encodes a 49 kDa protein with ADP-ribosylating activity and is present in invasive strains (Fleiszig *et al.*, 1997; Kaufman *et al.*, 2000; Pederson *et al.*, 2000); *exoU*, encoding a 70 kDa protein with cytotoxic activity present only in cytotoxic strains (Finck-Barbancon *et al.*, 1997), and *exoT*, encoding a 53 kDa GTPase-activating protein with anti-internalization properties (Fleiszig *et al.*, 1997; Garrity-Ryan *et al.*, 2000; Krall *et al.*, 2000), present in both cytotoxic and invasive strains. Although *exoS* and *exoT* are homologous, and have anti-internalization properties, strains possessing both *exoT* and *exoS* are more invasive than those that produce only *exoT* (Cowell *et al.*, 2000). Although distinctions in invasive and cytotoxic phenotypes may partly be attributable to variation to the *ExsA* regulon, more recent research suggests the cause of infectious phenotypes is multi-factorial, with issues such as host cell type and bacterial number influencing infectious mode (Zhu *et al.*, 2000). Mutations to the type III secretion system can also influence the infectious phenotype, allowing cytotoxic strains to be internalized (Hauser *et al.*, 1998). Some cytotoxic strains are still capable of invasion, especially when *exsA* is non-functional (Fleiszig *et al.*, 1996, 1997).

13.3.1 Proteomics of *P. aeruginosa* extracellular proteins

Investigations into bacterial surface proteins provide one means of appraising potential vaccine and diagnostic candidates, especially in conjunction with strain comparisons and immunoblotting. Extracellular proteins are often those involved in pathogenic processes and therefore the regulation of their expression and the method of secretion also provide new opportunities for novel therapies and drug targets. Bacterial secretion is generally growth-phase dependent and, therefore, the standard methodology is to grow cells in culture until mid-stationary phase is reached. Furthermore, many genes encoding extracellular proteins are regulated by quorum sensing (De Kievit and Iglewski, 2000), and hence cell density. Whole cells are removed via centrifugation and filtration and the free proteins in the culture supernatant are precipitated with ice-cold trichloroacetic acid, acetone, methanol or varying combinations of the above. It is virtually impossible to ensure that fractions derived from culture supernatants are pure due to the presence of contaminating, abundant cytosolic proteins (caused by cell autolysis during growth), membrane proteins and truly secreted proteins. Extracellular proteins are relatively easy to isolate for 2DGE analyses, as they are highly soluble, however few such analyses have been conducted on these proteins from *P. aeruginosa* to this time. We have recently described the stationary phase culture supernatant proteome (Figure 13.3) from two different strains of *P. aeruginosa*, and used these maps to investigate the role of the quorum-sensing (QS) regulators (*las* and *rhl*) on extracellular protein expression (Nouwens *et al.*, 2003). The major secreted protein of *P. aeruginosa* is elastase (LasB), a protease with a wide specificity allowing it to target host tissues. LasB is highly immunogenic and antibodies against this protein have also been found in cystic fibrosis patients (Klinger *et al.*, 1978). Another endopeptidase, LasA protease, is also a major component of the extracellular proteome of *P. aeruginosa*. This protease is specific for multiple glycine or gly–gly–ala sequences, providing it with staphylolytic activity due to the glycine cross-linkages in cell wall peptidoglycan. Several other minor constituents were observed in stationary phase culture supernatants (including proteins from major surface macromolecular structures such as flagella and pili); however, other highly abundant proteins that were identified by 2DGE/MS included an aminopeptidase (Cahan *et al.*, 2001), endoproteinase PrpL (Wilderman *et al.*, 2001), alkaline metalloproteinase (AprA) and chitin-binding protein. It is likely that many virulence factors including haemolysin, and several exotoxins, require host-specific interactions for their optimal expression.

The QS system in *P. aeruginosa* (encoded by *lasI*/*lasR* and *rhlI*/*rhlR*) is composed of two linked regulatory systems (*las* and *rhl*) that control the expression of approximately five per cent of the genome (Whiteley, Lee and Greenberg, 1999), most importantly including several virulence factors (Fugua and Greenberg, 1998; Pesci and Iglewski, 1997). Each system is composed of an autoinducer synthase (*lasI* and *rhlI*) and a regulatory protein (*lasR* and *rhlR*). The autoinducer synthase catalyses the production of acylhomoserine lactone (AHL) signalling molecules from S-adenosylmethionine and an acyl–acyl carrier protein, and these AHLs combine with the respective regulatory protein to form a complex that then binds to the promoters of genes containing an upstream '*lux*' recognition box. Many of these genes are controlled solely by

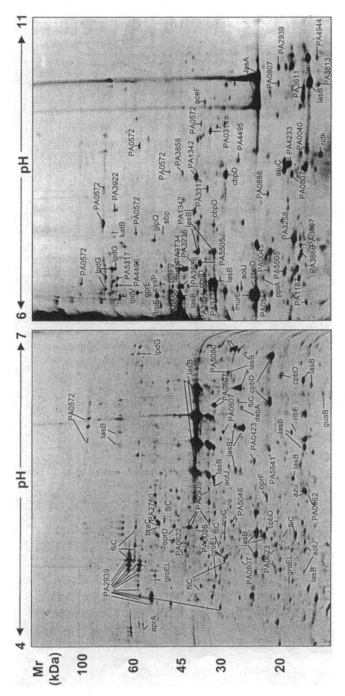

Figure 13.3 Reference map of *P. aeruginosa* extracellular proteins. Proteins were separated on pH 4–7 and pH 6–11 2D gels

one of the *las* or *rhl* operons, while the expression of others, including LasB itself, can be regulated by either system. Virulence factors regulated by both QS systems include elastase (*lasB*), LasA protease (*lasA*) and hemolysin. In our experiments, the effects on extracellular protein expression caused by multiple-deletion mutants within the *las* and *rhl* systems were analysed via 2DGE and MS (Nouwens *et al.*, 2003), revealing several previously unrecognized QS-regulated secreted proteins (Figure 13.4).

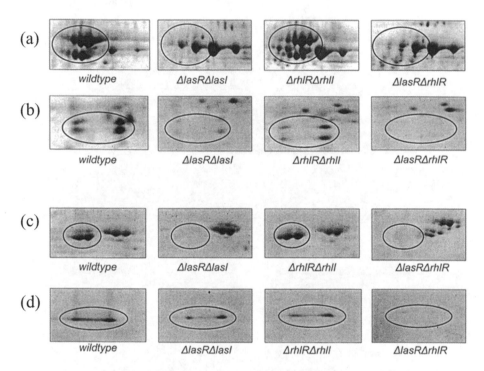

Figure 13.4 Influence of QS on extracellular protein expression in *P. aeruginosa*. Extracellular proteins from wild-type and QS mutants were separated on pH 4–7 and Ph 6–11 2D gels. (a) PA2939; (b) PrpL; (c) AprA; (d) LasA

13.4   Immunogenic proteins and vaccine discovery

The identification and further characterization of expressed proteins using proteomics, can be applied to the discovery of novel vaccine targets, antimicro-

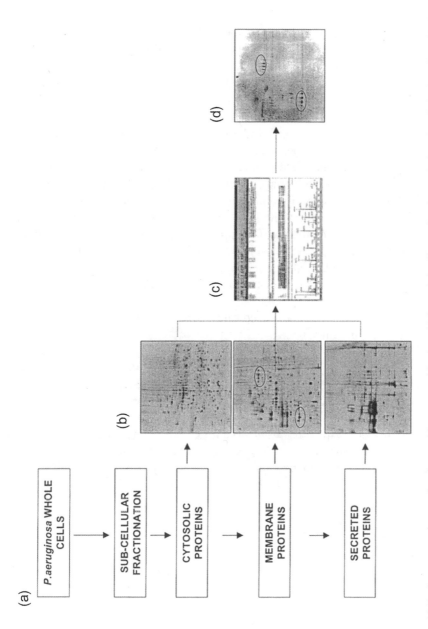

Figure 13.5 Immunogenic proteins detected by western blotting. (a) bacterial cells are fractionated via sub-cellular location. (b) Cytosolic, membrane and secreted protein enriched fractions are separated by 2DGE. (c) Proteins of interest are analysed by MS. (d) Replicate 2D gels of each fraction are screened by western blotting and the proteins mapped back to those identified in (c)

bial therapeutics and diagnostics, and to determine the effects of newly synthesized drugs. As described above, particularly useful information can be gained by examining proteins associated with a chosen sub-cellular location. This becomes even more relevant when attempting to determine which proteins from within a complex mixture elicit an immune response in the host. 2DGE combined with mass spectrometry and western blotting using patient sera is a highly effective method of predicting potential vaccine targets (Figure 13.5; Vytvytska *et al.*, 2002). the data generated from such an approach, however, should be examined with some caution, since highly abundant cytosolic proteins may often appear amongst the most immunogenic proteins when using western blotting. However, questions remain about whether such constituents are genuinely surface associated, and, therefore, about their suitability as vaccine targets. It seems likely that proteins such as the chaperone GroEL and glycolytic enzymes become immunogenic in the host perhaps only following cell autolysis; however, at least one study has shown convincing evidence that 'housekeeping' proteins elongation factor Tu and pyruvate dehydrogenase beta subunit are present as fibronectin-binding surface proteins in *Mycoplasma pneumoniae* (Dallo *et al.*, 2002). This suggests that, at least in this microbiological system, such proteins are capable of performing more than a sole function *in vivo*.

Pre-fractionation of surface-exposed and secreted proteins, combined with 2DE and western blotting, may provide much more focused information on potential vaccine targets. Furthermore, new methodology is also becoming available that utilizes surface display of genomic peptide libraries fused to *E. coli* OMPs for screening against sera with high antibody titre (Etz *et al.*, 2002). This method also allows the identification of specific epitopes that interact with antibody. Although this method has not yet been applied to *P. aeruginosa*, in *Staphylococcus aureus* 60 antigenic proteins were characterized and all were predicted surface or secreted proteins, suggesting that immunoreactive 'housekeeping' proteins may well be an artifact of the traditional western blotting process (Etz *et al.*, 2002).

13.5   Conclusions

Proteomics now provides many of the tools necessary for an in-depth analysis of *P. aeruginosa*. This has been aided significantly by the release of the complete genome sequence, coupled with recent improvements in protein separation, visualization and detection, as well as high sensitivity mass spectrometry. The ability to purify surface-associated and secreted proteins will also provide further capabilities for the development of novel therapies and vaccines against *P. aeruginosa*. Such analyses will allow us to appreciate a better understanding of *P. aeruginosa* pathogenicity and hopefully lead to better therapies with which to combat this organism.

References

Arora, S., Richings, B., Almira, E., Lory, S., and Ramphal, R. (1998) The *Pseudomonas aeruginosa* flagellar cap protein, fliD, is responsible for mucin adhesion. *Infect. Immun.* **66**, 1000–1007.

Cahan, R., Axelrad, I., Safrin, M., Ohaman, D. E., and Kessler, E. (2001) A secreted aminopeptidase of *Pseudomonas aeruginosa*: identification, primary structure and relationship to other aminopeptidases. *J. Biol. Chem.* **276**, 43 645–43 652.

Casadevall, A., and Pirofski, L.-A. (1999) Host–pathogen interactions: redefining the basic concepts of virulence and pathogenicity. *Infect. Immun.* **67**, 3703–3713.

Chalmers, M. J., and Gaskell, S. J. (2000) Advances in mass spectrometry for proteome analysis. *Curr. Opin. Biotechnol.* **11**, 384–390.

Chevallet, M., Santoni, V., Poinas, A., Rouquie, D., Fuchs, A., Kieffer, S., Rossignol, M., Lunardi, J., Garin, J., and Rabilloud, T. (1998) New zwitterionic detergents improve the analysis of membrane proteins by two-dimensional electrophoresis. *Electrophoresis* **19**, 1901–1909.

Comolli, J. C., Waite, L. L., Mostov, K. E., and Engel, J. N. (1999) Pili binding to Asialo-GM1 on epithelial cells can mediate cytotoxicity or bacterial internalization by *Pseudomonas aeruginosa*. *Infect. Immun.* **67**, 3207–3214.

Cornelis, P., Bouia, A., Belarbi, A., Guonvarch, A., Kammerer, B., Hannaert, V., and Hubert, J. C. (1989) Cloning and analysis of the gene for the major outer membrane lipoprotein from *Pseudomonas aeruginosa*. *Mol. Microbiol.* **3**, 421–428.

Cowell, B., Chen, D., Frank, D., Vallis, A., and Fleiszig, S. (2000) ExoT of cytotoxic *Pseudomonas aeruginosa* prevents uptake by corneal epithelial cells. *Infect. Immun.* **68**, 403–406.

Cowell, B. A., Willcox, M. D. P., Herbert, B., and Schneider, R. P. (1999) Effect of nutrient limitation on adhesion characteristics of *Pseudomonas aeruginosa*. *J. Appl. Microbiol.* **86**, 944–954.

Dallo, S. F., Kannan, T. R., Blaylock, M. W., and Baseman, J. B. (2002) Elongation factor Tu and E1 β subunit of pyruvate dehydrogenase complex act as fibronectin binding proteins in *Mycoplasma pneumoniae*. *Mol. Microbiol.* **46**, 1041–1051.

De Kievit, T. R., and Iglewsi, B. H. (2000) Bacterial quorum sensing in pathogenic relationships. *Infect. Immun.* **68**, 4839–4849.

Etz, H., Minh, D. B., Henics, T., Dryla, A., Winkler, B., Triska, C., Boyd, A. P., Sollner, J., Schmidt, W., von Ahsen, U., Buschle, M., Gill, S. R., Kolonay, J., Khalak, H., Fraser, C. M., von Gabain, A., Nagy, E., and Meinke, A. (2002) Identification of in vivo expressed vaccine candidate antigens from *Staphylococcus aureus*. *Proc. Nat. Acad. Sci. USA* **99**, 6573–6578.

Evans, K., Adewoye, L., and Poole, K. (2001) MexR repressor of the *mexAB-oprM* multidrug efflux operon of *Pseudomonas aeruginosa*: identification of Mex R binding sites in the *mexA–merR* intergenic region. *J. Bacteriol.* **183**, 807–812.

Evans, K., and Poole, K. (1999) The MexA–MexB–OrpM multidrug efflux system of *Pseudomonas aeruginosa* is growth-phase regulated. *FEMS Microbiol. Lett.* **173**, 35–39.

Finck-Barbancon, V., Goranson, J., Zhu, L., Sawa, T., Wiener-Kronish, J. P., Fleiszig, S. M., Wu, C., Mende-Mueller, L., and Frank, D. W. (1997) *ExoU* expression by *Pseudomonas aeruginosa* correlates with acute cytotoxicity and epithelial injury. *Mol. Microbiol.* **25**, 547–557.

Finlay, B. B., and Falkow, S. (1997) Common themes in microbial pathogenicity revisited. *Microbiol. Mol. Bio. Rev.* **61**, 136–169.

Fleiszig, S., Zaidi, T., Preston, M., Grout, M., Evans, D., and Pier, G. (1996) Relationship between cytotoxicity and corneal epithelial cell invasion by clinical isolates of *Pseudomonas aeruginosa*. *Infect. Immun.* **64**, 2288–2294.

Fleiszig, S. M. J., Wiener-Kronish, J. P., Miyazaki, H., Vallas, V., Mostov, K. E., Kanada, D., Sawa, T., Yen, T. S. B., and Frank, D. W. (1997) *Pseudomonas aeruginosa*-mediated cytotoxicity and invasion correlate with distinct genotypes at the loci encoding exoenzyme S. *Infect. Immun.* **65**, 579–586.

Fuqua, C., and Greenberg, E. P. (1998) Self perception in bacteria: quorum sensing with acylated homoserine lactones. *Curr. Opin. Microbiol.* **1**, 183–189.

Galloway, D. (1993) Role of exotonins in the pathogenesis of *P. aeruginosa* infections. In Campa, M., Bendinelli, M., and Friedman, H., (Eds.). *Pseudomonas aeruginosa as an Opportunistic Pathogen*. Plenum, New York, 107–127.

Garrity-Ryan, L., Kazmierczak, B., Kowal, R, Comolli, J., Hauser, A., and Engel, J. N. (2000) The arginine finger domain of ExoT contributes to actin cytoskeleton disruption and inhibition of internalization of *Pseudomonas aeruginosa* by epithelial cells and macrophages. *Infect. Immun.* **68**, 7100–7113.

Govan, J., and Deretic, V. (1996) Microbial pathogenesis in cystic fibrosis: mucoid *Pseudomonas aeruginosa* and *Burkholderia cepacia*. *Microbiol. Rev.* **60**, 539–574.

Gray, L. D., and Kreger, A. S. (1975) Rabbit corneal damage produced by *Pseudomonas aeruginosa* infection. *Infect. Immun.* **12**, 419–432.

Gygi, S. P., and Aebersold, R. (2000) Mass spectrometry and proteomics. *Curr. Opin. Chem. Biol.* **4**, 489–494.

Gygi, S. P., Rist, B., and Aebersold, R. (2000) Measuring gene expression by quantitative proteomics. *Curr. Opin. Biotechnol.* **11**, 396–401.

Hancock, R. E. (1998) Resistance mechanisms in *Pseudomonas aeruginosa* and other nonfermentative Gram-negative bacteria. *Clin. Infect. Dis.* Suppl. 1, S93–99.

Hauser, A., Fleiszig, S., Kang, P., Mostov, K., and Engel, J. (1998) Defects in Type III secretion correlate with internalization of *Pseudomonas aeruginosa* by epithelial cells. *Infect. Immun.* **66**, 1413–1420.

Hobot, J. A., Carlemalm, E., Villiger, W., and Kellenberger, E. (1984) Periplasmic gel:

new concept resulting from the reinvestigation of bacterial cell envelope ultrastructure by new methods. *J. Bacteriol.* **160**, 143–152.

Holden, M., Ram Chhabra, S., de Nys, R., Stead, P., Bainton, N., Hill, P., Manefield, M., Kumar, N., Labatte, M., England, D., Rice, S., Givskov, M., Salmond, G., Stewart, G., Bycroft, B., Kjelleberg, S., and Williams, P. (1999) Quorum-sensing cross talk: isolation and chemical characterization of cyclic dipeptides from *Pseudomonas aeruginosa* and other Gram-negative bacteria. *Mol. Microbiol.* **33**, 1254–1266.

Jang, I.-J., Kim, I.-S., Park, W., Yoo, K.-S., Yim, D.-S., Kim, H.-K., Shin, S.-G., Chang, O., Lee, N.-G., Jung, S., Ahn, D., Cho, Y., Ahn, B., Lee, Y., Kim, Y., Nam, S., and Kim, H.-S. (1999) Human immune response to a *Pseudomonas aeruginosa* outer membrane protein vaccine. *Vaccine* **17**, 158–168.

Kaufman, M. R., Jia, J., Zeng, L., Ha, U., Chow, M., and Jin, S. (2000) *Pseudomonas aeruginosa* mediated apoptosis requires the ADP-ribosylating activity of ExoS. *Microbiology* **146**, 2531–2541.

Klinger, J. D., Straus, D. C., Hilton, C. B., and Bass, J. A. (1978) Antibodies to proteases and exotoxin A of Pseudomonas aeruginosa in patients with cystic fibrosis: demonstration by radioimmunoassay. *J. Infect. Dis.* **138**, 49–58.

Koebnik, R., Locher, K. P., and Van Gelder, P. (2000) Structure and function of bacterial outer membrane proteins: barrels in a nutshell. *Mol. Microbiol.* **37** 239–253.

Krall, R., Schmidt, G., Aktories, K., and Barbieri, J. T. (2000) *Pseudomonas aeruginosa* ExoT is a Rho GTPase-activating protein. *Infect. Immun.* **68**, 6066–6068.

Kyte, J., and Doolittle, R. F. (1982) A simple method for displaying the hydropathic character of a protein. *J. Mol. Biol.* **157**, 105–132.

Lee, A., Mao, W., Warren, M. S., Mistry, A., Hoshino, K., Okumura, R., Ishida, H., and Lomovskaya, O. (2000) Interplay between efflux pumps may provide either additive or multiplicative effects on drug resistance. *J. Bacteriol.* **182**, 3142–3150.

Lopez, M. F., Berggren, K., Chernokalskaya, E., Lazarev, A., Robinson, M., and Patton, W. F. (2000) A comparison of silver stain and SYPRO Ruby protein gel stain with respect to protein detection in two-dimensional gels and identification by peptide mass profiling. *Electrophoresis* **21**, 3673–3683.

Lyczak, J. B., Cannon, C. L., and Pier, G. B (2000) Establishment of *Pseudomonas aeruginosa* infection: lessons from a versatile opportunist. *Microbes Infect.* **2**, 1051–1060.

Mansouri, E., Gabelsberger, J., Knapp, B., Hundt, E., Lenz, U., Hungerer, K.-D., Gilleland, H., Staczek, J., Domdey, H., and von Specht, B.-U. (1999) Safety and immunogenicity of a *Pseudomonas aeruginosa* hybrid outer membrane protein F-I vaccine in human volunteers. *Infect. Immun.* **67**, 1461–1470.

Meyer, J.-M. (2000) Pyoverdines: pigments, siderophores and potential taxonomic markers of fluorescent Pseudomonas species. *Arch. Microbiol.* **174**, 135–142.

Michea-Hamzehpour, M., Sanchez, J.-C., Epp, S., Paquet, N., Hughes, G., Hochstrasser, D., and Pechere, J.-C. (1993) Two-dimensional polyacrylamide gel electrophoresis isolation and microsequencing of *Pseudomonas aeruginosa* proteins. *Enzyme Protein* **47**, 1–8.

Morita, Y., Komori, Y., Mima, T., Kuroda, T., Mizushima, T., and Tsuchiya, T. (2001)

Construction of a series of mutants lacking all of the four major *mex* operons for multidrug efflux pumps or possessing each one of the operons from *Pseudomonas aeruginosa* PAO1: MexCD–OprJ is an inducible pump. *FEMS Microbiol. Lett.* **202**, 139–143.

Makae, T., Nakajima, A., Ono, T., Saito, K., and Yoneyama, H. (1999) Resistance to beta-lactam antibiotics in *Pseudomonas aeruginosa* due to interplay between the MexAB–OprM efflux pump and beta-lactamase. *Antimicrobial Agents Chemother.* **43**, 1301–1303.

Nicas, T. I., and Hancock, R. E. W. (1983) *Pseudomonas aeruginosa* outer membrane permeability: isolation of a porin protein F-deficient mutant. *J. Bacteriol.* **153**, 281–285.

Nouwens, A. S., Beaston, S. A., Whitchurch, C. B., Walsh, B. J., Schweizer, H. P., Mattick, J. S., and Cordwell, S. J. (2003) Proteome analysis of extracellular proteins regulated by the *las* and *rhl* quorum sensing systems in *Pseudomonas aeruginosa. Microbiol.* **149**, 1311–1322.

Nouwens, A. S., Cordwell, S. J., Larsen, M. R., Molloy, M. P., Gillings, M., Willcox, M. D. P., and Walsh, B. J. (2000) Complementing genomics with proteomics: the membrane subproteome of *Pseudomonas aeruginosa* PAO1. *Electrophoresis* **21**, 3797–3809.

Nouwens, A. S., Willcox, M. D. P., Walsh, B. J., and Cordwell, S. J. (2002) Proteomic comparison of membrane and extracellular proteins from invasive (PAO1) and cytotoxic (6294) strains of *Pseudomonas aeruginosa. Proteomics* **2**, 1325–1346.

Ochsner, U., Vasil, A., Johnson, Z., and Vasil, M. (1999) *Pseudomonas aeruginosa fur* overlaps with a gene encoding novel outer membrane lipoprotein, OmlA. *J. Bacteriol.* **181**, 1099–1109.

Oliver, D. B. (1996) Periplasm. In Neidhardt, F. C., Curtiss, III, R., Ingrahan, J. L., Lin, E. C. C., Low, K. B., Magasanki, B. W., Reznikoff, W. S., Riley, M., Schaechter, M., and Umbarger, H. E. (Eds.). ASM Press, Washington, DC.

Pavlovskis, O. R., and Wretlind, B. (1979) Assessment of protease (elastase) as a *Pseudomonas aeruginosa* virulence factor in experimental mouse burn infection. *Infect. Immun.* **24**, 181–187.

Pederson, K. J., Pal, S., Vallis, A. J., Frank, D. W., and Barbieri, J. T. (2000) Intracellular localization and processing of *Pseudomonas aeruginosa* ExoS in eukaryotic cells. *Mol. Microbiol.* **37**, 287–299.

Pesci, E. C., and Iglewski, B. H. (1997) The chain of command in *Pseudomonas* quorum sensing. *Trends Microbiol.* **5**, 132–134.

Pesci, E., Milbank, J., Pearson, J., McKnight, S., Kende, A., Greenberg, P., and Iglewski, B. (1999) Quinolone signaling in the cell-to cell communication system of *Pseudomonas aeruginosa. Proc. Natl. Acad. Sci. USA* **96**, 11 229–11 234.

Preston, M. J., Seed, P. C., Toder, D. S., Iglewski, B. H., Ohman, D. E., Gustin, J. K., Goldberg, J. B., and Pier, G. B. (1997) Contribution of proteases and LasR to the virulence of *Pseudomonas aeruginosa* during corneal infections. *Infect. Immun.* **65**, 3086–3090.

Price, B. M., Galloway, D. R., Baker, N. R., Gilleland, L. B., Staczed, J., and Gilleland,

H. E. Jr. (2001) Protection against *P. aeruginosa* chronic lung infection in mice by genetic immunization against outer membrane protein F (OprF) of *P. aeruginosa. Infect. Immun.* **69**, 3510–3515.

Ramphal, R., Koo, L., Ishimoto, K. S., Totten, P. A., Lara, J. C., and Lory, S. (1991) Adhesion of *P. aeruginosa* pilin-deficient mutants to mucin. *Infect. Immun.* **59**, 1307–1311.

Rawling, E. G., Brinkman, F. S. L., and Hancock, R. E. W. (1998) Roles of the carboxy-terminal half of *P. aeruginosa* major outer membrane protein OprF in cell shape, growth in low-osmolarity medium, and peptidoglycan association. *J. Bacteriol.* **180**, 3556–3562.

Santoni, V., Molloy, M., and Rabilloud, T. (2000) Membrane proteins and proteomics: *un amour impossible? Electrophoresis* **21**, 1054–1070.

Schirmer, T. (1998) General and specific porins from bacterial outer membranes. *J. Structural Biol.* **121**, 101–109.

Smolka, M. B., Zhou, H., Purkayastha, S., and Aebersold, R. (2001) Optimization of the isotope-coded affinity tag-labelling procedure for quantitative proteome analysis. *Anal. Biochem.* **297**, 25–31.

Srikumar, R., Paul, C. J., and Poole, K. (2000) Influence of mutations in the *mexR* repressor gene on expression of the MexA–MexB–OprM multidrug efflux system of *Pseudomonas aeruginosa. J. Bacteriol.* **182**, 1410–1414.

Staczek, R., Gilleland, H. J., Harty, R., Carcia-Sastre, A., Engelhardt, O., and Palese, P. (1998) A chimeric influenza virus expressing an epitope of outer membrane protein F of *Pseudomonas aeruginosa* affords protection against challenge with *P. aeruginosa* in a murine model of chronic pulmonary infection. *Infect. Immun.* **66**, 3990–3994.

Steadman, R., Heck, L., and Abrahamson, D. (1993) The role of proteases in the pathogenesis of *Pseudomonas aeruginosa* infections. In Campa, M., Bendinelli, M., and Friedman, H., (Eds.). *Pseudomonas aeruginosa as an Opportunistic Pathogen.* Plenum, New York, 129–143.

Tang, H. B., DiMango, E., Bryan, R., Gambello, M., Iglewski, B. H., Goldberg, J. B., and Prince, A. (1996) Contribution of specific *Pseudomonas aeruginosa* virulence factors to pathogenesis of pneumonia in a neonatal mouse model of infection. *Infect. Immun.* **64**, 37–43.

Urban, A., Leipelt, M., Eggert, T., and Jaeger, K.-E. (2001) DsbA and DsbC affect extracellular enzyme formation in *Pseudomonas aeruginosa. J. Bacteriol.* **183**, 587–596.

Vallis, A., Finck-Barbancon, V., Yahr, T., and Frank, D. (1999) Biological effects of *Pseudomonas aeruginosa* type III-secreted proteins on CHO cells. *Infect. Immun.* **67**, 2040–2044.

Van Delden, C., and Iglewski, B. (1998) Cell-to-cell signaling and *Pseudomonas aeruginosa* infections. *Emerging Infect. Dis.* **4**, 551–560.

Vytvytska, O., Nagy, E., Bluggel, M., Meyer, H. E., Kurzbauer, R., Huber, L. A., and Klade, C. S. (2002) Identification of vaccine candidates of *Staphylococcus aureus* by serological proteome analysis. *Proteomics* **2**, 580–590.

Washburn, M. P., Wolters, D., and Yates J. R. III (2001) Large-scale analysis of the yeast proteome by multidimensional protein identification technology. *Nature Biotechnol.* **19**, 242–247.

Westbrock-Wadman, S., Sherman, D. R., Hickey, M. J., Coulter, S. N., Zhu, Y. Q., Warrener, P., Nguyen, L. Y., Shawer, R. M., Folger, K. R., and Stover, C. K. (1999) Characterization of a *Pseudomonas aeruginosa* efflux pump contributing to amino-glycoside impermeability *Antimicrob. Agents Chemother.* **43**, 2975–2983.

Whitely, M., Lee, K. M., and Greenberg, E. P. (1999) Identification of genes controlled by quorum sensing in *Pseudomonas aeruginosa. Proc. Natl. Acad. Sci. USA* **96**, 13904–13909.

Wilderman, P. J., Vasil, A. L., Johnson, Z., Wilson, M. J., Cunliffe, H. E., Lamont, I. L., and Vasil, M. L. (2001) Characterization of an endoprotease (PrpL) encoded by a PvdS-regulated gene in *Pseudomonas aeruginosa. Infect. Immun.* **69**, 5383–5394.

Woodruff, W. A., and Hancock, R. E. W. (1989) *Pseudomonas aeruginosa* outer membrane protein F, structural role and relationship to the *Escherichia coli* OmpA protein. *J. Bacteriol.* **171**, 3304–3309.

Wu, H. C. (1996) Biosyntheiss of lipoproteins. In Neidhardt, F. C., Curtiss, R. III, Ingrahan, J. L., Lin, E. C. C., Low, K. B., Magasanki, B., Reznikoff, W. S., Riley, M., Schaechter, M., and Umbarger, H. E. (Eds.). *Escherichia coli and Salmonella typhimurium*, ASM Press, Washington, DC, 1005–1014.

Wu, X., Gupta, S., and Hazlett, L. (1995) Characterization of *P. aeruginosa* pili binding human corneal epithelial proteins. *Curr. Eye Res.* **14**, 969–977.

Zhao, Q., Li, X.-Z., Srikumar, R., and Poole, K. (1998) Contribution of outer membrane efflux protein OprM to antibiotic resistance in *Pseudomonas aeruginosa* independent of MexAB. *Antimicrob. Agents Chemother.* **42**, 1682–1688.

Index

Page numbers in italic, e.g. *78*, refer to figures. Page numbers in bold, e.g. **7**, signify entries in tables.

Genomics, Proteomics and Vaccines edited by Guido Grandi
© 2004 John Wiley & Sons, Ltd   ISBN 0 470 85616 5

Index compiled by John Holmes